Computational Atomic and Nuclear Physics

Proceedings of the Summer School of

Computational Atomic and Nuclear Physics

The University of the South
Sewanee, Tennessee
June 26 – July 7, 1989

Edited by
C Bottcher
M R Strayer
J B McGrory
Oak Ridge National Laboratory
Oak Ridge, TN 37831, USA

World Scientific
Singapore • New Jersey • London • Hong Kong

Published by

World Scientific Publishing Co Pte Ltd
5 Toh Tuck Link, Singapore 596224
USA office: 27 Warren Street, Suite 401-402, Hackensack, NJ 07601
UK office: 57 Shelton Street, Covent Garden, London WC2H 9HE

British Library Cataloguing-in-Publication Data
A catalogue record for this book is available from the British Library.

COMPUTATIONAL ATOMIC AND NUCLEAR PHYSIC

ISBN-13 978-981-02-0125-8
ISBN-10 981-02-0125-7

PREFACE

The evolution of parallel processor supercomputers in recent years provides opportunities to investigate in detail many complex problems, in many branches of physics, which were considered to be intractable only a few years ago. But to take advantage of these new machines, one must have a better understanding of how the computers organize their work than was necessary with previous single processor machines. Equally important, the scientist must have this understanding as well as a good understanding of the structure of the physics problem under study. In brief, a new field of computational physics is evolving, which will be led by investigators who are highly literate both computationally and physically.

A Center for Computationally Intensive Problems has been established with the collaboration of the University of Tennessee Science Alliance, Vanderbilt University, and the Oak Ridge National Laboratory. The objective of this Center is to carry out forefront research in computationally intensive areas of atomic, nuclear, particle, and condensed matter physics. An important part of this effort is the appropriate training of students. An early effort of this Center was to conduct a Summer School of Computational Atomic and Nuclear Physics. A distinguished faculty of scientists in atomic, nuclear, and particle physics gave lectures on the status of present understanding of a number of topics at the leading edge in these fields, and emphasized those areas where computational physics was in a position to make a major contribution. In addition, there were lectures on numerical techniques which are particularly appropriate for implementation on parallel processor computers and which are of wide applicability in many branches of science. In these proceedings are presented the lectures of all but three of the participating faculty at the School.

The School was held at the University of the South, Sewanee, Tennessee from June 26 to July 7, 1989. Over forty graduate students attended the School. They were students in atomic and nuclear physics and in computer sciences. Virtually all the students attended on full scholarships. This was possible because of the generous support of the Department of Energy through the Physics Division at the Oak Ridge National Laboratory, the state of Tennessee through the University of Tennessee Science Alliance, and of the Oak Ridge Associated Universities. We are particularly grateful to Ms. Barbara Everett and her staff at the University of the South for their generous and cheerful support in the local administration of the School. Finally, we acknowledge with great gratitude the considerable efforts of Ms. Althea Tate, the Summer School secretary.

C. Bottcher
M. R. Strayer
J. B. McGrory

CONTENTS

MONTE CARLO TECHNIQUES AND MANY-BODY PROBLEMS

M. H. Kalos

Laboratory of Atomic and Solid State Physics
and
Center for Theory and Simulation in Science and Engineering

Cornell University
Ithaca, New York 14853 USA

ABSTRACT

This paper covers some fundamentals of Monte Carlo methods as applied to many-body physics. A class of simple integral equations is introduced that correspond to elementary random walks and whose analytical solutions are given. They yield a deeper understanding of some issues of quantum Monte Carlo and suggest new approaches to the problem of antisymmetrization.

1. INTRODUCTION

These lectures are intended to give a brief survey of Monte Carlo methods and their application to quantum many-body problems through Green's function Monte Carlo. Rather than concentrating on the technical details, they will emphasize qualitative understanding through the use of model problems and model random walks. These models permit sufficient analytic treatment that the importance sampling transformation and "the fermion problem" become transparent. They suggest, in fact, a possible approach to the solution of the fermion problem. The disadvantage of this approach is that the physics is very much played down.

The material is divided into four parts: a brief introduction to Monte Carlo methods; ground state solutions of the Schroedinger equation; introduction of model problems; the fermion problem.

2. OVERVIEW OF MONTE CARLO METHODS

These are a form of numerical analysis useful for many-dimensional problems, such as the evaluation of many-dimensional integrals that arise in many-body physics.

$$\text{Let} \qquad G = \int f(x) g(x) dx$$

$$\text{where} \qquad f(x) \text{ is a probability distribution function}$$

$$f(x) \geq 0 \text{ and } \int_{\Omega_0} f(x) dx = 1$$

2.1 Sampling

"Sampling" a function $f(x)$ means finding an algorithm that gives a sequence $x_1, x_2, \ldots, x_m$ so that:

$$P\{x \epsilon \Omega, \Omega \subset \Omega_0\} = \int_{\Omega} f(x) dx \text{ for every } \Omega.$$

The x's may or may not be statistically independent.

We take M elements of this sequence and form the sum:

$$G_M \equiv \frac{1}{M} \sum_{k=1}^{M} g(x_k), \text{ where } [g_n] = G$$

Assume that the variance of G_M exists, then

$$\sigma_g^2 \equiv Var[g] = \int f(x) [g(x) - G]^2 dx = \int f(x) g^2(x) dx - G^2$$

An asymptotic condition on G_M as $M \rightarrow \infty$ is:

$$P\left\{|G_M - G| < \frac{s\sigma_g}{\sqrt{M}}\right\} = \frac{2}{\sqrt{2\pi}} \int_0^s e^{-\frac{1}{2}t^2} dt$$

Using $s = 1$, we can say that:

$$|G_M - G| < \frac{\sigma_g}{\sqrt{M}} \quad 68\% \text{ of the time}$$

The cost of the computation is governed by the variance σ_g. (The cost doesn't have to grow faster than the number of dimensions.)

2.2 Simple Examples

$$\int_0^1 x^2 dx = \int_0^1 1 \cdot x^2 \, dx = < x^2 >_1$$

$$\uparrow \quad \uparrow$$

$$\equiv f \quad \equiv g \qquad f = \text{ constant, uniform distribution.}$$

$$\text{or } = \int_0^1 2x \cdot \frac{1}{2} x \, dx = < \frac{1}{2} x >_{2x}$$

$$\uparrow \quad \uparrow$$

$$\equiv f \quad \equiv g$$

$$\text{or } \int_0^1 3x^2 \cdot \frac{1}{3} \, dx = < \frac{1}{3} >_{3x^2}$$

$$\uparrow \quad \uparrow$$

$$\equiv f \quad \equiv g \qquad \text{Note here } g \text{ is a constant}$$

$$or = \int_0^1 4x^3 \, \frac{1}{4x} \, dx = < \frac{1}{4x} >_{4x^3}$$

$$\uparrow \quad \uparrow$$

$$\equiv f \quad \equiv g$$

These are not all equivalent because the variance differs. For some choice, σ_g^2 may not even exist. And note that for g=constant, the variance is zero! Zero variance is useful in formal relations and as a guide to methods with low variance. That is, although a practical scheme with zero variance is usually impossible, the formal relations can help produce efficient schemes.

2.3 Sampling Algorithms

This is a long and interesting subject, to which we can give here only a brief introduction.

i) Composition
ii) Mapping
iii) Rejection; (Accept/reject random variables based on how close they are to a given distribution)
iv) Metropolis; (A random walk with a given asymptotic distribution)

There are also other random walk procedures. All of these transform random variables into other random variables.

We start with the basic distribution (0,1): ξ is uniform on (0,1), i.e.

$$f(\xi) = \begin{cases} 1 & \text{if } 0 < \xi < 1 \\ 0 & \text{otherwise} \end{cases}$$

To get this distribution, we use a suitable random number generator. We must test the generator in a way relating to the application; e.g., test it on a solved problem resembling the unknown.

For example, one can use:

$$\xi_n = \frac{mk}{N}; N = 2^P; m_k = am_{k+1} + b \bmod N$$

These are good generators, but they have correlated behavior on cycles of powers of two. If the calculation involves systematic use of random numbers in fixed patterns whose length is powers of two, systematic errors are likely.

Tausworthe generator:

$$m_k = m_{k-q} \otimes m_{k-p}$$

Here "$\otimes$" means "exclusive or".

2.3.1 Mapping

$$f(x) = 2x$$

$$\text{If } x = \sqrt{\xi} \text{ then } f(x) = 2\xi^2$$

Another example, solving for x as a function of ξ in:

$$\xi = \int_x^\infty f(y)dy \text{ samples } f(x)$$

Box-Muller; bivariate distribution of

$$\frac{1}{\sqrt{2\pi}} e^{-\frac{1}{2}x^2} \frac{1}{\sqrt{2\pi}} e^{-\frac{1}{2}y^2}$$

where

$$x = \sqrt{-2\log\xi_1}\cos 2\pi\xi_2$$
$$y = \sqrt{-2\log\xi_1}\sin 2\pi\xi_2$$

2.3.2 Composition

$$x = \max[\xi_1, \xi_2]; f(x) = 2x$$

2.3.3 Rejection. For example, let $x_1 = \xi$. If $\xi_2 \leq x$, then choose $x = x_1$, else repeat. Thus we sample the shaded points to sample $f(x) = 2x$,

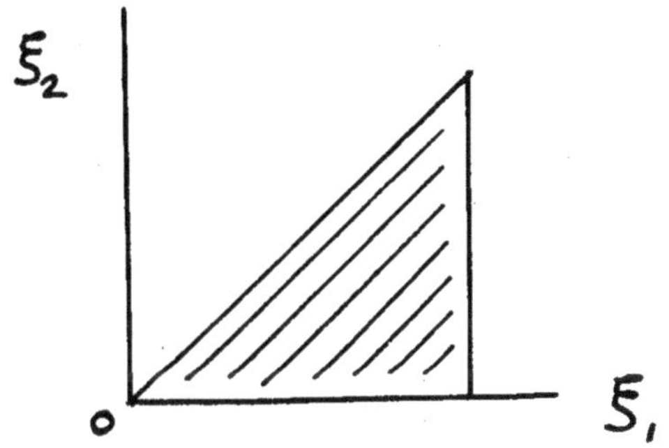

2.3.4 Metropolis. This is a most powerful algorithm from standpoints of simplicity and computing time. It can speed up multivariate computations enormously.

In this method, a pointer (walker) moves around the space on which the probability distribution is defined. For a simple example, consider sampling the distribution $f(x) = 2x, 0 \leq x \leq 1$.

Let x_0 = anything, say x_0=0.5. Then let $x'_{n+1} = \xi_n$, a trial position.

$$\text{If } x'_n + 1 > x_n \text{ then set } x_{n+1} = x'_{n+1}$$

$$\text{If } x'_{n+1} < x_n \text{ and } \xi'_n < \frac{x'_{n+1}}{x_n} \text{ then } x_{n+1} = x'_{n+1}$$

$$\text{Else } x_{n+1} = x_n$$

Thus we generate $x_0, x_1, x_2, \ldots$ Suppose we want to sample $f(x)$. Suppose at some stage the walker is at x_n. We sample x'_{n+1} from a stochastic kernel $T(x'_{n+1}|x_n)$. Then we will accept $x_{n+1} = x'_{n+1}$ with probability

$$\text{Prob } (x_n \to x'_{n+1}) = \min\left[1, \frac{T(x_n|x'_{n+1})f(x'_{n+1})}{T(x'_{n+1}|x_n)f(x_n)}\right]$$

and $x_{n+1} = x_n$ with probability $1 - \min\left[1, \frac{T(x_n|x'_{n+1})f(x'_{n+1})}{T(x'_{n+1}|x_n)f(x_n)}\right]$. The result gives the correct distribution asymptotically. The probability distribution of x_n goes to $f(x)$ for large n. Note that successive x's are correlated. If these correlations are large so that many steps m are required before the variables x_n and x_{n+m} are effectively independent then serious inefficiency can occur, since M-C quadrature is usually efficient when successive variants are uncorrelated. More serious is that the autocorrelation in Metropolis is unknown in advance and calculation of variance requires careful analysis of the results. But because of its simplicity, Metropolis is the most useful for sampling distribution in a high-dimensional space, e.g. to sample Boltzmann's distribution. Here, T is general. (Choosing T properly can minimize sequential correlations). For example, in Ising problems it is advantageous to optimize T to minimize autocorrelations.

General points

In some problems one must sample with respect to some non-obvious distribution. Recall $G = \int f(x)g(x)dx$. We noted that the separation into $f(x)$ and $g(x)$ was arbitrary, so generally

$$G = \int \tilde{f}(x)\frac{f(x)g(x)}{\int \tilde{f}(y)dy}dx$$

Is there an optimal $\tilde{f}(x)$? If $g(x) \geq 0$ then the choice $\tilde{f}(x) = \frac{f(x)g(x)}{\int f(y)g(y)dy}$ makes the variance zero.

One can show this with the calculus of variations to minimize σ_g^2 with the condition that $\int f dx = 1$. This happens because the average of a constant gives zero variance. But to use the optimal choice one must know the integral already! However, this observation is non-trivial. If we can make $\tilde{f}$ approximately equal to this function then we can reduce the variance with this choice to an arbitrary degree.

More general points

i) M-C methods permit us to approach many-body problems using explicit many-body integration.

ii) M-C is a kind of numerical analysis. It is not a simulation of the physics underlying the Schroedinger equation. However, it is a simulation of a related diffusion process.

M-C methods for the Schroedinger equation have the character of random walks. Consider a numerical approximation to the wavefunction ψ_T. One will generate a sequence of points R_k each having a numerical weight w_k that approximate to the wavefunction ψ in the following sense:

$$\psi(r_1, r_2, \ldots r_N) \equiv \psi(R) \cong \sum_k \delta(R - R_k) w_k$$

Thus we find an energy for the system by integration using a trial function $\psi_T(R)$

$$E = \frac{\int \psi_T H \psi dR}{\int \psi_T \psi dR} = \frac{\int \psi H \psi_T dR}{\int \psi \psi_T dR} \cong \frac{\sum_k w_k H \psi_T(R_k)}{\sum_k w_k \psi_T(R_k)}$$

The last relation is approximate in the sense that the integrals are replaced by sums over a finite sample of points. The errors are purely statistical.

iii) The M-C method of solution is <u>grid-free</u>. There are no truncation errors. It can be exact except for <u>Monte Carlo errors</u>, that is, statistical variance. Improvement of the calculation may require additional computing but the time for fixed accuracy need not grow faster than the number of particles. The Variational Principle gives a bound on the energy,

$$E \le E_v = \frac{\int \psi_T^*(R) H \psi_T(R) dR}{\int \psi_T^* \psi_T dR} = \frac{\int \psi_T^* \psi_T \frac{H\psi_T}{\psi_T} dR}{\int \psi_T^* \psi_T dR}$$

Let

$$f = \psi_T^*, \quad g = \frac{H\psi_T}{\psi_T}.$$

Then

$$E_v = \frac{\int f(R) g(R) dR}{\int f(R) dR} \cong \frac{\sum_k \left(\frac{H\psi_T}{\psi_T}\right)_{R=R_k}}{\sum_k 1}.$$

Again the last relation involves statistical errors in replacing the integrals by sums over a finite sample.

3. SOLUTION OF THE SCHROEDINGER EQUATION

We begin with $H = -\nabla^2 + V$ ($\hbar = 1 = m$, etc.), where $-\nabla^2 \equiv -\sum_i \nabla_i^2$. Green's function satisfies

$$\left[-\nabla^2 + \frac{\partial}{\partial\tau} + V(R)\right] G(R, R', \tau) = 0$$

$$G(R, R', 0) = \delta(R - R').$$

In this diffusion-like equation, V(R) plays the role of a probability per unit time that the walker disappears from sight at R.

Introduce eigenfunctions of H:

$$H\psi_k = E_k\psi_k$$

And the corresponding expansion of G:

$$G(R, R', \tau) = \sum_k \psi_k(R)\psi_k(R')e^{-E_k\tau}$$

$$G(R, R', \tau) \underset{\tau \to \infty}{\longrightarrow} \psi_0(R)\psi_0(R')e^{-E_0\tau}$$

The long-time behavior is dominated by the ground state.

G satisfies the convolution relation:

$$G(R, R_0, \tau_1 + \tau_2) = \int G(R, R', \tau_1)G(R', R_0, \tau_2)dR'$$

$$= \int \sum_k \psi_k(R)\psi_k(R')e^{-E_k\tau_1} \sum_\ell \psi_\ell(R')\psi_\ell(R_0)e^{-E_\ell\tau_2}dR'.$$

Use orthonormality to get

$$= \sum_k \psi_k(R)\psi_k(R_0)e^{-E_k(\tau_1+\tau_2)}$$

Applying this repeatably gives the "Trotter" form and path integrals for small values of τ:

$$G(R, R', \Delta\tau) \underset{\Delta\tau \to 0}{\cong} \frac{1}{(4\pi\Delta\tau)^{\frac{3}{2}}} \exp\left[-\frac{(R - R')^2}{4\Delta\tau}\right] e^{-V(R)\Delta\tau}$$

This goes to a δ-function as $\tau \to 0$. We can use it in path integral relations. But rather than form path integrals, we will do something different, as follows. $\psi_0(R)$ and G satisfy:

$$\psi_0(R) = E_0 \int G(R, R', \tau)\psi_0(R')dR'd\tau$$

$$= E_0 \int \sum_k \psi_k(R)\psi_k(R')e^{-E_k\tau}\psi_0(R')dR'd\tau$$

and

$$\psi_0(R) = e^{E_0\tau} \int G(R, R', \tau)\psi_0(R')dR' \text{ (for any } \tau)$$

These forms are the basis of "Green's function Monte Carlo".

Consider a series of functions $\varphi_0, \varphi_1, \varphi_2, \ldots$ where φ_0 is some given function and φ_n is defined through:

$$\varphi_n(R) = E_0 \int\int G(R, R', \tau)\varphi_{n-1}(R')dR'd\tau$$

We expand φ_0 in the eigenfunctions: $\varphi_0(R) = \sum_k a_k\psi_k(R)$. Then

$$\varphi_1(R) = E_0 \int \sum_\ell \psi_\ell(R)\psi_l(R')e^{-E_\ell\tau} a_k\psi_k(R')dR'd\tau$$

$$= \sum_k a_k \left(\frac{E_0}{E_k}\right) \psi_k(R)$$

and

$$\varphi_n(R) = \sum_k a_k \left(\frac{E_0}{E_k}\right)^n \psi_k(R)$$

$$\underset{n \to \infty}{\longrightarrow} A\psi_0(R), \quad \text{where A is some constant}$$

So the operation of getting φ_{n+1} from φ_n projects out the ground state.

3.1 Relation to Monte Carlo Methods

G is a solution to a diffusion problem; it is non-negative. It gives the behavior of a diffusing object or walker. A random walker following the appropriate rules has a chance of surviving proportional to $e^{-V(R)\Delta\tau}$. G can be sampled by M-C techniques with no approximations. The general idea involves convolution methods in which one generates G iteratively. So suppose we have a way of sampling G with a random walk procedure. And suppose our functions are positive everywhere, like the known function. $\varphi_0(R) \geq 0$ We can then treat it as a probability density and sample it by Metropolis methods. So $R_1, R_2, \ldots R_n$ are sampled from $\varphi_0(R)$. For each of these R_m, we sample $E_0G(R, R_m, \tau)$. We can ignore the role of τ since it is integrated out, that is, ignored in the Monte Carlo. The factor E_0 makes the population constant in size, on the average, even though some of the walkers have disappeared.

From this we get a distribution of new positions of walkers. We get the number of walkers at R within dR from the ones which started at R'. The probability for a walker to start at R' is proportional to φ_0:

$$\varphi_1(R)dR = E_0 \int\int G(R, R', \tau)\varphi_0(R')dR'dR\, d\tau$$

The iteration requires merely a random walk from old positions to new positions.

4. MODEL PROBLEMS

We will now introduce some integral identities that yield random walks like those that occur in Green's function Monte Carlo (GFMC). These will permit simple analysis and intuitive understanding of that technique.

These integral equations will have the form of

$$\varphi(y) = E_0 \int G(y, z)W(z)\varphi(z)dz$$

where $G(y, z)$ is a stochastic kernel such that

$$G(y, z) = G(z, y)$$

$$G(y, z) \geq 0 \text{ for all } y, z$$

$$\int G(y, z)dy = 1 \text{ for every } z$$

and

$$W(z) \geq 0$$

A random walk that is described by this equation is as follows. A walker at z becomes N walkers ($N \geq 0$) with the expected value of N, satisfying

$$< N >= E_0 W(z)$$

A simple implementation is

$$N = \lfloor E_0 W(z) + \xi \rfloor$$

where ξ is uniform on (0,1) and $\lfloor\ \rfloor$ denotes the floor function. Recall that:

$$G(R, R', \Delta\tau) \cong \frac{1}{(4\pi\Delta\tau)^{\frac{3N}{2}}} e^{-\frac{(R-R')^2}{4\Delta\tau}} e^{-V(R)\Delta\tau}$$

Now consider the relation (true for all β):

$$e^{-1/2y^2} = \int \sqrt{\frac{1+2\beta}{4\pi\beta}}\, e^{\frac{-(1+2\beta)}{4\beta}(y-z)^2} \sqrt{1+2\beta}\; e^{-\beta z^2}\; e^{-1/2z^2}\; dz$$

A normalized Gaussion sampling function!	Like the E_0 factor!	Like the $e^{-V\Delta\tau}$ factor.	Corresponds to the $\varphi_n(z)$ function.

In this example, a Gaussian transforms into another Gaussian. Here $E_0 W(z) = \sqrt{1+2\beta}e^{-\beta z^2}$. Consider the specific case $\beta = \frac{1}{2}$ and $\varphi_0(y) = e^{-(y-y^2)^2}/2\sigma_0^2$ in this formula. Then iteration give new Gaussions (centered at new y values and with new widths σ):

$$y_0 \rightarrow y_1 \rightarrow y_2 \ldots$$

$$\sigma_0 \rightarrow \sigma_1 \rightarrow \sigma_2 \ldots$$

We find

$$y_{n+1} = \frac{y_n}{1+\sigma_n^2}$$

$$\sigma_{n+1}^2 = \frac{1+3\sigma_n^2}{2(1+\sigma_n^2)}$$

Thus $\sigma_n \underset{n\rightarrow\infty}{\longrightarrow} 1$ and $y_{n+1} \underset{n\rightarrow\infty}{\longrightarrow} \frac{1}{2}y_n$. We can start with a delta function distribution centered at any position y_0; the iteration will converge to an equilibrium distribution $\exp(-\frac{1}{2}y^2)$:

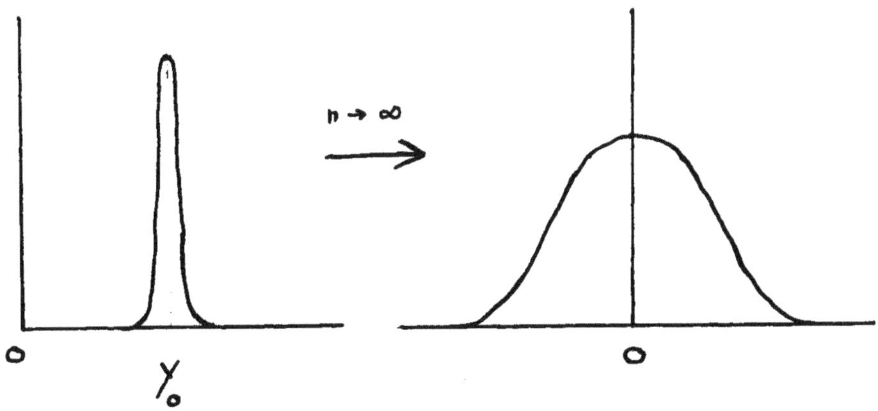

Diffusion makes the walkers wander off to infinity but the $W(y)$ function eliminates the ones which go too far. We always compute observables asymptotically, i.e. with respect to the distribution that approaches the correct one asymptotically.

We now introduce an importance function transformation. Consider the integral $\int f(x)g(x)dx$. As we saw, the optimal distribution is the entire integrand. General principles of Monte Carlo way that we should bias the distribution by the "importance" of a walker at any position.

Let us use the eigenfunctions of our model equation:

$$\varphi_k(y) = E_k \int G(y,z)W(z)\varphi_k(z)dz$$

These are not orthogonal in the usual sense. It is easy to see, however that the following generalization of the orthogonality relation holds.

$$\int \varphi_k(y)W(y)\varphi_\ell(y)dy = 0 \text{ if } E_\ell \neq E_k$$

Therefore, the functions $\sqrt{W(y)}\varphi_k(y)$ are orthogonal. An arbitrary function $v(y)$ may be expanded as:

$$v(y) = \sum_\ell c_\ell \sqrt{W(y)}\varphi_\ell(y)$$

$$E_0 \int G(y,z)v(z)dz = \sum_\ell c_\ell \left(\frac{E_0}{E_\ell}\right) \sqrt{W(y)}\varphi_\ell(y)$$

If one iterates many times,

$$\begin{aligned} E_0^n \int G(y,z_n) \int G(z_n,z_{n+1}) \ldots \int G(z_2,z_1)V(z_1)dz_1 \ldots dz_n \\ = \sum_\ell c_\ell \left(\frac{E_0}{E\ell}\right)^n \sqrt{W(y)}\varphi_\ell(y) \\ \to c_0\sqrt{W(y)}\varphi_0(y) \end{aligned}$$

$$c_0 = \int \sqrt{W(z)}\varphi_0(z)v(z)dz$$

and if

$$v(z) = \delta(z - y_0)$$

$$c_0 = \sqrt{W(y_0)}\varphi_0(y_0)$$

Thus the asymptotic population at y resulting from a walker that starts at y_0 is proportional to $W(y_0)\varphi_0(y_0)$. This is a plausible function to use as the importance. To use it, we multiply our integral equation through $W(y)\varphi_0(y)$ and multiply and divide the integrand on the right by $\varphi_0(z)$ to obtain:

$$\varphi_0(y)W(y)\varphi(y) = E_0 \int \left[W(y)\varphi_0(y)G(y,z)/\varphi_0(z)\right]\varphi_0(z)W(z)\varphi(z)dz$$

or

$$\tilde{\varphi}(y) = E_0 \int W(y)\tilde{G}(y,z)\tilde{\varphi}(z)$$

$$\text{with } \tilde{\varphi}(y) = \varphi_0(y)W(y)\varphi(y) \qquad \tilde{G}(y,z) = \varphi_0(y)G(y,z)/\varphi_0(z)$$

If we put our model equation in this form, we get:

$$e^{-(1+\beta)y^2} = \int \sqrt{\frac{(1+2\beta)^2}{4\pi\beta}} e^{\frac{-(1+2\beta)^2}{4\beta}(y-\frac{z}{1+2\beta})^2} e^{-(1+\beta)z^2} dz$$

$$\sqrt{\frac{(1+2\beta)^2}{4\pi\beta}} e^{\frac{-(1+2\beta)^2}{4\beta}(y-\frac{z}{1+2\beta})^2} \equiv \tilde{G}(y,z)$$

or

$$\tilde{\varphi}(y) = \tau_0 \int \tilde{G}(y,z)\tilde{\varphi}(z)dz$$

The transition kernel here satisfies

$$\int \tilde{G}(y,z)dz = 1 \text{ for all } y, z$$

so that N=1 and there is no multiplication (or termination) of the walkers. This circumstance, in which the population size has no variance, is clearly a more efficient random walk realization of the problem.

Also note that $\tilde{G}$ is not centered at z:

$$\tilde{G} \propto e^{-2(y-\frac{z}{2})^2} \text{ for } \beta = \frac{1}{2}$$

The walker is shifted to $\frac{z}{2}$ and the distribution is more narrow.

This $\tilde{G}$ was built using a knowledge of the exact eigenfunction. In practice we use an approximate trial function for φ_0, which reduces the fluctuations of the random walkers.

Importance sampling is a key element in getting accurate answers; one can get chemical accuracy for light atoms, one part in 10^5 or 10^6.

The analog of the new center-of-gravity shift is the drift that importance sampling produces.

What about E_0? That's what we are trying to find, so how can we use it without knowing its value in advance? We guess it; if guessed incorrectly the population will increase or decrease on the average. We fix the guess so that the population is stable. We can summarize our results for the model integral equation:

$$u_0(y) = e^{-\frac{1}{2}y^2} \qquad E_0 = 1$$

$$u_0(y) = \int_{-\infty}^{\infty} \frac{1}{\sqrt{\pi}} e^{-(y-z)^2} \sqrt{2}\, e^{-\frac{1}{2}z^2} u_0(z) dz$$

$$\sqrt{2} e^{-\frac{1}{2}z^2} \equiv W(z), \text{ plays role of potential}$$

$$u_k(y) = E_k \int_{-\infty}^{\infty} \frac{1}{\sqrt{\pi}} e^{-(y-z)^2} \sqrt{2}\, e^{-\frac{1}{2}z^2} u_k(z) dz$$

$$\tilde{u}_k(y) \equiv W(y) u_0(y) u_k(y)$$

$$\tilde{u}_k(y) = E_k \int_{-\infty}^{\infty} \sqrt{\frac{2}{\pi}}\; e^{-2(y-\frac{z}{2})^2} \tilde{u}(z) dz$$

$$u_1(y) = y e^{-\frac{1}{2}y^2} \qquad E_1 = 2$$

5. THE FERMION PROBLEM

The difficulty comes from anti-symmetric solutions to the Schroedinger equation and from the nature of the Monte Carlo method.

In the framework of Monte Carlo, u is represented by:

$$u_{MC} \cong \sum_{\ell=1}^{L} \delta(y - y_\ell)$$

We can always write:

$$u_1(y) = u^+(y) - u^-(y) \qquad \text{with } u^\pm \geq 0$$

with no loss of generality. In fact, this decomposition is not unique. We use in particular:

$$u^{\pm}(y) = \begin{cases} + & \max\,[u_1(y), 0] \\ - & \max\,[-u_1(y), 0] \end{cases}$$

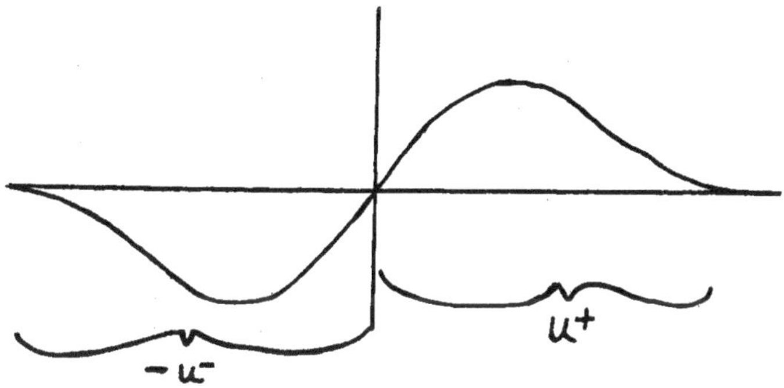

But one can always add a symmetric function:

$$u^{\pm} = u^{\pm} \;+\; u_{\text{symm}}$$

to get another choice.

We define a sequence:

$$u^{\pm}_m(y) = \int G(y,z)w(z)u^{\pm}_{m-1}(z)dz$$

These converge because the anti-symmetric part converges to the lowest anti-symmetric eigenfunction. In fact, for an anti-symmetric state we just need u^+ since the eigenvalue is then:

$$E = \frac{\int\; u^{(A)}_T(y)Hu^+(y)dy}{\int\; u^{(A)}_T(y)u^+(y)dy}$$

u^+ has symm and anti-symm parts, but the anti-symm part is projected out by $u^{(A)}_T$, some antisymmetric test (or trial) function.

$$E \cong \frac{\sum_\ell Hu^{(A)}_T(y_\ell)}{\sum_\ell u^{(A)}_T(y_\ell)}$$

The problem is that the distribution u^+ converges to u_0, which is orthogonal to any anti-symmetric test function. The signal will be lost in the Monte Carlo noise.

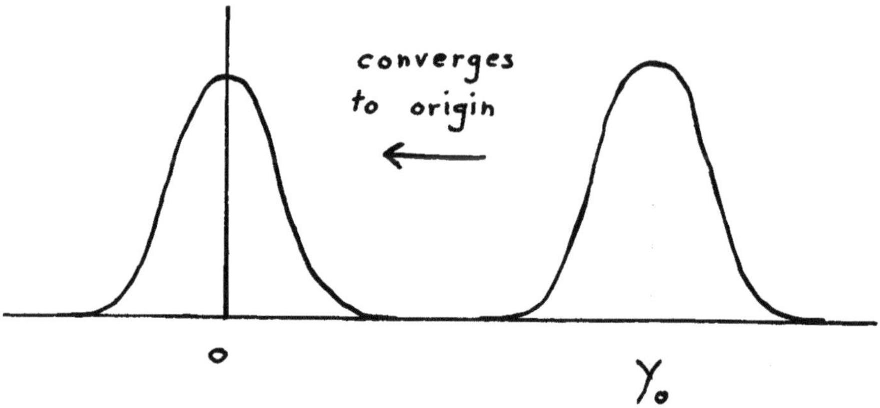

The eigenvalues converge but with large error bars:

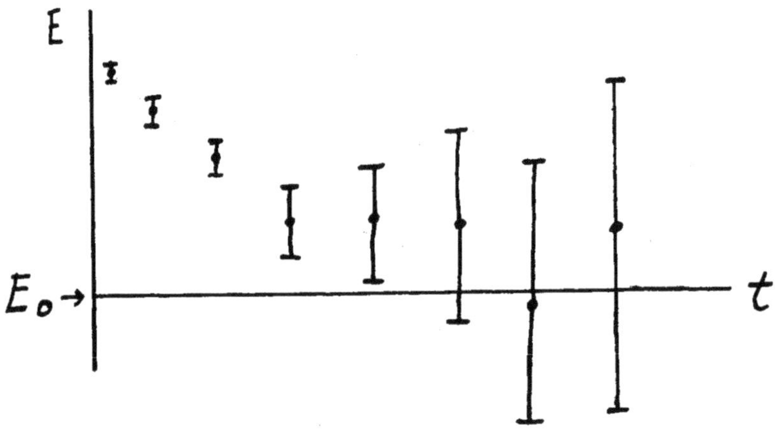

Energy converges in principle, but errors are large and growing for fixed computer time at each step. Terms associated with this are: "Transient Estimation" or "Node Relaxation".

5.1 Fixed Node Approximation

Anti-symmetric solutions of the Schroedinger equation must change sign somewhere.

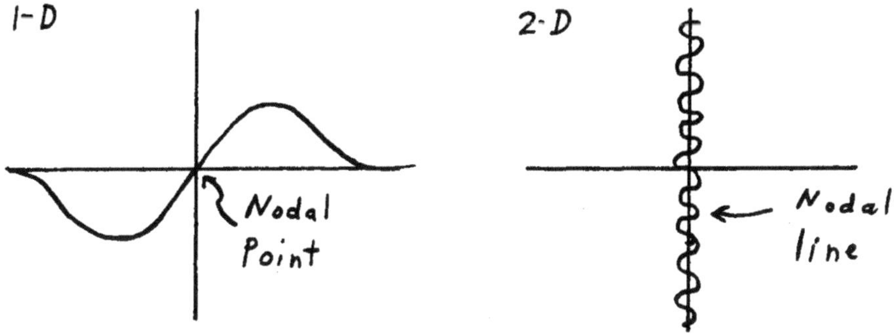

Symmetry often gives restrictions on the model surfaces. But generally we don't know where the nodes are.

One can solve the Schroedinger equation subject to the condition that the wavefunction vanish on some surface. This is called the "fixed-node approximation". It gives an upper bound for the energy, but it is not totally satisfactory. We want more objective methods whose accuracy can be determined from the calculation. Most of all we want exact methods. "Node relaxation" refers to starting with fixed nodes and then removing the constraint they create.

5.2 Mirror Potentials

Start with

$$\psi(R) = \psi^+(R) - \psi^-(R)$$

$$\left.\begin{aligned} H\psi^+ &= E\psi^+ \\ H\psi^- &= E\psi^- \end{aligned}\right\} \text{Difference of these two equations is the Schroedinger equation for } \psi$$

Now add $C\psi^+\psi^-$ to the last two equations:

$$H\psi^+(R) + C(R)\psi^-(R)\psi^+(R) = E\psi^+(R)$$

$$H\psi^-(R) + C(R)\psi^+(R)\psi^-(R) = E\psi^-(R)$$

where $C(R)$ is any symmetric function. The difference of these two equations still gives the Schroedinger equation for ψ. [But the solutions for ψ^+ and ψ^- are changed from before].

The equations for $\psi^{\pm}$ are now:

$$[H + C(R)\psi^{\mp}(R)]\psi^{\pm}(R) = E\psi^{\pm}(R)$$

Thus ψ^- acts as a potential for ψ^+; ψ^- is small where ψ^- is large and vice versa. This stabilizes the antisymmetric part of each function. In practice it proved hard to get an algorithm where the walkers themselves would determine such a "mirror potential" in a completely accurate way.

We solve:

$$u^{\pm} = \int G(y,z)[w(z) - u^{\pm}(z)]u^{\pm}(z)dz$$

So we are solving a Schroedinger equation for a stochastic potential. Kalos and Carlson were unable to make an exact method based on this, but found that a "mirror potential" derived from a trial function gave very good results.

Arnows method:

$$\text{Choose} \quad u^+(y) = \begin{cases} u(y) & \text{for } u > 0 \\ 0 & \text{otherwise} \end{cases}$$

$$u^-(y) = \begin{cases} -u(y) & \text{for } u < 0 \\ 0 & \text{otherwise} \end{cases}$$

$$u(y) = u^+(y) - u^-(y)$$

so

$$u(y) = E_1 \int G(y,z)w(z)[u^+(z) - u^-(z)]dz$$

The normalization of u^+ and u^- must be the same so that there must be the same number of walkers representing each function.

Then we have:

$$u(y) = E_1 \sum_{\ell} [G(y, z_\ell^+)W(z_\ell^+) - G(y, z^-)W(z_\ell^-)]$$

$$\text{Define} \quad G^+(y, z^+, z^-) = \begin{cases} [G(y,z^+) - G(y,z^-)] & \text{if this is } \geq 0 \\ 0 \quad \text{otherwise} \end{cases}$$

$$G^-(y, z^+, z^-) = \begin{cases} [G(y,z^+) - G(y,z^-)] & \text{if this is } \leq 0 \\ 0 \quad \text{otherwise} \end{cases}$$

This procedure should also help stabilize the solution. It worked well on a model problem but did not look promising for many-body problems and has not been pursued.

In one dimension, it is easy to use this method efficiently since a dense population of walkers can easily be obtained. A walker in a population representing u_1 +

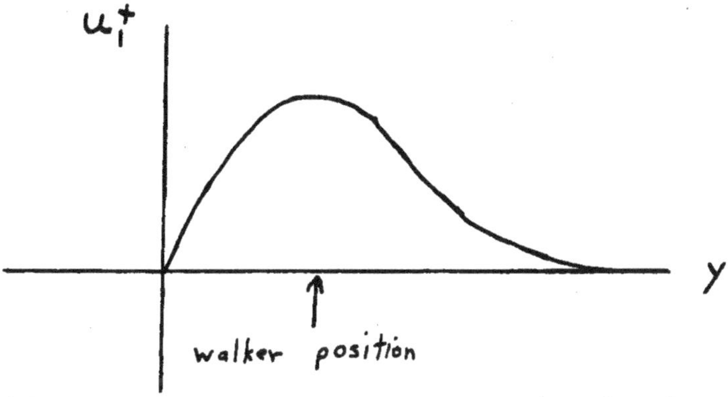

is not prevented from crossing the node at $x = 0$. It can be reflected in the origin so that $x > 0$ again at the expense of giving it a negative weight. For our model walk in one dimension it is easy to generalize the Arnow's method in an interesting way. Let this walker have weight -1 and be at $x_2 > 0$. Suppose we can find two walkers with weights of +1 at x_1 and x_3 with $0 < x_1 < x_2 < x_3$. Then if $x_3 - x_1 \leq 2\sqrt{\log 2} = 1.67$ a composite Green's function using all three coordinates is positive everywhere. Thus the negative-weight walker is absorbed into the other pair, and the resulting walker has positive weight. This will not work so well in higher dimensions. But the importance sampling transformation discussed above plays an interesting role in enhancing the effect. Recall that (for $\beta = \frac{1}{2}$ in our model problem) the kernel has the form

$$\exp\left[-2\left(y - \frac{z}{2}\right)^2\right]$$

in any number of dimensions. The fact that this is centered at $\frac{z}{2}$ (rather than z) has the effect of improving the cancellation of a negative weight walker by a set of nearby positive walkers. This acceleration is strengthened at each iterate. Now it is possible to move all of the walkers together in a correlated way so as to retain the chance of cancellation until they move close enough so that the aggregate kernel is everywhere positive. This appears to be a promising possibility for future work.

ACKNOWLEDGEMENTS

This article is based on excellent and invaluable notes taken by David P. Murdock during the lectures at Sewanee College. It is a pleasure to acknowledge his help and that of Liane Strantz and Karl Runge in the preparation of the manuscript. The work was supported by Oakridge National Laboratory and the Cornell Theory Center, funded in part by the NSF, New York State, and IBM.

REFERENCES

1. P.A. Whitlock and M.H. Kalos, *Monte Carlo Methods, Volume 1: Basics*, John Wiley. New York (1986).
2. J. Carlson and M.H. Kalos, "Mirror Potential and the Fermion Problems", Physical Review C, Vol. 32, 1735 (1985).
3. M.H. Kalos, Monte Carlo Methods in Quantum Problems, Proceedings of a NATO Advanced Research Workshop. Reidel, Dordrecht (1984).
4. M.H. Kalos, "Optimization and the Many-Fermion Problem" in Monte Carlo Methods in Quantum Problems, Reidel, Dordrecht (1984).
5. K.E. Schmidt and M.H. Kalos, "Few-and Many-Fermion Problems", Monte Carlo Methods in Statistical Physics, Vol. 2, K. Binder (Ed.), Springer-Verlag, 1984.
6. D.M. Arnow, M.H. Kalos, M.A. Lee, and K.E. Schmidt, "Green's function Monte Carlo for few fermion problems", J. Chem. Phys., 77, 5562 (1982).

RELATIVISTIC HEAVY ION REACTIONS

David M. Brink

Department of Theoretical Physics, 1 Keble Rd., Oxford OX1 3NP, UK

ABSTRACT: The theory of quantum chromodynamics predicts that if nuclear matter is heated to a sufficiently high temperature then quarks might become deconfined and a quark-gluon plasma could be produced. One of the aims of relativistic heavy ion experiments is to search for this new state of matter. These lectures survey some of the new experimental results and give an introduction to the theories used to interpret them.

I. INTRODUCTION

According to the conventional model of nuclear structure normal nuclear matter is made of neutrons and protons. Each nucleon is composed of three quarks. A proton consists of the combination (uud) of two u-quarks each with charge number 2/3 and one d-quark with charge number -1/3. A neutron is a (udd) combination. Free quarks have not been found in nature, and no reaction has been observed in which a nucleon is broken up into its constituent quarks. Quantum chromodynamics explain this by the confinement hypothesis. Quarks are confined in nucleons because of colour symmetry. The quarks carry a colour quantum number which takes three values, but only the colour singlet combination can occur in nature. This symmetry restriction does not allow free quarks but it does allow the quark combinations which correspond to the known hadrons.

The colour symmetry restriction should also allow quarks to form larger clusters. There is no reason, in principle, why the 12 quarks in the 4 nucleons of a ^{4}He nucleus should not clump together rather than remain confined inside the individual neutrons and protons. It just does not seem to happen in normal nuclei. The theory of QCD predicts that at a sufficiently high temperature there should be a deconfinement phase transition. Above the transition point the quarks would leave the nucleons and move in a common region of space to form some kind of quark matter. The deconfinement transition is somewhat analogous to the ionization of ordinary matter at a high enough temperature. Ordinary ionized matter is called a plasma and its properties are governed by the balance between the free charges which produce electromagnetic fields and the forces on the charges due to the fields. Deconfined quark matter is often called a quark-gluon plasma and there are interactions between colour charges and colour fields analogous to those in the electrical plasma. Gluons in QCD are the analogues of photons in quantum electrodynamics. They are the quanta of the QCD field.

One of the aims of relativistic heavy ion reactions is to produce the quark-gluon plasma. In a high energy reaction the nucleons in the nuclei involved interact very strongly and a lot of energy of relative motion is transformed into excitation energy. The temperature could rise to a high enough value for the quark deconfinement transition to occur. The deconfined phase is expected to last only for a limited time. The quark-gluon plasma would expand and cool and eventually transform back into ordinary hadronic matter. At the hadronization or "freeze-out" stage many particles would be produced, nucleons, pions, kaons, deltas etc. The experimental problem is to find some signal of the formation of the deconfined phase. How can an intermediate state, which consists of some sort of deconfined quark matter, be distinguished from one consisting of ordinary hadronic matter? One possibility is that the energy and momentum distribution of the fragments could be changed by the formation of the plasma, another is that there might be an enhancement or suppression of the formation of certain products, a third is that the spacial or temporal extension of the intermediate stage might be affected.

The current experimental status of relativistic heavy ion reactions was reviewed at the Quarkmatter '88' conference by Braun-Munzinger and Stachel [1]. At present there are six large experimental set-ups at the CERN SPS accelerator and three at the BNL AGS complex. The CERN accelerator produces beams of ^{16}O and ^{32}S nuclei with energy per nucleon $E_{lab}/A = 60$ GeV and 200 GeV. At Brookhaven the projectiles are the same as at CERN and the energies are $E_{lab}/A = 14.5$ GeV. Because of the high energies the results of experiments measure variables which are expressed in terms of relativistic quantities like light cone variables, rapidity and transverse momentum. These are defined in section 2 of these notes and some of the experiments are described in section 3.

The detectors in an experiment see only the final products of a reaction. How is it possible to find out if a quark-gluon plasma was produced as an intermediate stage? One way to proceed is to make the hypothesis that there is no exotic quark matter stage in a relativistic heavy ion reaction, but that the collision is just a superposition of independent nucleon-nucleon collisions. If a model constructed with this hypothesis fits the experimental data then the experiment gives no evidence for the formation of a quark-gluon plasma. Several phenomenological models of multiparticle production based on the hypothesis of independent nucleon-nucleon collisions have been developed and their predictions have been compared with experimental data. There is an introduction to one of these models, the LUND model, in section 4 of these notes.

If a quark-gluon plasma is formed then models are needed to predict the consequences of its formation. The models used at present assume that the reaction proceeds in three stages. In the first stage the plasma is formed. Then it expands and cools. Finally it breaks up into the reaction products. The expansion and cooling are normally studied using a hydrodynamic model. We need an equation of state and a set of initial conditions in order to solve the hydrodynamic equations. An introduction to the equation of state of the plasma is given in section 5 and the hydrodynamical equations together with the initial conditions are discussed in section 6. The various factors which influence the hadronization stage are considered in section 7.

I will introduce two more specialized topics at the end of the course. One is the production of J/ψ particles in relativistic heavy-ion reactions and the other is the investigation of the spacial and temporal extension of the intermediate stage of the reaction by pion interferometry.

2. SOME DEFINITIONS

This section contains definitions of some quantities which are used in studies of relativistic heavy ion collisions. When discussing this topic it is the custom to use units where the velocity of light is $c = 1$. A particle with rest mass m moving with speed v has a total energy $E = \gamma m$ where $\gamma = 1/\sqrt{1-v^2}$. The z-axis is chosen to be parallel to the direction of the incident beam. The components of momentum parallel and perpendicular to this direction (p_z, p_T) are related to the corresponding velocity components by $p_z = \gamma m v_z$ and $p_T = \gamma m v_T$. The relation between the energy and momentum can be written as

$$E^2 = m^2 + p_T^2 + p_z^2 = m_T^2 + p_z^2. \tag{1}$$

The transverse mass m_T of a particle is related to its rest mass m by $m_T^2 = m^2 + p_T^2$. Both m_T and p_T are invariant for Lorentz transformations parallel to the z-axis.

The results of many relativistic heavy-ion experiments are expressed in terms of rapidities. The rapidity y of a particle is defined by

$$y = \tfrac{1}{2} \ln \frac{E + p_z}{E - p_z} = \ln \frac{E + p_z}{m_T} = \tfrac{1}{2} \ln \frac{1 + v_z}{1 - v_z} \tag{2}$$

It is related to the z-component of the particle velocity by

$$v_z = \tanh y.$$

Eq.(2) leads to some simple relations between E, p_z, m_T and y

$$E + p_z = m_T e^{y}, \qquad E - p_z = m_T e^{-y}, \tag{3}$$

$$E = m_T \cosh y, \qquad p_z = m_T \sinh y. \tag{4}$$

Equation (4) shows that the rapidity corresponds to an imaginary rotation angle in the space of 4-vectors. The relativistic velocity addition formula has a very simple form when written in terms of rapidity. A frame of reference S' moves with velocity V in the z-direction relative to a frame S. The rapidity y' of a particle in the frame S' is related to its rapidity y in S by

$$y' = y + Y$$

where $V = \tanh Y$. The analogy with rotations clarifies this addition property of rapidities. The Brookhaven AGS accelerator produces heavy-ion beams with momentum per nucleon $p_{lab}/A \approx 14.5$ GeV/c. The rapidity of ions in such a beam is $y \approx 3.4$. Beams with $p_{lab}/A = 60$ or 200 GeV/c at the CERN SPS have rapidities $y = 4.8$ or 6.0 respectively. The rapidity and other relativistic quantities are discussed in more detail in the book on relativistic kinematics by Byckling and Kajante [2].

The momentum components p_z ant p_T of a charged particle can be measured by deflection in a magnetic field. But it is also necessary to know either its energy or its rest mass in order to evaluate its rapidity. The mass is known if the particle can be identified but particle identification in an experiment is not always possible. For this reason experimental results are often expressed in terms of the pseudo-rapidity η defined by

$$\eta = \ln(\cot\theta/2) \tag{5}$$

where $\theta = \tan^{-1}(v_T/v_z)$ is the angle between the direction of the particle velocity and the z-axis. The rapidity and pseudo rapidity are almost equal provided the particle is highly relativistic ($E >> m$) and θ is not too small. We have $E \approx p$ and $p_T \approx m_T$ provided $\gamma \sin\theta >> 1$. When these conditions are satisfied

$$y \approx \ln\frac{p + p_z}{p_T} = \ln\frac{1 + \cos\theta}{\sin\theta} = \ln(\cot\theta/2) = \eta.$$

Light-cone variables are sometimes used in calculations (eg. the Lund model). The light cone momenta (p_+, p_-) are defined as

$$p_\pm = E \pm p_z \tag{6}$$

and are related by $p_+ p_- = m_T^2$. In a nucleon-nucleon collision

$$a + b \rightarrow c + X \tag{7}$$

where b is the beam particle and a is the target particle , the nucleons a and b produce a particle c together with other fragments X. We use the notation $E = a_0, p_z = a_z$ for the energy and z-component of momentum for the particle a and similarly for b and c. The light-cone variable for projectile fragmentation x_+ is defined as

$$x_+ = \frac{c_0 + c_z}{b_0 + b_z} \tag{8a}$$

and the light-cone variable for target fragmentation x_- is defined as

$$x_- = \frac{c_0 - c_z}{a_0 - a_z}. \tag{8b}$$

Both x_+ and x_- are relativisticly invariant quantities. They are related to the rapidities of the particles a, b and c by

$$x_+ = \frac{m_{cT}}{m_N} \exp(y_c - y_b) \tag{9}$$

$$x_- = \frac{m_{cT}}{m_N} \exp(y_a - y_c) \tag{10}.$$

In these equations m_{cT} is the transverse mass of c and m_N is the nucleon mass.

The energy of a reaction like (7) is sometimes specified by the invariant quantity s defined by

$$s = (E_a + E_b)^2 - (p_a + p_b)^2.$$

It is equal to the square of the total energy in the centre-of-mass system because then $(p_a + p_b) = 0$. In the laboratory system $p_b = 0$, $E_b = m_b$ and $E_a^2 - p_a^2 = m_a^2$. Hence

$$s = m_a^2 + m_b^2 + 2m_b E_{lab}.$$

A 200 GeV proton has $\sqrt{s} = 19.6$ GeV.

Relativistic heavy ion experiments often measure a quantity called the transverse energy defined as

$$E_T = \sum_i E_i \sin\theta_i. \tag{11}$$

where E_i is the energy of the ith reaction product and θ_i is the angle between its path and the z-axis. The measured E_i for a meson is its total energy, while for a baryon it does not include the nucleon rest mass. If the products are highly relativistic then the transverse energy is approximately the same as the transverse momentum, $E_i \sin\theta_i \approx p_{Ti}$. In this limit E_T is approximately invariant for Lorentz transformations parallel to the z-axis. The transverse energy gives a measure of the energy deposition in relativistic collisions. There is a kinematical limit [3]

$$\begin{aligned} E_{Tmax} &= \sqrt{s} - (\tilde{A} + \tilde{B})m_n \\ &= [(\tilde{A}^2 + \tilde{B}^2 + 2E_b\tilde{A}\tilde{B}/m_n)^{\frac{1}{2}} - \tilde{A} - \tilde{B}]m_n. \end{aligned} \tag{12}$$

where E_b is the laboratory energy of the projectile, m_n is the nucleon rest mass and $\tilde{A}$ and $\tilde{B}$ are the number of participants in the target and projectile respectively. This limit corresponds to all the reaction products moving in the transverse direction in the centre-of-mass system. In the case of an incident beam with an energy of 200 GeV per nucleon the kinematical limit is $E_{Tmax} = 17.5$ GeV for protons on a proton target, 47 GeV for protons on a lead target and 500 GeV for a central collision of an oxygen beam on a lead target (taking $\tilde{A} = 56$. CERN experiments measure values of E_T up to 60% of E_{Tmax} while some BNL experiments measure $E_T \approx E_{Tmax}$ [3].

3. EXPERIMENTS

At the present time there are several large experiments running at CERN and at the BNL AGS complex which aim to study relativistic heavy-ion collisions. A recent survey of these experiments was given at the Quark Matter 1988 Conference by Braun-Munzinger and Stachel [1]. The information in this section is taken from their paper and from more detailed reports of individual groups. The experimental set-ups include large detectors to measure the overall energy flow of reaction products together with more specialized detectors for particle identification, correlation measurements, and so on. As an example I will discuss the CERN WA80 experiment. Details of the experiment together with some early results are given in the proceedings of the Quark Matter 1987 Conference [4,5].

The WA80 set-up includes two calorimeters, the Mid-Rapidity Calorimeter (MIRAC) and the Zero-Degree Calorimeter (ZDC). MIRAC measures the transverse energy produced in the reaction on an event-by-event basis and covers the pseudo-rapidity range $2.4 < \eta < 5.5$. The ZDC is located 11m from the target and measures the total energy of particles which pass through the beam hole in MIRAC. It collects charged reaction products with $\theta_i < 0.3$ deg corresponding to a pseudo-rapidity $\eta > 6$. An important aspect of a relativistic heavy-ion collision is that the collision geometry is determined by the sizes of the target and projectile nuclei, the impact parameter and the overlap volume. The ZDC energy spectrum of an event can be used to give an indication of the impact parameter. Consider for example a $^{16}O + ^{197}Au$ collision. Events with low ZDC energies result from central collisions where the oxygen projectile is stopped by the gold target nucleus and produces only a few product particles at angles less than 0.3 deg. Events with large ZDC energies are due to peripheral collisions. In such a collision only a few nucleons in the oxygen projectile interact with the target. The remainder continue in the forward direction and are collected by the ZDC [4].

A unique feature of WA80 is the nearly 4π sr coverage for the measurement of the charged particle multiplicity. There are several detectors which measure the number of charged particles produced in an event in different angular regions. The target is surrounded by a Plastic Ball which detects charged particles in the angular range $\theta > 30$ deg ($\eta < 1.3$). A Large-Angle Multiplicity detector (LAM) covers the angular range $30 > \theta > 10$ deg, while the Mid-Rapidity Multiplicity detector (MIRAM) covers the range $13 > \theta > 1$ deg.

The beam rapidity in a CERN experiment at 200 Gev per nucleon is $y = 6$. Events with high rapidity ($y > 4.5$) are likely to be associated with remanants of the projectile. This is called the projectile fragmentation region. In a reaction with a light projectile and a heavy target some of the target nucleons will be participants in the reaction and some will be spectators. The spectators and some of the participants will give rise to reaction products with low rapidity ($y < 1.5$). These products belong to the target fragmentation region. The mid-rapidity region ($1.5 < \eta < 4.5$) or central region is considered to be the most interesting one. If a quark-gluon plasma or some other exotic form of matter is formed in the reaction then the clearest signals are likely to appear in the central region because the particles with intermediate rapidities have probably had the strongest interactions.

The typical shape of a transverse energy distribution is shown in fig. 1. The curves correspond to preliminary data from the NA35 collaboration [6] for S+S and S+Au collisions for a beam energy of 200 GeV per nucleon. The pseudo-rapidity range is $2.2 < \eta < 3.6$ which corresponds to the central region as defined in the last paragraph. Similar E_T distributions for different target and projectile distributions and different pseudo-rapidity ranges have been obtained by other groups [6,7,8,9,10].

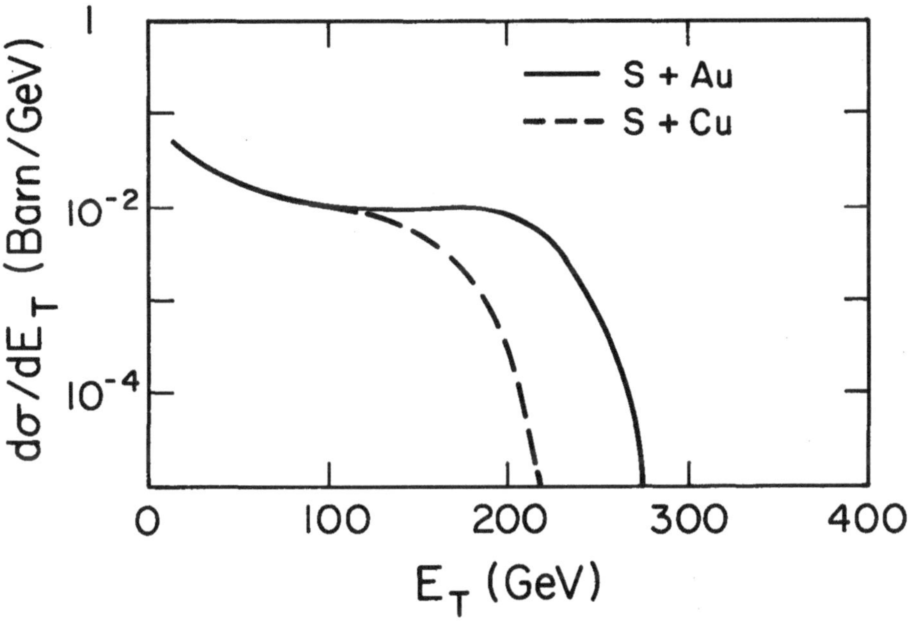

Fig.1 Sketch of typical transverse energy distributions for two reactions. The scales correspond to NA35 collaboration expreiments at 200 GeV/u for a rapidity interval $2.2 < \eta < 3.6$.

To a first approximation the shape of the E_T distribution corresponds to that expected if the nucleons in the projectile interact with the target independently. The transverse energy differential cross-section can be estimated by folding the transverse energy distributions of the individual nucleon-nucleus collisions [11,12]. The low E_T part of the distribution corresponds to glancing collisions with large impact parameter, where only a few nucleons enter the target. The large E_T limit corresponds to head-on collisions. The agreement between the experimental results and the folded distributions is reasonably good. There are, however, discrepancies in the details which might be significant [13].

In his opening talk at the Quark Matter '88' conference Jacob [14] pointed out that the first aim of the relativistic heavy-ion experiments was to obtain large excitation energies. As discussed in section 2 the transverse energy E_T gives a measure of the energy deposited. For 200 GeV/n O + Au and S + Au in the pseudo-rapidity interval $2.2 < \eta < 3.6$ the NA35 collaboration measures E_T at one percent of the plateau cross section to be $E_T^{tail} = 150$ GeV and 275 GeV. Using other data they estimate the corresponding numbers for the larger interval $-.1 < \eta < 3.6$ to be $E_T^{tail} = 245$ Gev and 405 GeV respectively. These values correspond to 0.63 and 0.59 of the kinematical limit discussed in the last section.

One striking result of the WA80 experiment is the similarity of the transverse energy (E_T) and charged particle multiplicity distributions for reaction products in the central rapidity region. The value of E_T per charged particle lies between 0.5 and 0.6 GeV for interactions between an oxygen projectile and various targets at 60 and 200 A GeV [15].

Each of the CERN experiments NA34(Helios),NA35, NA38 and WA80 has calorimeters operating in certain pseudo-rapidity ranges and at the same time measures some more specialized aspects of the reaction. For example Na35 measures transverse momentum (p_T) distributions of pions, and strange particles, and studies two-pion interferometry. The Na38 experiment has a di-muon detector to study the production of J/ψ particles. The WA80 experiment also contains a more specialized detector SAPHIR. It is an electromagnetic calorimeter consisting of 1278 lead glass modules that can detect photons which are produced by direct processes or by the decay of short-lived particles such as neutral pions. This detector has been used to measure transverse momentum (p_T) distributions of neutral pions which are identified by their decay photons ($\pi^0 \to 2\gamma$) [16]. The BNL accelerator operates at an energy lower than CERN but measures similar observables. The BNL E802 experiment has a powerful time of flight particle identification system. References to some of the recent results of these collaborations will be given later in the lecture notes.

4. INDEPENDENT NUCLEON-NUCLEON COLLISION MODELS

A benchmark for the interpretation of relativistic heavy-ion experiments is the comparison of experimental results with predictions of independent nucleon-nucleon collision models. The Lund Model [17,18,19] and the Dual Parton Model [20] are examples of extrapolations from the phenomenology of pp collisions to predictions about nucleus-nucleus collisions. The dual parton model has been extended and applied to the analysis of experimental results in a number of papers by Werner [21]. A simplified version of these multiple-collision models has been published recently by Wong and Lu [22]. These are all examples of "wounded nucleon string" models. The number of interactions is calculated using classical Glauber (straight-line) geometry. Interactions are assumed to lead to excitation of the nucleons into string-like configurations. Because of time dilation the strings fragment into hadrons on a long time scale and the excitation and fragmentation are treated as independent processes. The details of the fragmentation are adjusted to fit the properties of pp data.

Before discussing some of the details of the Lund model I will give some simple geometrical considerations which apply to all the multiple collision models. A nucleon-nucleus collision is described using straight line Glauber geometry. When the projectile passes through the nucleus it collides with and excites ν nucleons. If the incident nucleon has an impact parameter b a thickness function can be defined by

$$T_t(b) = \int \rho_t(b,z)\mathrm{d}z \tag{13}$$

where the nucleon density is normalized to unity. The average number of nucleons in the target participating in the reaction is

$$<\nu>= \sigma_{nn} A_t T(b) \tag{14}$$

where $\sigma_{nn} = 29.4$ mb [23] is the cross-section for a nucleon-nucleon collision and A_t is the number of nucleons in the target.

For the case of a nucleus-nucleus collision the average number of participants in the target is

$$W_t = A_t \int \mathrm{d}^2 s T_t(s)[1 - (1 - T_p(b-s)\sigma_{nn})^{A_p}]. \tag{15}$$

There is a similar expression for W_p, the average number of participants in the projectile. If the absorption is strong enough and the nuclear densities are assumed to be uniform with sharp surfaces these expressions can be simplified to give an estimate for the total number of participants

$$W = W_t + W_p = \int_{S_{eff}} \mathrm{d}^2 s[A_t T_t(s) + A_p T_p(s-b)] \tag{16}$$

where S_{eff} is the transverse area of overlap of the two nuclei. For a central collision with $A_p < A_t$ this expression gives

$$W(0) = A_p + A_t(1 - (1 - (A_p/A_t)^{2/3})^{3/2}). \tag{17}$$

Hadrons are treated as vortex lines in a superconducting vacuum in the Lund model. A vortex line consists of a hard core surrounded by a damped field. This field is equivalent to a field formed by coloured dipoles, lined up along the vortex line. In a nucleon-nucleon collision the fields overlap and the dipole links are treated as partons which may exchange momenta. Many incoherent momentum transfers will lead to a substantial excitation of the strings. The total transverse momentum exchange is assumed to be small and the important part is the longitudinal transferred momentum. After a collision the nucleons are assumed to be stretched out into excited strings with the same colour structure as the original strings. The Lund model assumes no colour exchange. In this it differs from the Dual Parton Model.

At the parton level the momentum transfer is calculated using the Feynman wee parton spectrum [18]. In terms of light-cone variables $p_+ = E + p_z$ and $p_- = E - p_z$ the probability distribution in the final state can be written as

$$P = \frac{dp_+^l}{p_+^l}\frac{dp_-^r}{p_-^r}$$

The superscripts indicate whether the particle is moving to the left or to the right. The corresponding p_+^r and p_-^l are given by momentum conservation. After the collision the masses of the two objects are calculated using $p_+p_- = m_T^2$.

The following simple physical picture is useful. An excited meson in the string model consists of a quark-antiquark pair connected by a coloured string. The total excitation energy oscillates between kinetic energy of the quarks and potential energy of the string. The string tension is a constant κ which is independent of the extension. A string of length l has a potential energy κl. The numerical value of the tension is $\kappa \approx 1$ GeV/fm or $\kappa \approx 0.2$ GeV^2. The conversion factor between these two numbers is $\hbar c = 197$ MeV fm ≈ 0.2 GeV fm. An excited nucleon is similar except that there are three quarks, a di-quark pair at one end of the string and a single quark at the other end. The di-quark pair has the properties of an antiquark.

The Lund hadronization model [17,27] is used to describe the fragmentation of the excited strings. Consider an excited meson represented by a string with a quark q at one end and an antiquark $\bar{q}$ at the other. The meson decays into a pair of mesons by breaking the string with the creation of a $\bar{q}_1 q_1$ pair at the point of scission. The two fragments consist of the strings $q\bar{q}_1$ and $q_1\bar{q}$. The meson can also decay into a baryon-antibaryon pair. In that case a quark pair and an antiquark pair are created at the scission point. String breaking is a quantum tunnelling process and the probability of scission is proportional to $\exp(-\pi\mu^2/\kappa)$ where μ is the meson mass and κ is the string tension [17]. Transverse momentum distributions are obtained by replacing μ^2 by the square of the transverse mass $m_T^2 = \mu^2 + p_T^2$ in this formula. Thus the transverse momentum has a universal form $\exp(-\pi p_T^2/\kappa)$ which depends on the string tension parameter κ. Predictions made with this formula do not agree well with the experimental data of the NA35 collaboration [6] and the model for the p_T distribution is clearly too simple.

String breaking is a dynamical process and momentum and energy can shared between the fragments in different ways. The Lund model allows a string to break into several fragments, but assumes that the scission points have a space-like separation. Then there is no causal connection between the breaks amd they can be considered as independent events. As no production point is more fundamental than another the fragmentation can be calculated by starting at either end of the string ang working along it. Starting with the meson produced by the break nearest the end of the string which contains the original quark q plus the antiquark $\bar{q}_1$ the fragmentation is discussed in terms of a light cone variable

$$z = \frac{(E + p_z)_{frag}}{(E + p_z)_{init}}$$

which is invariant for Lorentz transformations parallel to the z-axis. It is assumed to have a probability distribution $f(z)$. After the first break the remaining system is very similar. The leading quark q is replaced by q_1 and the process is repeated. This procedure leads to an integral equation for the fragmentation [17,27]. The form of the probability distribution $f(z)$ is determined by symmetry arguments to be

$$f(z) = N\frac{(1-z)^a}{z}\exp(-b\mu^2/z)$$

where μ is the meson mass and the three parameters N, a and b are related by a normalization condition. The implementation of this procedure is discussed in ref.[27].

The Lund model gives a good description of the charged particle multiplicity, transverse energy production and pion transverse momentum distributions in high energy nucleon-nucleon collisions. One simple feature of multiplicity distributions $P(n)$ in high energy nucleon-nucleon collisions is Koba-Nielsen-Olesen (KNO) scaling [25,26]. This states

$$< n > P(n) = \psi(n/ < n >)$$

where ψ is an energy independent scaling function. The Lund model reproduces this scaling behaviour [18]. Experimental evidence for this scaling law has been presented by Arnison et al [26]. They also point out that the average E_T per secondary particle is about 0.5 GeV, quite similar to the value in relativistic nucleus-nucleus collisions. KNO scaling in string models has been discussed recently by Werner [21]. He shows that a certain scaling of string masses leads to KNO scaling.

The Lund model describes a nucleon-nucleus collision as a succession of nucleon-nucleon collisions. When the projectile passes through the target nucleus it collides with ν nucleons. The projectile does not have time to fragment between subcollisions and after each subcollision the mass of the projectile increases. After the interaction we are left with $(\nu + 1)$ excited strings, ν from the target and one from the projectile. These strings fragment independently. The distribution of participating nucleons, ν, is calculated by the procedure described in ref [18]. It is obtained by assuming a Woods-Saxon distribution for the nuclear density and by using a straight line Glauber geometry.

In the Lund model the step to a nucleus-nucleus collision is straightforward. Both the projectile and target constituents are distributed according to Woods-Saxon density distributions and the two nuclei are made to interact at a random impact parameter. Using a straight line geometry the number of subcollisions is recorded and the excitation of the participating nuclei is calculated. The excited nucleons then decay according to the Lund Monte Carlo program for jet fragmentation [27]. The Lund model is implemented in a Monte Carlo programme called FRITIOF. (Fritiof is the hero in a Scandanaviain saga. His adventures are recorded in a poem by Esias Tenger. Various subroutines in the programme are named after other characters in the programme. For example INGEBORG is his fiancee and ANGURVADEL is his sword.)

FRITIOF is an event generator.The output of each run records the results of one collision. It gives the number of participants and the impact parameter of the collision and then goes on to list the particles produced together with their energy and the components of their momenta. One example of a peripheral collision given in ref [19] has three participants and produces 38 secondary particles. The event produces 13 pions, 14 direct gamma rays as well as kaons,deltas, and rho and eta mesons. The short lived particles will decay before they are detected and the program has an option to allow the decay. Photon production in the Lund model has been discussed in a recent article by Anderson et al [28].

Wounded nucleon string models like FRITIOF provide baselines to which new data can be compared. It is hoped that new physics can be identified by systematic deviations of the experimental data from such baselines. Predictions of the Lund model have been compared with experimental data in a number of papers. Sorenson et al. [4] show that the predictions of the transverse energy spectra are qualitatively correct but that there are systematic deviations at the high energy end. Werner [21] finds a similar effect in his calculations.

5. THE QUARK-GLUON PLASMA

In normal nuclear matter quarks are confined inside hadrons because of restrictions on the colour degree of freedom. Calculations based on quantum chromodynamics suggest that colour should no longer be confined at sufficiently high temperatures or at high enough quark density. Quark deconfinement probably requires temperatures in excess of 200 MeV at low quark density or densities in excess of five times the nuclear density at low temperature. It is hoped that such high temperatures or densities can be reached in high energy heavy-ion collisions.

The most reliable information about deconfinement parameters comes from lattice gauge theory calculations. The situation in 1987 was reviewed by Karsch at the Quark Matter '87' conference [29]. These calculations show a sharp phase transition when the QCD parameter $\beta = 6/g^2 \approx 5.5$. At this point the ratio ε/T^4, where ε is the energy density and T is the temperature, changes discontinuously by an order of magnitude. The critical temperature is found to be of the order of 200 Mev. This corresponds to an energy density needed for plasma formation $\varepsilon \approx 2\text{Gev/fm}^3$.

Theoretical studies of the possibility of forming a quark-gluon plasma in high-energy heavy-ion collisions require an equation of state. In the absence of something more fundamental we can choose some simple phenomenological form for this equation. The theory of quarks and gluons in QCD contains two fundamental features, asymptotic freedom and confinement. The two features are incorporated in a simple way in the MIT bag model and in 1982 Satz [30] was able to show that the SU(2) Yang-Mills system, above the critical temperature for colour deconfinement, approaches asymptotically free behaviour in accord with the bag model.

We assume that hadronic matter at a temperature near the critical temperature $T_c \approx 200$ MeV for the quark gluon phase transition consists mainly of pions and that at those temperatures the pion is highly relativistic so that its rest mass can be neglected. If the pion gas is electrically neutral then its entropy density is given by the black body radiation formula

$$s_\pi = c_\pi T^3 = 4\frac{g_\pi \pi^2}{90}(T/\hbar c)^3. \tag{18}$$

Often theoretical papers use units with $\hbar c = 1$. The constant $g_\pi = 3$ refers to the number of kinds of pions (π^+, π^0 and π^-). Eq.(18) is obtained from the formula for the entropy of a system of non-interacting bosons

$$S = \sum[(n_i + 1)\ln(n_i + 1) - n_i \ln n_i]$$

where n_i is the number of bosons in the ith quantum state. Using a Bose-Einstein distribution for n_i and assuming a zero chemical potential this can be written as

$$S = \int \rho(\varepsilon)\mathrm{d}\varepsilon[\frac{\varepsilon/T}{\mathrm{e}^{\varepsilon/T} - 1} - \ln(1 - \mathrm{e}^{-\varepsilon/T}].$$

For bosons with zero rest mass moving in a volume V the density of states

$$\rho(\varepsilon) = \frac{Vg}{2\pi^2}\frac{\varepsilon^2}{(\hbar c)^3}.$$

Evaluating the integral gives the entropy density

$$s = S/V = \frac{2g}{3\pi^2}(T/\hbar c)^3 I_3.$$

The integral

$$I_n = \int \frac{x^n \mathrm{d}x}{\mathrm{e}^x - 1} = n!\zeta(n+1)$$

where ζ is the Riemann zeta function($\zeta(2) = \pi^2/6 = 1.645$, $\zeta(3) = 1.202$ and $\zeta(4) = \pi^4/90 = 1.082$).

The number of bosons in a system is

$$N = \int \frac{\rho(\varepsilon)\mathrm{d}\varepsilon}{\mathrm{e}^{\varepsilon/T} - 1}$$

which gives a number density

$$n = N/V = \frac{g}{2\pi^2}(T/\hbar c)^3 I_2.$$

Hence the number density of pions is related to the entropy density by

$$n_\pi = \frac{\zeta(3)}{4\zeta(4)} s_\pi = s_\pi/3.6 \tag{19}$$

while the pressure and the energy density are

$$p_\pi = \varepsilon_\pi/3 = s_\pi T/4. \tag{20}$$

These relations fail badly when the temperature falls below the pion rest mass ($\approx$ 140 MeV). This equation of state was used by Landau in his 1953 paper on relativistic hydrodynamics [31].

The above relations are changed in the quark-gluon plasma phase in two ways. The coefficient in eq.(18) for the entropy is changed because the fundamental particles are now quarks and gluons rather than pions. The factor g_π is replaced by

$$g_{qg} = g_g + (7/8)g_q \tag{21}$$

where g_g and g_q are the numbers of kinds of quarks and gluons. The factor (7/8) is due to the Fermi statistics of the quarks. For quarks with zero rest mass and zero chemical potential the integral I_3 in the expression for the entropy density is replaced by

$$J_3 = \int x^3 \mathrm{d}x/(\mathrm{e}^x + 1) = (7/8)I_3.$$

If there are two quark flavours and eight gluon colours then $g_{qg} = 37$ ($g_g = 16$ and $g_q = 24$). Hence

$$s_{qg} = c_{qg}T^3 = \frac{148\pi^2}{90}(T/\hbar)^3. \tag{22}$$

Confinement in the bag model is achieved by giving the QCD vacuum a constant energy density B which restricts the quarks and gluons to small regions of space. The new expression for the energy density is

$$\varepsilon_{qg} = (3/4)c_{qg}T^4 + B \tag{23}$$

and the pressure is

$$p_{qg} = (1/4)c_{qg}T^4 - B.$$

These two expressions can be derived from a free energy

$$F = -(1/4)c_{qg}T^4V + BV$$

where V is the plasma volume by using the thermodynamic relations

$$S = -(\partial F/\partial T)_V, \qquad P = -(\partial F/\partial V)_T.$$

A droplet of plasma in equilibrium at ($P = 0$ has a temperature

$$T = T_c = (4B/c_{qg})^{1/4}. \tag{24}$$

Such a droplet has an entropy $S = c_{qg}VT_c^3$, an energy $E = ST_c$ and a volume $V = E/4B$. It could decay by evaporating pions. As it evaporates it shrinks in size but its temperature remains constant. Numerical estimates can easily be obtained from these formulae. A quark-gluon plasma with a temperature $T = T_c \approx 200$ MeV has $T/\hbar c \approx 1\ \text{fm}^{-3}$. In this case the entropy density is $s \approx 16\ \text{mf}^{-3}$ and the energy density is $\epsilon \approx 3.2\ \text{fm}^{-3}$. The entropy density of a pion gas at the same temperature is $s \approx 1.3\ \text{fm}^{-3}$ and is much less than for the plasma because the degeneracy factor g is much smaller.

The expressions for F, S, and E for small dropets of quark-gluon plasma are modified by the inclusion of surface terms. For example the expression for the free energy for a finite drop can be expanded as

$$F(T) = F_0(T)V + F_1(T)A + F_2(T)C + \dots \tag{25}$$

where A is the surface area of the droplet and C is a measure of the average curvature of the surface. These surface corrections could have important consequences for the behaviour of droplets of quark-gluon plasma. For example the temperature of a droplet at zero pressure would depend on its size. Thus the temperature of droplet would change, rather than remaining constant, as it loses energy by evaporating hadrons.

The contributions to F_1 and F_2 could have an intrinsic or a dynamic origin. Farhi and Jaffe [32] argue that the intrinsic contribution to F_1 should be small at $T = 0$. The dynamic contributions to F_1 and F_2 are due to the surface corrections to the density of states of quarks and gluons in a finite droplet. It turns out that the term proportional to A in the density of states of massless quarks and gluons is zero and that the curvature term is the leading correction. The density of states of non-interacting gluons with boundary conditions corresponding to a perfect chromomagnetic conductor is the same as for the density of states of the electromagnetic field in a cavity with conducting walls. Balian and Bloch give a formula for the electromagnetic case [33]. The expression for the density of states with surface corrections for massless quarks is quoted in a convenient form by Farhi and Jaffe [32]. The density of states in wave number k is

$$\rho(k) = g\left(\frac{Vk^2}{2\pi^2} + C_s kA + \frac{C_R}{8\pi}\int d^2s\left(\frac{1}{R_1} + \frac{1}{R_2}\right) + ..\right).$$

Here V, A, R_1 and R_2 are the volume, surface area and radii of curvature and g is the degeneracy factor. For zero mass non-interacting quarks $C_s = 0$, $C_R = -\pi/6$ [32], while for massless gluons $C_s = 0$, $C_R = 2/3\pi$ [33]. The leading correction to the density of states for massive quarks is proportional to A and a formula is given by Berger and Jaffe [34]. Thus if quarks have a non-zero mass there will be a dynamic contribution to the surface tension of a quark matter droplet.

6. HYDRODYNAMICS

In 1953 Landau [31] proposed a theory of relativistic hydrodynamics to account for phenomena occuring in collisions initiated by high energy cosmic ray particles. He argued that at the moment of collision a large number of particles are created which are concentrated in a small volume. The mean free path of the particles in such a system is very small compared with its dimensions. In the course of time the system expands but the mean free path in the early stages remains small. Each element of the system remains in local thermal equilibrium and this part of the expansion has a hydrodynamic character. Since the velocities in the system are comparable with the velocity of light relativistic hydrodynamics have to be used.

The total number of particles is not constant during the hydrodynamic stage of the expansion. Hence the number of particles emerging from the collision is determined not by the number created at the instant of the collision but by the number produced at the second stage of the expansion; that is the stage of free separation of the particles. This is the hadronization or the "freeze-out" stage of the reaction. At the time when he developed his theory Landau had in mind a fluid consisting mainly of pions. The freeze-out stage occurs when the temperature of the fluid drops below the pion rest mass. Then the equilibrium density of pions falls off like $\exp(-m_\pi/T)$, the mean free path becomes relatively large and hydrodynamics are no longer applicable. Landau pointed out that the calculation of the number of pions appearing at the break-up is simplified if the motion of the fluid is assumed to be adiabatic. Then the entropy in each individual region of the fluid remains constant during the hydrodynamic expansion and the number of pions produced can be calculated from the relation (19) between the pion density and the entropy density.

Now we are interested in the dynamics of the quark-gluon plasma rather than a pionic fluid. We can still use hydrodynamics provided the mean-free path of quarks and gluons is small enough so that the plasma is in local thermal equilibrium. The hydrodynamic expansion of the quark matter fluid depends not only on the equations of motion but also on the initial conditions. Two forms of the initial condition have been proposed. One of these corresponds to the initial condition used by Landau in his 1953 paper and is called the Landau stopping condition. An extensive discussion of this initial condition was given by Feinberg at the Quarkmatter '87 conference [35]. This boundary condition may apply to the lower energy BNL experiments [36]. The second is known as the Bjorken scaling condition and was proposed by Bjorken [37] as being more appropriate for highly relativistic nucleus-nucleus collisions.

In the Landau stopping domain the participating nucleons form a highly flattened disk at the instant of the collision. The matter in the disk is initially at rest but is under high pressure and expands rapidly. The disk shape is maintained during the first stage of the hydrodynamic flow and the expansion is at first one dimensional along the beam direction. It may become three dimensional at a later stage.

Bjorken has a different scenario. The two nuclei have the shape of highly

Lorentz contracted pancakes in the centre-of-mass frame of the participating nucleons. The two pancakes interpenetrate so that after a short time we have two pancakes which recede from the collision points with a speed near the speed of light. These pancakes contain most of the baryon number of the initial participating nucleons. The region between the pancakes is called the central rapidity region. It contains a fluid consisting of the other products of the nucleon-nucleon collisions. The fluid in the central region is assumed to have a boost invariant flow pattern. Exact boost invariance would require the fluid to extend from $-\infty$ to $+\infty$ in both the space cordinate in the z-direction as well as in rapidity. The fluid is finite in extent so boost invariance can only be an approximation which should be best in the central rapidity region. In a situation where boost invariance is exact there is a relation between the transverse energy per unit rapidity range dE_T/dy and the initial energy density. For a finite system this leads to an approximate relation [37]

$$\varepsilon_0 \approx (dE_T/dy)_{max}(\tau_0 \pi R^2)^{-1}$$

where τ_0 is the equilibration time for the plasma and R is its radius. A typical measured value for $(dE_T/dy)_{max}$ for the reaction $^{16}O + {}^{197}Au$ at CERN energies is 60 GeV. Estimates for the other parameters are $\tau_0 = 1$ fm/c and $R = 3$ fm. These numbers give $\varepsilon = 2.1$ fm^{-3}. This is less than the estimate $\varepsilon = 3.2$ in the equilibrium plasma at $T_c = 200$ MeV, but these numbers are estimates and in any case in a finite system the phase transition would not be sharp. Thus deconfinement could already be occuring in the present experiments.

Now we write the hydrodynamic equations assuming that there is no dissipation. We follow the argument in Cleymans et al [38] and details can be found there. We also specialize to the case of a baryon-neutral fluid so that the baryon chemical potential is zero. The form of the energy momentum tensor

$$T^{\sigma\tau} = (\varepsilon + p)u^\sigma u^\tau - g^{\sigma\tau} p \tag{26}$$

follows from the assumption that it is diagonal and spatially isotropic in a local rest frame. In eq.(26) ε and p are the energy density and pressure, and u^σ is the fluid velocity 4-vector. The equations of motion are just the energy and momentum conservation equations

$$\partial_\mu T^{\mu\nu} = 0. \tag{27}$$

The entropy conservation equation is obtained by contracting eq.(26) with the velocity 4-vector

$$u_\nu \partial_\mu T^{\mu\nu} = 0. \tag{28}$$

Using the fact that $u_\mu u^\mu = 1$ and $u_\nu \partial_\mu u^\nu = 0$ one obtains

$$u^\nu \partial_\nu \varepsilon + (\varepsilon + p)\partial_\nu u^\nu = 0. \tag{29}$$

The energy density, pressure, entropy density s, temperature T, baryon chemical potential μ_B and the baryon density n_B are related by

$$\varepsilon + p = Ts + \mu_B n_B.$$

When the baryon chemical potential is zero this reduces to

$$\varepsilon + p = Ts \tag{30}$$

and we have $d\varepsilon = Tds$. Using these equations to eliminate ε and p in favour of s we get the entropy conservation relation

$$\partial_\nu (u^\nu s) = 0. \tag{31}$$

Cleymans et al. [38] show how to obtain the entropy conservation equation when the baryon chemical potential is non-zero. A second set of hydrodynamic equations is obtained by contracting eq.(26) with the tensor $g_{\mu\nu} - u_\mu u_\nu$. This leads to the following set of equations

$$(\varepsilon + p)u^\mu \partial_\mu u_\nu - \partial_\nu p + u_\nu u^\mu \partial_\mu p = 0.$$

Again eliminating ε and p in favour of s and T we get

$$u^\mu \partial_\mu (u_\nu T) - \partial_\nu T = 0.$$

There is a big simplification if we specialize to the case where the flow is one dimensional in the z-direction. We put

$$u^\mu = (u^0, u^1) = (\cosh\theta, \sinh\theta). \tag{32}$$

Here θ is essentially the fluid rapidity. The hydrodynamic equations can be written in terms of the new variables as

$$\cosh\theta\partial\varepsilon/\partial t + \sinh\theta\partial\varepsilon/\partial z + (\varepsilon + p)(\sinh\theta\partial\theta/\partial t + \cosh\theta\partial\theta/\partial z) = 0$$

$$\sinh\theta\partial p/\partial t + \cosh\theta\partial p/\partial z + (\varepsilon + p)(\cosh\theta\partial\theta/\partial t + \sinh\theta\partial\theta/\partial z) = 0.$$

Next we change from the variables (t, z) to a light cone variable Y and the proper time τ

$$Y = \tfrac{1}{2}\ln\frac{t+z}{t-z}, \qquad \tau = \sqrt{t^2 - z^2}, \tag{33}$$

or equivalently

$$t = \tau\cosh Y, \qquad z = \tau\sinh Y. \tag{34}$$

The partial derivatives with respect to t and z can be transformed as

$$\partial/\partial t = \cosh Y\partial/\partial\tau - (1/\tau)\sinh Y\partial/\partial Y$$

$$\partial/\partial z = -\sinh Y\partial/\partial\tau + (1/\tau)\cosh Y\partial/\partial Y$$

Then the hydrodynamical equations can be written in the form [38]

$$\tau\partial\varepsilon/\partial\tau + \tanh(\theta - Y)\partial\varepsilon/\partial Y + (\varepsilon + p)[\partial\theta/\partial Y + \tanh(\theta - Y)\tau\partial\theta/\partial\tau] = 0, \tag{35}$$

$$\partial p/\partial Y + \tanh(\theta - Y)\tau\partial p/\partial\tau + (\varepsilon + p)[\tau\partial\theta/\partial\tau + \tanh(\theta - Y)\partial\theta/\partial Y]. \qquad (36)$$

The important input from physics is the initial condition. Bjorken's scaling solution corresponds to the initial condition

$$\theta(Y, \tau_0) = Y \qquad (37)$$

at the initial proper time τ_0. It is boost invariant in the sense that $\theta(Y + Y_0, \tau_0) = Y + Y_0$. These relations are preserved at later proper times and the hydrodynamical equations become very simple

$$\tau\partial\varepsilon/\partial\tau + \varepsilon + p = 0, \qquad \partial p/\partial Y = 0. \qquad (38)$$

The entropy equation takes a similar form

$$\tau\partial s/\partial\tau + s = 0$$

and has the solution

$$s = s_0\tau_0/\tau. \qquad (39)$$

Bjorken's solution can hold exactly for all z only if the fluid is infinite in extent and fills the whole region $-\infty < z < +\infty$. It is assumed to give a good approximation to the flow pattern in the central rapidity region.

7. HADRONIZATION

The equations of hydrodynamics are used to model the complex behaviour of the central region in a high energy heavy-ion collision. The material in this region might consist of a dense hadron gas or something more exotic like quark matter, but it is assumed to evolve like an ideal relativistic fluid until hadronization occurs. The fluid can break up in a number of ways. One possibility is that as it expands its temperature drops and the mean free path of the constituent particles becomes comparable with the dimensions of the system. Another is that the fluid becomes unstable and breaks up into droplets which then evaporate hadrons.

The first of these possibilities was the one discussed by Landau in his 1953 paper. The freeze-out of hadrons occurs at a particular point in the fluid when the local temperature drops below some critical value T_c. Landau had in mind a fluid consisting of pions and nucleons. He argued that when the temperature drops below the pion rest mass the pion density decreases dramatically and the mean free path becomes large enough for the constituent hadrons to be essentially free. This argument leads to a critical temperature comparable with the pion rest mass, $T_c \approx 140$ MeV. Hadronization does not occur at a definite time in the laboratory frame of reference because the freeze-out conditions are defined in the local rest frame at each point in the fluid. For example in a fluid expanding according to the Bjorken scaling law freeze-out would occur at the same proper time τ for each element of the fluid.

The possibility that a quark-gluon plasma fluid becomes unstable as it expands was considered by Bertsch et al. [39]. They assume that the plasma breaks up into droplets which they call globs. The instability occurs when the pressure of the fluid drops to $P \approx 0$. If the equation of state corresponds to the bag model discussed in section 5 then the critical temperature T_c is given by eq.(24). Each droplet has a collective velocity determined by the local velocity of the fluid at the point of its origin. Superimposed on this is a thermal motion at a temperature T_c. The droplets evaporate by emitting pions. As they evaporate they shrink in size and eventually disappear. They can also absorb pions. There are two limiting cases.

a. The rate at which the droplets separate is so slow that any given droplet can emit and absorb many pions and the conversion into pions is adiabatic. The number of pions emitted is determined by entropy conservation and is given by eq.(19). Using the relation between the internal energy of a droplet and its entropy, the total number of pions evaporated is $N_\pi = U/3.6T_c$, where U is the total internal energy of the droplets.

b. Evaporation takes place with little reabsorption. Then energy conservation determines the number of pions. Each pion is emitted with an average energy 2.7 T_c and $N_\pi = U/2.7T_c$.

Bertsch et al. assume a transition temperature $T_c = 200$ MeV. Then a droplet with a radius of 1 fm has an energy of 13 GeV. Such a droplet would evaporate a net total of about 18 pions if it emits and reabsorbs many pions (case a) and 24 if there is little reabsorption (case b). In case (a) a smaller number of pions evaporate but the droplets interact through the emission and reabsorption process. The extra energy goes into collective motion of the droplets.

The evaporation of pions from an uncharged droplet with a temperature T can be calculated using the black-body radiation law. The rate dW for pions to evaporate into a volume element d^3p of momentum space is given by the detailed balance formula

$$dW = \frac{3\sigma v}{(2\pi\hbar)^3}\frac{d^3p}{e^{E/T}-1}. \tag{40}$$

Here $\sigma \approx \pi R^2$ is the cross-section for a pion with speed v to be absorbed by the droplet of radius R. The factor 3 in eq.(40) is due to the three charge states of the pions. Integrating this formula and neglecting the pion rest mass ($m_\pi << T$) gives an emission rate

$$W = 0.366\sigma c(T_c/\hbar c)^3.$$

Bertsch et al. [39] use this formula to estimate the rate at which globs emit pions. They assume that a droplet with radius r_g absorbs any incident pion coming within a distance $r_g + d$ so that the cross section $\sigma = \pi(r_g + d)^2$ and $T_c = 200$ MeV. The numerical value of the evaporation rate is $W = 1.2$ c/fm for $r_g + d = 1$ fm. The total time for a glob of this size to evaporate would be about 15 fm/c or 5×10^{-23} s.

The distribution of transverse momentum p_T as a function of rapidity can be obtained by changing variables in the volume element

$$d^3p = 2\pi p_T dp_T dp_z = \pi dp_T^2 dp_z = \pi E dp_T^2 dy \tag{41}$$

where we have calculated

$$(\mathrm{d}p_z/\mathrm{d}y)_{p_t} = m_T \cosh y = E$$

from eq.(5). For a fixed y the distribution of p_T is

$$\mathrm{d}W = \frac{N p_T m_T \cosh y \mathrm{d}p_T}{e^{(m_T \cosh y)/T} - 1} \tag{42}$$

where N is a normalization constant. If $T >> m_\pi$ we can neglect the pion rest mass so that $p_T \approx m_T$.Then the average transverse momentum of the pions is

$$< p_T >= \frac{I_3}{I_2} \frac{T}{\cosh y} = \frac{2.7T}{\cosh y}. \tag{43}$$

The NA35 collaboration measured p_T distributions of negatively charged particles (mainly negative pions) in ^{16}O + Au collisions at 200 GeV/n and compared the results with those obtained from p + p collisions at the same energy. The p_T distributon (rapidity range $2 < y < 4$) for the p + p case is well fitted by a thermal source (42) with a temperature $T = 135 \pm 2$ MeV. The p_T spectrum in the O + Au case (rapidity range $2 < y < 3$) has a more complicated shape (Strobele [40]) with an excess of particles with low p_T. The spectrum has been fitted by assuming two thermal sources with different temperatures and intensities N_1 and N_2. The temperatures of the two sources are $T_1 = 43 \pm 6$ MeV and $T_2 = 153 \pm 5$ Mev with $N_1/N_2 = 0.23 \pm 0.07$. The origin of the low energy component is not well understood and could be a signal of some kind of interesting phenomenon. The NA35 collaboration has also investigated the p_T distributions of K_0, Λ and anti-Λ particles by measuring the charged decay products [6]. The p_T distributions of neutral pions has been studied by the WA80 collaboration [41]. They observe the pions by detecting their decay gamma rays in the SAPHIR detector. The p_T distribution of the neutral pions in the WA80 experiment is similar to the distribution for charged pions from p + p collisions but the data do not extend to low enough p_T values and cannot give any information about the low temperature source seen in the NA35 experiment.

8. J/ψ SUPPRESSION

The J/ψ particle is a bound state of c and $\bar{c}$ quarks. It is produced in high energy nucleon-nucleon collisions and can be detected by the decay branch J/$\psi \rightarrow \mu^+ + \mu^-$. Under normal conditions the $c\bar{c}$ are held together by a confining QCD interaction. Matsui and Satz [42] predicted that the $c\bar{c}$ interaction in a quark-gluon plasma would be reduced by the Debye screening of the colour charge and that under such conditions the J/ψ particle might not be bound. In a nucleon-nucleon collision parton-parton interactions produce $c\bar{c}$ pairs. The resonant interaction between the c and $\bar{c}$ leads to J/ψ production. If, however the $c\bar{c}$ production occurs in a quark-gluon plasma then the pair finds itself in a deconfining environment. If the temperature is high then the interaction is not strong enough to bind the pair and the J/ψ will not be produced. The c and $\bar{c}$ will follow separate trajectories and will eventually combine with normal quarks to form charmed mesons.

A number of conditions have to be satisfied for the formation of J/ψ to be suppressed. i) The J/ψ should not escape from the production region before the plasma is formed. ii) The temperature and density of the plasma should be high enough. At the same time the possibility of other non-plasma mechanisms of J/ψ suppression should be considered .

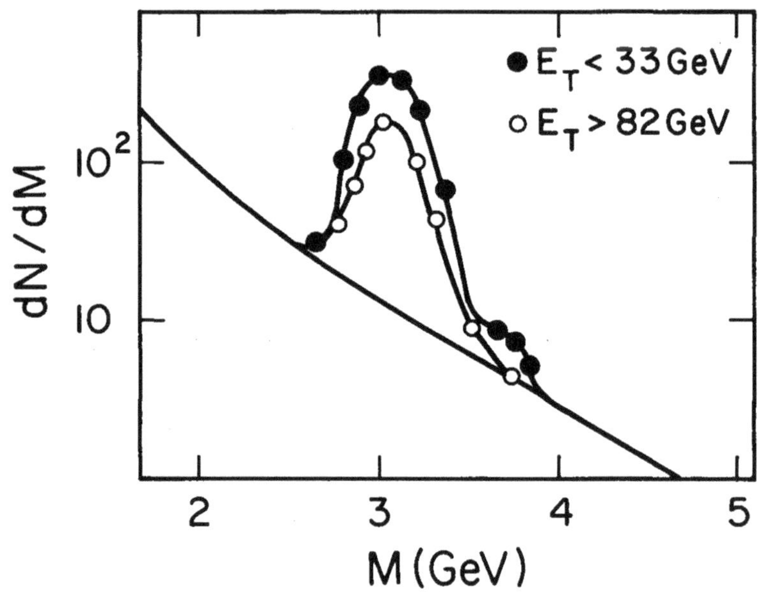

Fig. 2. Sketch of J/ψ production cross section. The open points correspond to central collisions (large E_T) and the solid points to peripheral collisions (small E_T). The data are normalized to the same Drell-Yan background.

The CERN NA38 experiment [43]was designed to investigate J/ψ suppression. The detector in the experiment consists mainly of a muon spectrometer and an electromagnetic calorimeter. The low probability of producing a muon pair in a hadronic collision requires a high luminosity,achieved by an intense beam and a thick target. The electromagnetic calorimeter measures the total transverse energy of the charged reaction products in the pseudo-rapidity range $1.7 < \eta < 4.1$. A quark-gluon plasma is more likely to be formed in a central collision rather than in a peripheral collision. Central collisions are characterized by large values of E_T. Hence the Debye screening argument of Matsui and Satz suggests a suppression of J/ψ production in events with large E_T as compared to those with small E_T.

A $\mu^+\mu^-$ pair detected in coincidence in a relativistic heavy-ion collision could have been produced in one of three different ways.

a) The pair could result from the decay of a J/ψ particle formed in the collision.

b) The two muons could have been produced by some other pair creation mechanism. One example is the Drell-Yan process in which the $\mu^+\mu^-$ pair is formed by a $(q, \bar{q})$ annihilation. Drell-Yan pairs with a large invariant mass would be formed in the initial stages of a collision and there is no reason why their formation should be affected by the formation of a quark-gluon plasma.

c) The μ^+ and μ^- might have been formed independently from the weak decays of pions or kaons. The cross section for this process can be calculated from measured $\mu^+\mu^+$ and $\mu^-\mu^-$ cross-sections.

Fig. 2 is a sketch of the $\mu^+\mu^-$ invariant mass spectrum for events selected according to the value of E_T. The cross section due to pairs produced independently (process (c)) has been subtracted out. The smooth background is assumed to be due to the Drell-Yan process and in the figure the curves for the two ranges of E_T have been normalized to give the same background. The peak in the distribution has the correct invariant mass (3 GeV) of a pair produced by the decay of the J/ψ. The width of the peak is due to the instrumental resolution of the detector. The results of the experiment show clearly that J/ψ particles are produced and that the production cross-section is smaller, by a factor of two, relative to the smooth background for the central collisions as compared with the peripheral collisions.

The J/ψ suppresion has also been measured as a function of p_T of the $\mu^+\mu^-$ pair. The suppression effect disappears gradually as the p_T value of the J/ψ increases. The experiment also shows that the Drell-Yan background has no such p_T dependence. This effect was predicted by Matsui and Satz [42]. A $c\bar{c}$ pair with large p_T escape from the plasma and bind to form a J/ψ when they get outside.

The results of the NA38 experiment are clear. There is a suppression of J/ψ relative to the Drell-Yan background for central collisions. No such effect is seen in proton-nucleus collisions. But do the results really demonstrate the existence of the quark-gluon plasma? It is too early to reach a definite conclusion. More theoretical work needs to be done to test alternative explanations. Several such explanations have been proposed [44] and there is even a suggestion that the plasma should enhance the J/ψ production [45].

9. PION INTERFEROMETRY

A high energy heavy ion collision goes through an intermediate state which might be a collection of excited colour strings or a quark-gluon plasma or a hadronic fluid. In each of these models there is a point where the intermediate state breaks up into independent hadrons. The study of $\pi^-\pi^-$ interferometry gives a way to measure the time scale of the intermediate state and the size of the system at freeze-out. Pion interferometry is a generalization of the Hanbury-Brown and Twiss method and uses the Bose nature of the pion to probe the spacial and temporal extension of the intermediate stage of a heavy ion collision. Some recent theoretical references are Pratt [46], Kolehmainen and Gyulassy [47] and Bertsch et al. [48].

The basic idea of pion interferometry can be explained simply in the case of a static monochromatic source. Then it can easily be extended to a time-dependent source. Pion radiation from an extended monochromatic source is observed with two detectors at points D_1 and D_2 a long way from the source and coincidences between pairs of negative pions are measured. The vectors $\mathbf{p}_1$ and $\mathbf{p}_2$ are wave vectors directed from the centre of the source to the detectors at D_1 and D_2. They correspond to the momenta of the pions detected ta D_1 and D_2. Pions are produced at points $\mathbf{s}_a$ and $\mathbf{s}_b$ in the source with amplitudes $A(\mathbf{s}_a)$ and $A(\mathbf{s}_b)$. The total amplitude for detection of two pions, one at D_1 and the other at D_2 is proportional to

$$A(1,2) = \frac{N}{\sqrt{2}} \int \mathrm{d}^3\mathbf{s}_a \mathrm{d}^3\mathbf{s}_b A(\mathbf{s}_a) A(\mathbf{s}_b) [e^{i(\mathbf{p}_1 . \mathbf{s}_a + \mathbf{p}_2 . \mathbf{s}_b)} + e^{i(\mathbf{p}_1 . \mathbf{s}_b + \mathbf{p}_2 . \mathbf{s}_a)}]. \quad (44)$$

where N is a normalization constant. The two terms in the square brackets in this equation arise because of the Bose symmetry of the pions. In the first term the source at $\mathbf{s}_a$ produces the pion detected at D_1 and the source at $\mathbf{s}_b$ produces the pion detected at D_2. In the second term it is the other way around. In order to calculate the average intensity $< |A(1,2)|^2 >$ we assume that the source amplitudes are gaussian random variables so that the average

$$< A^*(\mathbf{r}) A^*(\mathbf{s}) A(\mathbf{t}) A(\mathbf{u}) > = < A^*(\mathbf{r}) A(\mathbf{t}) >< A^*(\mathbf{s}) A(\mathbf{u}) > + \\ < A^*(\mathbf{r}) A(\mathbf{u}) >< A^*(\mathbf{s}) A(\mathbf{t}) > . \quad (45)$$

We also assume that the average $< A(\mathbf{t}) A(\mathbf{u}) > = 0$. The average $< A^*(\mathbf{s}_a) A(\mathbf{s}_b) >$ is a kind of density matrix for the source. A source density can be defined by taking the Wigner transform

$$D(\mathbf{s}, \mathbf{p}) = \int \mathrm{d}^3\mathbf{t} < A^*(\mathbf{s} + \tfrac{1}{2}\mathbf{t}) A(\mathbf{s} - \tfrac{1}{2}\mathbf{t}) > e^{i\mathbf{p}.\mathbf{t}}. \quad (46)$$

Semi-classically $D(\mathbf{s}, \mathbf{p})$ is proportional to the probability of producing a pion at the point $\mathbf{s}$ with momentum $\mathbf{p}$.

The pion intensity $I(\mathbf{p}_1)$ at the detector D_1 is proportional to $\int \mathrm{d}^3\mathbf{s} D(\mathbf{s}, \mathbf{p}_1)$ with a corresponding expression for $I(\mathbf{p}_2)$. If the amplitudes at different points in the source are uncorrelated then

$$< A^*(\mathbf{s}_a) A(\mathbf{s}_b) > = D(\mathbf{s}_a) \delta(\mathbf{s}_a - \mathbf{s}_b). \quad (47)$$

In this case the source density is independent of $\mathbf{p}$ and using the definition (44) of $A(1,2)$ and the property (47) we obtain

$$< |A(1,2)|^2 > = |N|^2 \int\int \mathrm{d}^3\mathbf{s}_a \mathrm{d}^3\mathbf{s}_b D(\mathbf{s}_a) D(\mathbf{s}_b) [1 + \cos(\mathbf{p}_1 - \mathbf{p}_2).(\mathbf{s}_a - \mathbf{s}_b)]. \quad (48)$$

Equation (48) can also be written as

$$|A(1,2)|^2 = I(\mathbf{p_1})I(\mathbf{p_2})R(\mathbf{p_1},\mathbf{p_2}) \tag{49}$$

where

$$R(\mathbf{p_1},\mathbf{p_2}) = 1 + \frac{\int\int \mathrm{d}^3\mathbf{s_a}\mathrm{d}^3\mathbf{s_b}D(\mathbf{s_a})D(\mathbf{s_b})\cos(\mathbf{p_1}-\mathbf{p_2}).(\mathbf{s_a}-\mathbf{s_b})}{[\int \mathrm{d}^3\mathbf{s}D(\mathbf{s})]^2}. \tag{50}$$

The correlation function $R(\mathbf{p_1},\mathbf{p_2})$ is a result of the interference between the two waves in the amplitude (44). In the case of a general source density $D(\mathbf{s},\mathbf{p})$ eq.(49) still holds but the correlation function is replaced by

$$R(\mathbf{p_1},\mathbf{p_2}) = 1 + \frac{\int\int \mathrm{d}^3\mathbf{s_a}\mathrm{d}^3\mathbf{s_b}D(\mathbf{s_a},\mathbf{P})D(\mathbf{s_b},\mathbf{P})\cos(\mathbf{p_1}-\mathbf{p_2}).(\mathbf{s_a}-\mathbf{s_b})}{\int \mathrm{d}^3\mathbf{s_a}D(\mathbf{s_a},\mathbf{p_1})\int \mathrm{d}^3\mathbf{s_b}D(\mathbf{s_b},\mathbf{p_2})} \tag{51}$$

where $\mathbf{P} = \frac{1}{2}(\mathbf{p_1}+\mathbf{p_2})$. When the source is time dependent then $\mathbf{s}$ is the position-time 4-vector and $\mathbf{p}$ is the momentum-energy 4-vector. The integrals in eqs.(50) are four dimensional integrals over space-time variables.

Experimental results on $\pi^-\pi^-$ correlations are often analysed in terms of a three parameter gaussian fit

$$R(Q_T,Q_L) = 1 + \lambda \mathrm{e}^{-\frac{1}{2}(Q_T^2R_T^2+Q_L^2R_L^2)}$$

where $Q_T = p_{1T} - p_{2T}$ is the transverse component of the momentum difference, $Q_L = p_{1z} - p_{2z}$ is the longitudinal component and λ is a chaoticity parameter. This form woud arise from an uncorrelated source with a gaussian intensity distribution. The correlation function $R(\mathbf{p_1},\mathbf{p_2})$ is a function only of $(\mathbf{p_1}-\mathbf{p_2})$ for an uncorrelated source and the parameters R_T and R_L give a measure of the transverse and longitudinal dimensions of the source. If $\mathbf{p_1} = \mathbf{p_2}$ then eq.(50) shows that $\lambda = 1$. Correlations between pions with different energies would give $\lambda < 1$. Pratt [46] and Kolehmainen and Gyulassy [47] argue that the actual correlation function should be rather different and should depend on the reaction time as well as the source size. Bertsch et al. [48] have proposed a parametrization which is Lorentz covariant and incorporates some aspects of a realistic flow pattern and time dependence. The form they use is

$$D(\mathbf{r},\mathbf{P}) = f(p_T)\tau \mathrm{e}^{-(\tau/\tau_0)^2}\mathrm{e}^{-(r_T/R)^2}\mathrm{e}^{-(Y-y)^2/y_0^2}. \tag{52}$$

Here y is the rapidity associated with the energy and momentum of a particle (eq.2) and Y is the rapidity variable defined in eq.(33) associated with the position and time. The parameters τ_0, R and y_0 characterize the source and should be determined by fitting the experimental data. Bertsch et al. show that the correlation in Q_L is sensitive to the longitudinal dimension of the source but point out that Q_T has two components if the average momentum of the pions has a transverse component . One of these is parallel to P_T. They call this the *outward* component and denote it by Q_{out}. They call the component perpendicular to P_T the *sideways* component and denote it by Q_{side}. There are some fits to data obtained by the NA35 group using this method of analysis [6] but more experiments are needed to obtain good information about the characteristics of the intermediate state of the reaction.

10. CONCLUSION

Much progress in our understanding of relativistic heavy ion physics has be made over the past few years. Many very interesting experiments are in progress and results are already available which provide insight and which will act as a guide to future theoretical work. Independent nucleon string models like the Lund model have been developed to the stage where detailed predictions can be made. These models are reasonably successful at describing E_T and multiplicity distributions but are not so successful for p_T distributions. At present they do not say anything about J/ψ suppression or about pion interferometry. More theoretical work is needed on quark-gluon plasma models. At the most fundamental level it is very important to have more detailed predictions about the deconfining phase transition. The hadronization stage of a reaction needs to be understood better in order to be able to make precise predictions about the distributions of the reaction products and to identify places where there are significant differences between the predictions of string models and quark-gluon plasma models.

11. ACKNOWLEDGEMENTS

These lectures were prepared during a period of sabbatical leave spent at the NSCL/Cyclotron Laboratory at MSU. I would like to thank G. Bertsch and the members of the NSCL Laboratory for their warm hospitality for helpful discussions during my stay there. I am also grateful for the hospitality of the TRIUMF laboratory where the lecture notes were completed.

REFERENCES

1. Braun -Munzinger, P. and Stachel, J., "Quark Matter '88'", to be published in Nucl. Phys.
2. Byckling, E. and Kajnate, K. Particle Kinematics (Wiley 1972)
3. Baym, G. Proceedings of IX Autumn School, The Physics of the Quark-Gluon Plasma, Dias de Deus, J. and Costa Ramos, S. eds.(World-Scientific, Singapore,1988), 209.
4. Sorenson, S.P. et al, "Quark Matter 1987" Z.Phys. C38, 3 (1988).
5. Lund, I. et al , "Quark Matter 87", Z.Phys. C38, 51 (1988).
6. Harris, J.W., "Quark Matter '88'", to be published in Nucl. Phys.
7. Albrecht, R. et al., Phys. Lett. B119, 297 (1987).
8. Akesson et al. Phys. Lett. B214, 295 (1988).
9. Remsberg, L.P., Z.Phys. C38, 35 (1988).
10. Braun-Munzinger, P., Z.Phys. C38, 45 (1988).
11. Jackson, A.D. and Boggild, H., Nucl.Phys. A470, 669 (1987).

12. Abbott, T. et al., Z.Phys. C38, 35 (1988).
13. Baym, G., Friedman, G. and Sarcevic, I., Phys.Lett. B219, 205 (1989).
14. Jacob, M., "Quark Matter '88'", to be published in Nucl. Phys.
15. Albrecht, R. et al , Phys. Lett. B202, 596 (1988).
16. Albrecht, R. et al, Phys. Lett. B201, 390 (1988)
17. Anderson, B. et al , Phys.Rep. 97, 31 (1983)
18. Anderson, B., Nucl.Phys. B281, 289 (1987).
19. Nilsson-Almquist, E. and Stenlunl, E., Comp.Phys.Commun.43, 387 (1987)
20. Capella et al, Z.Phys. C3, 68 (1980); C10, 249 (1981); C33, 541 (1987).
21. Werner, K., Z.Phys. C42, 85 (1989); Phys.Lett. B219, 111 (1989); Phys.lett. B220, 243 (1989); Phys. Rev. D39,780 (1989).
22. Wong, C.Y.and Lu, Z.D., Phys.Rev. D39, (1989).
23. Wong, C.Y., Phys.Rev. D30, 972 (1984); Phys.Rev. D32, 94 (1985).
24. Koba, Z., Nielsen, B. and Olesen, P. Nucl.Phys. B40, 317 (1972).
25. Arnison et al., Phys.lett. B107, 320 (1981).
26. Werner, K., Phys.Rev.Lett. 61, 1050 (1988).
27. Sjostrand, T., Comp.Phys.Commun. 39, 347 (1986).
28. Anderson, B., Dahlqvist, P. and Gustafson, G., Nucl. Phys. B317, 635 (1989).
29. Karsch, F., Z.Phys. C38, 147 (1988).
30. Satz, H. Phys.Lett. B113, 245 (1982).
31. Landau, L.D. Collected Papers of L.D.Landau 569
32. Farhi, E. and Jaffe, R.L., Phys. Rev. D30, 2379 (1984).
33. Balian, R. and Bloch, C., Ann. Phys. 64, 271 (1971).
34. Berger,M.S. and Jaffe, R.L., Phys. Rev. C35, 213 (1987).
35. Feinberg, E.L., Z.Phys. C38, 229 (1988).
36. Stachel, J. and Braun-Munzinger, P., Phys. Lett. B216, 1 (1989).
37. Bjorken, J.D., Phys.Rev. D27, 140 (1983).
38. Cleymans, J., Gavai, R.V. and Suhonen, E. Phys.Rep. 130, 217 (1986).
39. Bertsch, G., Gong, M., McLerran, L., Ruuskanen, V. and Sarkkinen, E. Phys.Rev. D37, 1202 (1988)
40. Strobele, H., Z.Phys. C38, 89 (1988).
41. Lohner, H. Z.Phys. C38, 97 (1988).
42. Matsui, T. and Satz, H., Phys.Lett. B178, 416 (1986).
43. Baglin, et al Phys.Lett. B220, 471 (1989).
44. Blaizot, J.P. and Ollitrault, J.Y., Phys. Lett. B217, 386, 392 (1989).
45. Milana, J., Phys. Rev. Lett. 62, 2921 (1989).
46. Pratt, S., Phys. Rev. D33, 1314 (1986).
47. Kolehmainen, K. and Gyulassy, M., Phys. Lett. B180,203 (1986).
48. Bertsch, G., Gong, M. and Tohyama, M., Phys.Rev. C37, 1896 (1988).

COMPUTATIONAL METHODS IN RELATIVISTIC ATOMIC PHYSICS

W. R. Johnson
Department of Physics
University of Notre Dame
Notre Dame, IN 46556

I. INTRODUCTION

These lectures describe some of the numerical techniques that are used to work out the structure of atoms and ions of high nuclear charge where relativistic effects are significant. Relativistic atomic physics calculations are important in many practical applications. For example, relativistic transition rate calculations provide input data for determining thermodynamic conditions in plasmas such as those inside a TOKAMAK, or those inside a star. Calculations of energy levels and transition amplitudes are used in interpreting experiments on high-Z, few-electron ions designed to test QED in strong fields. Relativistic atomic structure calculations are essential in interpreting experiments on the parity-violating effects of weak interactions in atoms and in analyzing measurements designed to detect a possible atomic electric dipole moment in high Z atoms.

The inner (K-shell) electrons in an atom with nuclear charge Z move with a velocity $v = \alpha Z c$, where c is the speed of light and α is the fine-structure constant, $\alpha = 1/137.03598\cdots$. For an atom such as tungsten, Z = 74, one finds relativistic corrections of order $v^2/c^2 \simeq 0.25$, and for heavier atoms even larger corrections. One effect of relativity is to produce a radial Lorentz contraction of the inner orbitals. The contracted inner orbitals screen the nucleus more effectively than nonrelativistic orbitals; causing the outer orbitals of an atom with high nuclear charge to expand beyond what one would expect nonrelativistically, and leading to a relativistic redistribution of charge.

In addition to the relativistic charge redistribution, there is the effect of the spin-orbit interaction which splits each atomic subshell into fine-structure components, depending on the combined orbital and spin angular momenta. The spin-orbit interaction is usually introduced into nonrelativistic calculations in an ad-hoc manner; it is automatically and correctly included in relativistic calculations. The largest part of the spin-orbit interaction, that arising from interaction with the nuclear Coulomb field, is contained in the one-electron Dirac hamiltonian; the remaining part is in the Breit interaction, a correction to the electron-electron Coulomb interaction that arises from the exchange of transverse photons between the atomic electrons. In the present calculations, we adopt the static form of the Breit interaction:

$$b_{12} = -\frac{e^2}{2r_{12}} \left(\boldsymbol{\alpha}_1 \cdot \boldsymbol{\alpha}_2 + \boldsymbol{\alpha}_1 \cdot \hat{r}_{12} \, \boldsymbol{\alpha}_2 \cdot \hat{r}_{12} \right) \quad . \tag{1}$$

There are frequency dependent corrections to b_{12} that can be associated with the radiative corrections and evaluated by perturbation theory. The treatment of these small corrections is beyond the scope of the present lectures.

The Dirac hamiltonian describing the electron's rest energy, its kinetic energy, and its intraction with the nucleus is written:

$$h_0 = c\boldsymbol{\alpha}.\mathbf{p} + \beta mc^2 + V_{nuc}(r) + U(r) \quad , \tag{2}$$

where $\boldsymbol{\alpha}$ and β are 4×4 Dirac matrices,

$$\boldsymbol{\alpha} = \begin{bmatrix} 0 & \boldsymbol{\sigma} \\ \boldsymbol{\sigma} & 0 \end{bmatrix} , \quad \beta = \begin{bmatrix} 1 & 0 \\ 0 & -1 \end{bmatrix} ,$$

and where $V_{nuc}(r)$ is the nuclear potential,

$$V_{nuc}(r) = -\frac{Z\,e^2}{r} \quad .$$

The potential U(r) in Eq.(2) is introduced to account for the electron-electron interaction approximately. In many-body calculations, U(r) is artificially added to the lowest-order hamiltonian to establish an accurate zeroth-order wavefunction; an equal and opposite counter term is added to the interaction hamiltonian.

The many-electron hamiltonian describing an atom or ion is written

$$H = H_0 + V \quad , \tag{3}$$

$$H_0 = \sum_i h_0(r_i) \quad . \tag{4}$$

where $h_0(r)$ is given by Eq.(2), and where

$$V = \frac{1}{2}\sum_{ij}{}' [\frac{e^2}{r_{ij}} + b_{ij}] - \sum_i U(r_i) \quad . \tag{5}$$

In relativistic calculations, the two-body interaction given in Eq.(5) leads to so called "continuum dissolution" difficulties. These difficulties (wavefunctions based on the potential given in Eq.(5) collapse into the negative energy continuum) and their resolution have been described by Brown and Ravenhall[1], and more recently by Sucher[2] and by Mittleman[3]. These authors find that a proper treatment of the electron-electron interaction based on quantum electrodynamics leads to a modified version of the interaction given in Eq.(5) that is free of the continuum dissoluion difficulties. This modified interaction is given by Eq.(5) with the right-hand side placed between positive-energy projection operators, Λ_+:

$$V = \Lambda_+ \left[\frac{1}{2}\sum_{ij}{}' [\frac{e^2}{r_{ij}} + b_{ij}] - \sum_i U(r_i) \right] \Lambda_+ \quad . \tag{6}$$

The many-electron projection operators in Eq.(6) are products of one-electron positive-energy projection operators determined from the single-electron Dirac hamiltonian in Eq.(2). These projection operators depend on the choice of the initial potential, U(r); the hamiltonian governing relativistic atomic structure, therefore, depends on the (arbitrary) choice of zeroth-order potential. QED corrections to the atomic structure also depend on U(r); and indeed, in such a way that the atomic structure calculations when combined with the corresponding QED corrections must be independent of the choice of U(r). It might be added here that the dependence of the atomic structure part of the calculation alone on the choice of the initial potential, U(r), was found to be less than one ppm for the case of He where the ground state energy was calculated starting from two different initial potentials[4]. The dependence on U(r) is expected to be very slight for other neutral atoms as well.

In lowest approximation, $H = H_0$, a single-particle operator,

$$H_0 = h_0(r_1) + h_0(r_2) + \cdots + h_0(r_N) \quad . \tag{7}$$

and the corresponding lowest-order wavefunction is a product of one-electron orbitals:

$$\Psi_0(\mathbf{r}_1,\mathbf{r}_2,\cdots,\mathbf{r}_N) = u_a(\mathbf{r}_1)\, u_b(\mathbf{r}_2)\, u_c(\mathbf{r}_3) \cdots u_n(\mathbf{r}_N) \;, \tag{8}$$

where $u_a(\mathbf{r})$ is an eigenfunction of the one-electron Dirac hamiltonian with eigenvalue ϵ_a,

$$h_0(\mathbf{r})\, u_a(\mathbf{r}) = \epsilon_a\, u_a(\mathbf{r}) \quad . \tag{9}$$

The lowest-order energy of the many-electron system is

$$E_0 = \epsilon_a + \epsilon_b + \cdots + \epsilon_n \;. \tag{10}$$

A many-electron eigenfunction of H_0 with energy E_0, satisfying the Pauli exclusion principle, can be obtained by replacing Eq.(8) with a Slater determinant of one-electron orbitals:

$$\Psi_0 \;=\; |a,b,c,\cdots,n\rangle \;=\; \frac{1}{\sqrt{N!}} \begin{vmatrix} u_a(\mathbf{r}_1) & u_b(\mathbf{r}_1) & \cdots & u_n(\mathbf{r}_1) \\ u_a(\mathbf{r}_2) & u_b(\mathbf{r}_2) & \cdots & u_n(\mathbf{r}_2) \\ \cdots & & & \\ u_a(\mathbf{r}_N) & u_b(\mathbf{r}_N) & \cdots & u_n(\mathbf{r}_N) \end{vmatrix} \quad . \tag{11}$$

This determinental wavefunction vanishes if any of the single-paticle indices, a,b, $\cdots$, are equal. Changing the order of any pair of orbital indices mearly changes the sign of the wavefunction. Equation (11) gives the lowest-order wavefunction appropriate to the description of the ground-state of a closed-shell N-electron atom. If we wish to describe an atom with unfilled shells in its ground-state or to treat excited states, then we must combine together several Slater determinants to give the lowest-order wavefunction. For simplicity, we consider the non-degenerate case only in the remainder of these lectures.

To proceed beyond the zeroth-order approximation, we employ perturbation theory. We expand the wavefunction and energy in powers of the potential V as:

$$\Psi = \Psi_0 + \Psi_1 + \cdots \tag{12}$$

$$E = E_0 + E_1 + \cdots \tag{13}$$

where

$$(H_0 - E_0)\Psi_0 = 0 \quad , \tag{14a}$$

$$(H_0 - E_0)\Psi_1 = (E_1 - V)\Psi_0 \quad , \tag{14b}$$

$$(H_0 - E_0)\Psi_2 = (E_1 - V)\Psi_1 + E_2\,\Psi_0 \quad , \text{etc.} \tag{14c}$$

If we neglect the Breit interaction, temporarily, then we may easily find the first-order correction to the energy from Eq.(14b):

$$E_1 = \langle \Psi_0 | \, V \, | \Psi_0 \rangle =$$

$$\langle \Psi_0 | \, \frac{1}{2} \sum_{ij}{}' \frac{e^2}{r_{ij}} \, | \Psi_0 \rangle - \langle \Psi_0 | \, \sum_i U(r_i) \, | \Psi_0 \rangle \quad . \tag{15}$$

This expression for the energy is a sum of matrix elements of a two-particle operator, $1/2 \; \Sigma' e^2/r_{ij}$, and a one-particle operator, $\Sigma \; U(r_i)$, where

$$F = \sum_i f(\mathbf{r}_i) \tag{16}$$

is a typical one-particle operator and where

$$G = \frac{1}{2} \sum_{ij}{}' g(r_{ij}) \tag{17}$$

is a typical two-particle operator.

The following theorems will help evaluate matrix-elements of one- and two-particle operators between Slater-determinant wavefunctions.

Theorem 1.

$\langle a' \; b' \; c' \cdots n' | \; F \; | \; a \; b \; c \cdots n \rangle =$

$= 0$	if the sets $\{a' \; b' \; c' \cdots n'\}$ and $\{a \; b \; c \cdots n\}$ differ in more than one place,
$= f_{a'a}$	if only a' and a are different, and
$= \sum_a f_{aa}$	if all indices are identical in the two sets.

Here we use the notation $f_{ab} = \langle a | \; f \; | b \rangle$ for single-particle matrix elements of $f(\mathbf{r})$.

Theorem 2.

$\langle a' \; b' \; c' \cdots n' | \; G \; | \; a \; b \; c \cdots n \rangle =$

$= 0$	if the sets $\{a' \; b' \; c' \cdots n'\}$ and $\{a \; b \; c \cdots n\}$ differ in more than two places,
$= g_{a'b'ab} - g_{a'b'ba}$	if only the pair a', b' and a, b differ,
$= \sum_b [g_{a'bab} - g_{a'bba}]$	if only a' and a differ, and

$$= \frac{1}{2}\sum_{ab} [g_{abab} - g_{abba}] \quad \text{if all indices are identical.}$$

Here we use the notation $g_{abcd} = \langle ab|\ g(r_{12})\ |cd\rangle$, for the unsymmetrized two-particle matrix element of $g(r_{12})$.

With the aid of these two theorems, we obtain the following expression for the first-order energy:

$$E_1 = \frac{1}{2}\sum_{ab} [g_{abab} - g_{abba}] - \sum_{a} U_{aa} \quad , \tag{18}$$

where $g(r_{ij}) = e^2/r_{ij}$.

The potential U(r) in the one-electron hamiltonian h_0 in Eq.(2) is chosen to approximate in some way the real electron-electron interaction. One particularly important choice for many-body calculations is the Hartree-Fock potential, $U = V_{HF}$, defined by:

$$V_{HF}\ u(r) = \sum_{b} \int d^3r' \frac{e^2}{|\mathbf{r}' - \mathbf{r}|} [(u_b^\dagger u_b)'\ u(\mathbf{r}) - (u_b^\dagger u)'\ u_b(\mathbf{r})\ ]. \tag{19}$$

Matrix elements of the Hartree-Fock potential are given by

$$\langle c\ |V_{HF}|\ d\rangle = (V_{HF})_{cd} = \sum_{a} [g_{cada} - g_{caad}] \quad . \tag{20}$$

from which it follows that

$$E_1 = \sum_{a} [\tfrac{1}{2}\ (V_{HF})_{aa} - (U)_{aa}] \quad . \tag{21}$$

If we start our atomic structure calculations using the Hartree-Fock potential, then we have

$$E_0(HF) = \sum_{a} \epsilon_a = \sum_{a} [\ (c\boldsymbol{\alpha}\cdot\mathbf{p} + \beta mc^2 + V_{nuc})_{aa} + (V_{HF})_{aa}\ ] \ , \tag{22a}$$

$$E_1(HF) = -\frac{1}{2}\sum_{a} (V_{HF})_{aa} \quad . \tag{22b}$$

$$E_0(HF) + E_1(HF) = \sum_{a} [\ (c\boldsymbol{\alpha}\cdot\mathbf{p} + \beta mc^2 + V_{nuc})_{aa} + \frac{1}{2}\ (V_{HF})_{aa}\ ]$$

$$= \sum_{a} (c\boldsymbol{\alpha}\cdot\mathbf{p} + \beta mc^2 + V_{nuc})_{aa} + \frac{1}{2}\sum_{ab} [g_{abab} - g_{abba}] \quad . \tag{22c}$$

It is this later expression which, when considered from a variational point of view, leads to the Hartree-Fock equations. From Eq.(22c) it follows that the matrix element of H calculated with HF orbitals (orbitals determined self-consistently in the HF potential) are correct through first-order in perturbation theory.

To proceed, we calculate the first-order wavefunction from Eq.(14b):

$$\Psi_1 = \frac{1}{4}\sum_{abnm} \frac{g_{nmab}}{\epsilon_a + \epsilon_b - \epsilon_n - \epsilon_m} \mid ab \rightarrow nm \rangle$$
$$+ \sum_{an} \frac{(V_{HF} - U)_{na}}{\epsilon_a - \epsilon_n} \mid a \rightarrow n \rangle \quad , \tag{23}$$

where $\mid ab \rightarrow nm \rangle$ is the ground-state Slater determinant with the occupied orbitals a and b replaced by excited orbitals n and m, respectively, and where $\mid a \rightarrow n \rangle$ is a Slater determinant with occupied orbital a replaced by excited orbital n. It is clear that n and m cannot themselves be occupied orbitals; otherwise the Slater determinants would vanish. The first term in Eq.(23) is a sum over doubly excited states and the second over singly excited states. The effect of the positive-energy projection operators in Eq.(6) is to restrict the excited orbitals in Eq.(23) to the positive-energy part of the spectrum of the Dirac equation. With this restriction, the denominators in Eq.(23) are negative definite numbers; without the projection operators there would be an infinity of zeros in the denominators of the first term of Eq.(23).

It is interesting to note that the singly excited state contributions to the first-order wavefunction vanish precisely when $U(r) = V_{HF}$, the Hartree-Fock potential. It is, therefore, particularly convenient to choose V_{HF} as a starting potential for many-body calclations.

If we choose $U = V_{HF}$ and write out the expression for the second-order energy from Eq.(14c), then we find

$$E_2 = \frac{1}{4}\sum_{abnm} \frac{(g_{abnm} - g_{abmn})(g_{nmab} - g_{nmab})}{\epsilon_a + \epsilon_b - \epsilon_n - \epsilon_m} \quad . \tag{24}$$

Despite its benign appearance, this double sum is extremely complicated to deal with computationally, since each index, n and m, ranges over all positive-energy bound-state and continuum solutions to the single-particle Dirac equation.

Let me end this first lecture with a summary of what is required from a computational point of view to carry out relativistic atomic structure calculations of the type outlined above:

1. We need to determine bound-state solutions to the one-electron Dirac equation in a given central potential U(r).
2. We require self-consistent field methods to treat the important special case where U(r) is the Hartree-Fock potential.
3. We need to determine all of the positive-energy excited state solutions, bound and continuum, of the Dirac equation in order to evaluate the sums over n and m in the expression for the second-order energy, Eq.(24).

We will discuss how these problems are solved in the following lectures.

II. THE BOUND-STATE EIGENVALUE PROBLEM FOR THE DIRAC EQUATION

We write the Dirac equation as

$$h_0\, u(\mathbf{r}) = [\, c\boldsymbol{\alpha}\cdot\mathbf{p} + \beta mc^2 + V(r)\,]\, u(\mathbf{r}) = \epsilon\, u(\mathbf{r}) \quad , \tag{25}$$

where $V(r) = -e^2/r + U(r)$. It is not difficult to show that the total angular momentum vector, $\mathbf{J} = \mathbf{L} + \mathbf{S}$, where $\mathbf{L}$ is the orbital angular momentum,

$$\mathbf{L} = [\, \mathbf{r} \times \mathbf{p}\,] \; , \tag{26}$$

and where $\mathbf{S}$ is the 4×4 spin angular momentum matrix,

$$\mathbf{S} = \frac{1}{2}\begin{bmatrix} \boldsymbol{\sigma} & 0 \\ 0 & \boldsymbol{\sigma} \end{bmatrix} \; , \tag{27}$$

commutes with the single-particle hamiltionian. We may, therefore, classify the eigenstates of h_0 according to the eigenvalues of energy, J^2 , and J_z .The eigenstates of J^2 and J_z are easily constructed using the two-component representation of $\mathbf{S}$. They are

$$\Omega_{j\ell m}(\hat{r}) = \sum_{\mu} \langle \ell, m-\mu, 1/2, \mu \mid j, m \rangle \, Y_{\ell\, m-\mu}(\hat{r}) \, \chi_{\mu} \quad . \tag{28}$$

In Eq.(28), $Y_{\ell\, m_\ell}(\hat{r})$ is a spherical harmonic and χ_μ is a two-component eigenfunction of σ_z. The quantities $\langle \ell, m_\ell, 1/2, \mu \mid j, m \rangle$ are Clebsch-Gordon coefficients. The resulting two-component angular functions are referred to as spherical spinors; they satisfy the relations

$$J^2 \, \Omega_{j\ell m} = j(j+1) \, \Omega_{j\ell m} \quad , \quad \text{and} \tag{29a}$$

$$J_z \, \Omega_{j\ell m} = m \, \Omega_{j\ell m} \quad . \tag{29b}$$

There are two possible values of ℓ in Eqs.(28) and (29a,b) for each j, $\ell = j+1/2$ and $\ell = j-1/2$. The corresponding spherical spinors have opposite parity. We introduce the operator

$$K = -\,2\,\mathbf{L}\cdot\mathbf{S} - 1 \quad , \tag{30}$$

and note that

$$K \, \Omega_{j\ell m} = \kappa \, \Omega_{j\ell m} \quad , \tag{31}$$

where $\kappa = -(j+1/2)$ for $j = \ell+1/2$, and $\kappa = j+1/2$ for $j = \ell-1/2$. The operator K has integer eigenvalues, κ; the absolute value of κ determines j and the sign of κ determines ℓ (or the parity). We may, therefore, introduce the shorter and more convenient notation

$$\Omega_{\kappa m}(\hat{r}) \equiv \Omega_{j\ell m}(\hat{r}) \quad . \tag{32}$$

The spherical spinors satisfy the orthogonality relations

$$\int \Omega^{\dagger}_{\kappa m}(\hat{r}) \, \Omega_{\kappa' m'}(\hat{r}) \, d\Omega = \delta_{\kappa\kappa'} \, \delta_{mm'} \quad . \tag{33}$$

The following identities are often useful:

$$\boldsymbol{\sigma}\cdot\hat{r} \, \Omega_{\kappa m} = -\, \Omega_{-\kappa m} \quad , \quad \text{and} \tag{34a}$$

$$\boldsymbol{\sigma}\cdot\mathbf{p} \, f(r) \, \Omega_{\kappa m} = i\left[\frac{df}{dr} + \frac{1+\kappa}{r} f \right] \Omega_{-\kappa m} \quad . \tag{34b}$$

If we seek a solution to the Dirac equation (25) in the form

$$u_{n\kappa m}(\mathbf{r}) = \frac{1}{r} \begin{bmatrix} iP_{n\kappa}(r) \, \Omega_{\kappa m}(\hat{r}) \\ Q_{n\kappa}(r) \, \Omega_{-\kappa m}(\hat{r}) \end{bmatrix} \quad , \tag{35}$$

then we find, with the help of the identities (34a,b), that the radial functions P and Q satisfy a pair of coupled first-order differential equations:

$$(V + mc^2)P_{n\kappa} + c(\frac{d}{dr} - \frac{\kappa}{r})Q_{n\kappa} = \epsilon P_{n\kappa} , \quad (36a)$$

$$- c(\frac{d}{dr} + \frac{\kappa}{r})P_{n\kappa} + (V - mc^2)Q_{n\kappa} = \epsilon Q_{n\kappa} . \quad (36b)$$

The normalization condition for the orbital $u_{n\kappa m}(\mathbf{r})$,

$$\int u^{\dagger}_{n\kappa m}(\mathbf{r}) \, u_{n\kappa m}(\mathbf{r}) \, d^3r = 1 , \quad (37a)$$

can be written

$$\int_0^{\infty} [P^2_{n\kappa}(r) + Q^2_{n\kappa}(r)] \, dr = 1 , \quad (37b)$$

when expressed in terms of the radial functions $P_{n\kappa}$ and $Q_{n\kappa}$. The first numerical problem that we face is to determine the eigenvalues, ϵ, and the corresponding eigenfunctions P and Q for the radial Dirac equations (36a,b) corresponding to bound states in a central potential V(r).

Before starting the discussion of techniques to solve the eigenvalue problem for the radial Dirac equation, let me say a few words about the choice of an appropriate radial grid. We expect the solutions to Eqs.(36a,b) to vary rapidly near r = 0, like r^{γ} with $\gamma = \sqrt{(\kappa^2 - \alpha^2 Z^2)}$, and more gradually for large r, like $e^{-\lambda r}$, with $\lambda = \sqrt{(m^2c^4 - \epsilon^2)}$. To accommodate the rapid growth at small r and the gradual decrease at large r, we adopt a radial grid of the form

$$r_i = r_0 (e^{t_i} - 1) , \quad (38)$$

where

$$t_i = (i-1) h \quad , \quad i = 1,N , \quad (39)$$

is a uniformly-spaced grid. Typically we take N = 500 points, the parameter r_0 = 0.0001(a.u.) and the grid spacing h = 0.02 - 0.03. These choices permit the radial Dirac equations to be integrated with high accuracy (parts in 10^9) for binding energies as low as 0.01 a.u. We store the arrays r_i and $r'_i = (dr/dt)_i$ in a common block in our programs. The later quantity is the Jacobian of the transformation from the r grid to the uniformly-spaced t grid and is needed in integration algorithms.

For small r, we have

$$P(r) = a \, r^{\gamma} , \quad (40a)$$

$$Q(r) = b\, r^{\gamma} \quad , \quad \text{with} \tag{40b}$$

$$\frac{b}{a} = \frac{\alpha Z}{(\gamma - \kappa)} \quad . \tag{40c}$$

We can choose a arbitrarily and determine b from Eq.(40c). The leading terms in the power series solution to the radial equations can be determined once U(r) is chosen, and the resulting truncated power series can be used to obtain accurate numerical solutions for P(r) qnd Q(r) at the first few grid points. For large r, the radial functions behave as

$$P(r) = \surd(mc^2 + \epsilon)\; e^{-\lambda r} \tag{41a}$$

$$Q(r) = \surd(mc^2 - \epsilon)\; e^{-\lambda r} \quad , \tag{41b}$$

and we can use the first few terms in the asymptotic series to determine an accurate numerical solution on the last few points of the radial grid. Actually, it is sufficient to consider the radial functions only out to a "practical infinity", a point where the wavefunction is negligibly small. We choose this point to be the point where $\lambda r = 20$; then the exponential factor in Eqs.(41a,b) is reduced to about 10^{-10} .

The difficult region is the one at intermediate r where neither the power series, nor the asymptotic series is adequate. In this intermediate region, we use step-by-step numerical integration based on an Adams Bashforth - Adams-Moulton predict-correct scheme. We integrate the radial equations outward from the origin to the classical turning point, r_c, and inward from the practical infinity to r_c. We determine the eigenvalue of the Dirac equation by matching our radial solutions at r_c. The significance of the classical turning point is that for $r > r_c$ the functions P and Q are monotonically decreasing, while for $r < r_c$ the radial functions oscillate, except near the origin where they grow like a power of r.

We integrate inward in the asymptotic region since the desired solution in this region grows as r decreases while the complementary solution, which governs the error propagation, decreases with decreasing r. Similarly, we integrate the radial equations outward from the origin, since the desired solutions grow with increasing r, while the complementary solutions decrease with increasing r.

If we represent the pair of radial Dirac equations on the uniformly-spaced t grid as a two-component radial equation

$$\frac{dy}{dt} = f(y,t) \ , \tag{42}$$

with

$$y(t) = \begin{bmatrix} y_1(t) \\ y_2(t) \end{bmatrix} = \begin{bmatrix} P(r(t)) \\ Q(r(t)) \end{bmatrix} ,$$

then we can formally integrate from one grid point, t_n, to the next, t_{n+1}:

$$y_{n+1} = y_n + \int_{t_n}^{t_{n+1}} f(y(t),t)dt \ . \tag{43}$$

To carry out the integration, we introduce the backward difference operator, ∇:

$$\nabla f_n = f_n - f_{n-1} \ . \tag{44}$$

From Eq.(44) one may write:

$$f_{n+1} = (1 - \nabla)^{-1} f_n \ , \tag{45}$$

$$f_{n+2} = (1 - \nabla)^{-2} f_n \ , \tag{46}$$

or more generally

$$f_{n+x} = (1 - \nabla)^{-x} f_n \ . \tag{47}$$

In these last three expressions, it is understood that the terms in parentheses are to be expanded in a power series in ∇, and Eq.(44) is to be used iteratively to determine ∇^k. Using Eq.(47) for the integrand in Eq.(43), we may carry out the integration analytically with the result:

(Adams-Bashforth - predictor formula)

$$y_{n+1} = y_n - \frac{h\,\nabla}{(1-\nabla)\,\log(1-\nabla)} f_n \ , \tag{48a}$$

$$= y_n + h\,(1 + \tfrac{1}{2}\nabla + \tfrac{5}{12}\nabla^2 + \tfrac{9}{24}\nabla^3 + \cdots)f_n \ . \tag{48b}$$

Equations (48a,b) may be rewritten in the form:

(Adams-Moulton - corrector formula)

$$y_{n+1} = y_n - \frac{h\nabla}{\log(1-\nabla)} f_{n+1} \ , \tag{49a}$$

$$= y_n + h\,(1 - \tfrac{1}{2}\nabla - \tfrac{1}{12}\nabla^2 - \tfrac{1}{24}\nabla^3 + \cdots)f_{n+1} \ . \tag{49b}$$

Keeping terms only to third-order and using Eq.(44), we may rewrite Eq.(48b) as: (predictor formula)

$$y_{n+1} = y_n + \frac{h}{24}[55f_n - 59f_{n-1} + 37f_{n-2} - 9f_{n-3}] + \frac{19}{720}h^5y_n^{(5)} \quad . \quad (50)$$

where the last term is the error estimated from the neglected fourth-order term in Eq.(48b). From Eq.(49b) we find: (corrector formula)

$$y_{n+1} = y_n + \frac{h}{24}[9f_{n+1} + 19f_n - 5f_{n-1} + f_{n-2}] - \frac{1}{240}h^5y_n^{(5)} \quad . \quad (51)$$

The error in Eq.(51) is smaller by a substantial factor than that in Eq.(50) since interpolation is being used in Eq.(51) and extrapolation in Eq.(50). In practice, the less accurate predictor formula, Eq.(50), is used to advance from point t_n, where y_n and f_n (and presumably also f_{n-1}, f_{n-2} and f_{n-3}) are known, to t_{n+1} . Using the predicted value of y_{n+1}, one evaluates f_{n+1} . The resulting value of f_{n+1} can be used in the corrector formula (51) to give a more accurate value of y_{n+1} .

The scheme for integrating the system of equations (41a,b) is to go from one grid point to the next, first using the predictor formula to give a value of the dependent variable at the new grid point, and then using the corrector formula to obtain a more accurate value. It is an elementary exercise to show that the predictor step can be eliminated entirely and the corrector formula alone used when the system of equations (41a,b) is linear. It is left as an exercise for the reader to write out the computer code to implement the resulting Adams-Moulton corrector formulas for the specific case of the radial Dirac equations.

There are other algorithms of high accuracy, such as Milne's method, that could be used instead of the Adam's predict-correct methods to solve the system of equations (42). One should be cautious in applying such methods since they often lead to unsuspected numerical instabilities.

We use the Adams-Moulton formula to integrate Eq.(42) outward from the point r = 0 to the classical turning point, and to integrate inward from the practical infinity to the classical turning point. To start the solution we must supply values for y_k at the first three points by alternative methods, since function values at points t_1, t_2, and t_3 are required in determining the values at point t_4. Typically, the power series solution is used for the first few points. Similarly, the asymptotic series is used to supply the values of the dependent variable at the last three grid points, preparatory to inward integration.

The step-by-step integration is carried out from the origin to the classical turning point using the algorithms just described. A second integration is carried out from the practical infinity to the classical turning point. The radial Dirac functions P(r) and Q(r) determined in the region beyond the classical turning point are multiplied by a scale factor (this is permissible since the equations are linear and homogeneous) so that the large component, $P_<(r)$, of the solution determined by the outward integration agrees with the large component of the solution determined by inward integration, $P_>(r)$, at the classical turning point. Generally, unless the energy parameter, ϵ, in the radial Dirac equation (36) is an eigenvalue, the small components of the two solutions will be different at $r = r_c$; varying the energy systematically until this difference vanishes is the basis for determining eigenvalues of the Dirac equation numerically.

There are two steps in finding accurate eigenvalues of the radial Dirac equations: a preliminary node counting step to determine an approximate energy interval for a particular eigenvalue, followed by an iteration procedure in which a precise value of the eigenvalue is determined.

Node counting:

The number of internal nodes (zeros) for the large component of the radial Dirac function, $P_{n\kappa}(r)$, is $n-\ell-1$, where n is the principal quantum number and ℓ is the orbital angular momentum quantum number. The large component of the radial Dirac equation has the same number of nodes as the radial Schrodinger wavefunction. There are also $n-\ell-1$ nodes for the small component when $\kappa < 0$, and there are $n-\ell$ nodes for the small component when $\kappa > 0$. We proceed as follows: guess an energy value, ϵ, and solve the radial Dirac equations. Count nodes. If the number of nodes is too great, decrease the value of ϵ; if it is too small, increase ϵ. Keep track of the energy bounds determined at each try and "trap" the energy in an interval where the radial functions have the correct number of nodes. At this point you have a rough approximation to ϵ suitable for iterative refinement.

Perturbation step:

One can show from the radial Dirac equations that if the discontinuity in the small component of the radial wavefunction at the point r_c is $\Delta Q = Q_<(r_c)-Q_>(r_c)$, for $\epsilon = \epsilon_1$, then for $\epsilon = \epsilon_2$, with

$$\epsilon_2 = \epsilon_1 + \frac{c\ \Delta Q}{\int dr(P^2+ Q^2)} \quad , \tag{52}$$

the disconinuity in Q at r_c will vanish. This is an approximate expression accurate to second order in the energy difference, so a perturbation loop based on Eq.(52) converges in only a few steps.

Now you are in a position to write your own eigenvalue routine to solve Dirac equation in a central potential. If you undertake this project, I strongly recommend that you use the well-known analytic solutions to the Dirac equation in a Coulomb field to check out your routine. I'll end this second lecture here and take up the problem of determining the Hartree-Fock potential and orbitals in the next lecture.

III. DIRAC-HARTREE-FOCK EQUATIONS FOR A CLOSED-SHELL ATOM

Our aim is to write out the Dirac Hartree-Fock(DHF) equations as a set of radial differential equations suitable for numerical solution. For a closed-shell atom the HF potential, defined by Eq.(19), is a spherically symmetrical non-local potential acting in the space of one-electron Dirac orbitals, $u_a(r)$. As a preliminary step in our analysis, we express the HF potential as a non-local potential acting on two-component radial Dirac functions:

$$R_a(r) = \begin{bmatrix} P_a(r) \\ Q_a(r) \end{bmatrix} \quad .$$

For this purpose, we need the identity,

$$\frac{1}{|\mathbf{r}'-\mathbf{r}|} = \sum_{LM} (r_<^L / r_>^{L+1})\ C^*_{LM}(\hat{r})\ C_{LM}(\hat{r}') \quad . \tag{53}$$

where $r_>$ is the larger of r and r', and $r_<$ is the smaller of the two.

The quantities $C_{LM}(\hat{r})$ in Eq.(53) are just normalized spherical harmonics, given by

$$C_{LM}(\hat{r}) = \sqrt{\frac{4\pi}{2L+1}}\, Y_{LM}(\hat{r}) \ . \tag{54}$$

Using the relation (53), the integral in the first (direct) term in Eq.(19) becomes,

$$\int \frac{d^3r'}{|r'-r|} u_b^{\dagger}(r')u_b(r') = \sum_{LM} C^{*}_{LM}(\hat{r}) \int d\Omega'\, \Omega^{\dagger}_{\kappa_b m_b}(\hat{r}')C_{LM}(\hat{r}')\Omega_{\kappa_b m_b}(\hat{r}') \times$$
$$\int dr' \,(r_<^L/r_>^{L+1})\,(P_b(r')^2 + Q_b(r')^2) \ . \tag{55}$$

In writing Eq.(55) we have used the fact that the angular integral is invariant under the replacement $\kappa_b \rightarrow -\kappa_b$. We may use the Wigner-Ekhart theorem to express the m-dependence of the angular matrix element:

$$\langle \kappa_b, m_b | C_{LM} | \kappa_a, m_a \rangle = \int d\Omega\, \Omega^{\dagger}_{\kappa_b m_b}(\hat{r})C_{LM}(\hat{r})\Omega_{\kappa_a m_a}(\hat{r}) =$$
$$(-1)^{j_b - m_b} \begin{bmatrix} j_b & L & j_a \\ -m_b & M & m_a \end{bmatrix} \langle \kappa_b \| C_L \| \kappa_a \rangle \ , \tag{56}$$

where the square bracket is a Wigner 3j symbol and where the reduced matrix element is given by

$$\langle \kappa_b \| C_L \| \kappa_a \rangle = \sqrt{([ja][jb])}\ (-1)^{j_b+1/2} \begin{bmatrix} j_b & j_a & L \\ -1/2 & 1/2 & 0 \end{bmatrix} \Pi(\ell_b, \ell_a, L) \ . \tag{57}$$

In Eq.(57), $\Pi(\ell_b, \ell_a, L)$ is a parity factor defined by:

$$\begin{aligned} \Pi(\ell_1, \ell_2, \ell_3) &= 1 \text{ if } \ell_1+\ell_2+\ell_3 \text{ is even} \\ &= 0 \text{ if } \ell_1+\ell_2+\ell_3 \text{ is odd} \ , \end{aligned} \tag{58}$$

and $[j_b] = 2j_b+1$ is the occupation number of subshell b. If we sum the integral in Eq.(55) over a complete subshell, $m_b = -j_b \rightarrow j_b$, and use the identity

$$\sum_{m_b} (-1)^{j_b - m_b} \begin{bmatrix} j_b & L & j_b \\ -m_b & M & m_b \end{bmatrix} = \sqrt{[j_b]}\ \delta_{L0}\ \delta_{M0} \ , \tag{59}$$

to simplify the resulting sum, then we find

$$\sum_b \int \frac{d^3r'}{|r - r'|} u_b^{\dagger}(r')\ u_b(r') = \sum_{\kappa_b} \sqrt{[j_b]}\ \langle \kappa_b || C_0 || \kappa_b \rangle \times$$

$$\int_0^\infty \frac{1}{r_>} [\, P_b(r)^2 + Q_b(r)^2 \,]\, dr = \sum_{\kappa_b} [j_b]\, v_0(b,b,r) \quad . \tag{60}$$

Here we have introduced the function:

$$v_0(b,b,r) = \int_0^\infty \frac{1}{r_>} [\, P_b(r')^2 + Q_b(r')^2 \,]\, dr' \ . \tag{61}$$

This function is a special case of the Hartree screening functions

$$v_k(a,b,r) = \int_0^\infty (r_<^k / r_>^{k+1})\, [\, P_a(r')P_b(r') + Q_a(r')Q_b(r')]\, dr' \ . \tag{62}$$

With the aid of Eq.(61), the direct part of the HF potential can be written

$$V_{HF}^{dir}\, R_a(r) = \sum_b [j_b]\, v_0(b,b,r)\, R_a(r) \ , \tag{63}$$

where $R_a(r)$ is the two component radial function defined previously. The exchange part is slightly more complicated to analyze; one finds

$$V_{HF}^{exc}\, R_a(r) = \sum_b [j_b] \sum_L \Lambda_{\kappa_b L \kappa_a}\, v_L(b,a,r)\, R_b(r) \ , \tag{64}$$

where the angular factor $\Lambda_{\kappa_b L\, \kappa_a}$ is given by

$$\Lambda_{\kappa_b L \kappa_a} = \langle\, \kappa_b \| C_L \| \kappa_a \,\rangle^2 / [j_a][j_b] \ . \tag{65}$$

Combining the expressions, (63) and (64), we have

$$V_{HF}\, R_a(r) = \sum_b [j_b] \Big\{\, v_0(b,b,r) R_a(r) + \sum_L \Lambda_{\kappa_b L \kappa_a} v_L(b,a,r) R_b(r) \,\Big\} \ , \tag{66}$$

where the sum over b extends over all of the subshells of the atom.

Let us assume that subshells a, b, c, •••, n are occupied in the atomic ground state. Then for a given subshell, say subshell a, we have

$$(\, V + mc^2 \,)P_a + c(\, \frac{d}{dr} - \frac{\kappa_a}{r})Q_a = \epsilon_a P_a \quad . \tag{67a}$$

$$-\, c(\, \frac{d}{dr} + \frac{\kappa_a}{r})P_a + (\, V - mc^2 \,)Q_a = \epsilon_a Q_a \quad . \tag{67b}$$

where

$$V\, R_a(r) = -\, \frac{e^2}{r}\, R_a(r) + V_{HF}\, R_a(r) \ . \tag{68}$$

The equations (66,67) are solved self-consistently to give the radial functions, $R_a(r)$, for each occupied level, a, and the Hartree-Fock potential, $V_{HF}(r)$.

For inner electrons of heavy atoms, it is important to account for the finite size of the nucleus. This is done by replacing the nuclear potential, $V_{nuc}(r) = -e^2Z/r$, by

$$V_{nuc}(r) \rightarrow -\frac{e^2}{r}\int_0^r dr'r'^2\rho_{nuc}(r') - e^2\int_r^\infty dr'r'\rho_{nuc}(r') \quad , \tag{69}$$

where $\rho_{nuc}(r)$ is the normalized nuclear charge form factor,

$$\int_0^\infty dr\ r^2\ \rho_{nuc}(r) = Z \ . \tag{70}$$

For Z > 9, the nuclear form factor can be approximated by a uniform ball of charge of radius c_{nuc}, with

$$c_{nuc} = 1.079\ A^{1/3} + 0.745\quad (0.065)\quad \text{fm}\ . \tag{71}$$

A precise evaluation of nuclear size corrections is made using nuclear form factors determined from electron-nucleus scattering experiments [5] or from x-ray measurements in muonic atoms [6].

The most time consuming step in the numerical solution of the DHF equations is the evaluation of the Hartree screening potentials $v_k(a,b,r)$ defined in Eq.(62). These potentials may be written out explicitly as the sum of two indefinite integrals

$$v_k(a,b,r) = \frac{1}{r^{k+1}}\int_0^r dr'\ r'^k\ [\ P_a(r')P_b(r')+Q_a(r')Q_b(r')\ ]$$
$$+\ r^k\int_r^\infty \frac{dr'}{r'^{k+1}}[\ P_a(r')P_b(r')+Q_a(r')Q_b(r')\ ]\ . \tag{72}$$

One method that can be used to evaluate the functions $v_k(a,b,r)$ is to accumulate the two indefinite integrals in Eq.(72), segment-by-segment, using Lagrangian integration formulas to evaluate individual segments,

$$\int_{t_m}^{t_{m+1}} dt\ f(t) = \frac{h}{D}\sum_k a_k(m)\ f(t_k) \quad . \tag{73}$$

The Lagrangian coefficients $a_k(m)$ and D are given in [7]. For example, a Lagrangian integration scheme with an error of order h^6 per segment is given by choosing D = 720 and using the integration expansion coefficients, $a_k(m)$, given in Table I.

Table I. Lagrangian coefficients, $a_k(m)$, D = 720.

m\k	-2	-1	0	1	2
-2	251	646	-264	106	-19
-1	-19	346	456	-74	11
0	11	-74	456	346	-19
1	-19	106	-264	646	251

The first segment (from t_1 to t_2) of the first integral in Eq.(72) is evaluated using the formulas from the first row of the table, the second and successive segments are evaluated using the second row formulas, which are slightly more accurate. Each seqment is added to the sum of the previous segments in order to accumulate the indefinite integral. The second integral in Eq.(72) is accumulated by inward integration, starting from the fourth row coefficients to evaluate the last segment (t_{max} to t_{max-1}) and using the next to last row coefficients for the remaining intervals. The integrand in Eq.(72) must be multiplied by the Jacobian, r'(t), to carry out the integration correctly on the uniformly-spaced t grid. Alternatively, one can convert Eq.(72) into a second-order linear inhomogeneous differential equation, that can be solved by finite difference techniques similar to those described in the previous lecture. My personal preference is the Lagrangian scheme.

Both the Lagrangian and the finite difference schemes require an iteration procedure to be carried out; i.e., the value of an unknown function must be determined at point t_n before it can be found at point t_{n+1}. Such iteration schemes are not "vectorizable" on supercomputers such as the CRAY. The evaluation of these Hartree screening functions limits the speed of atomic structure calculations.

The definite integrals required for normalization of the wave-function and other purposes are vectorizable. My favorite scheme for evaluating a definite integral is the trapezoidal rule with endpoint corrections. This rule is based on the Euler-Maclaurin sum formula .

$$\int_{t_1}^{t_n} dt\ f(t) = h\sum_{k=1}^{n} f_k - \frac{1}{2}h(f_n + f_1) - \frac{B_2}{2!}h^2(f_n' - f_1')$$
$$- \frac{B_4}{4!}h^4(f_n''' - f_1''') \cdots - \frac{B_{2k}}{2k!}h^{2k}(f_n^{(2k-1)} - f_1^{(2k-1)}) + R_k \quad . \tag{74}$$

where B_{2k} are Bernoulli numbers and where the prime symbol is used to represent differentiation. The error, R_k, in Eq.(75) is of order h^{2k+3}. If derivatives are written as linear combinations of functional values near the endpoints, then formula (74) can be rewritten in the form

$$\int_{t_1}^{t_n} dt\ f(t) = \frac{h}{D}\sum_{\ell=1}^{m} c_\ell\ (f_\ell + f_{n-\ell+1}) + h\sum_{k=m+1}^{n-m} f_k \quad , \tag{75}$$

where c_ℓ and D are determined from Eq.(74) and from Lagrangian differentiation formulas. The first term in Eq.(75) is the endpoint correction term; the second is the contribution from the remainder of the interval. This second term is the largest contribution to the integral and is written in form perfectly suited to high-speed evaluation on a vector machine. To carry out a definite integral with an error of order h^7, for example, one would choose:

$$D = 1440 \ ,$$
$$(c_\ell) = (475, 1902, 1104, 1586, 1413) \ ,$$

in Eq.(75).

Now let us outline a two-step scheme for solving the DHF equations:

Step 1. Local Potential Approximation

Start with a good local approximation, U(r), to the DHF potential and solve the radial Dirac equations for each occupied orbital using the eigenvalue scheme developed in the last lecture. For example, in neon the $1s_{1/2}$, $2s_{1/2}$, $2p_{1/2}$, and $2p_{3/2}$ levels are all occupied, so we solve radial Dirac equations for the four corresponding orbitals. I use a local potential proportional to the direct part of the HF potential,

$$U(r) = (1 - \frac{1}{N})\ V_{HF}^{dir}(r) \ .$$

Here N is the total number of electrons in the atom. This is an approximation to the DHF equations that is adequate as a starting point for the iterative solution to the complete DHF equations with exchange. Evaluation of U(r) requires iteration as well, since the direct part of the HF potential depends on the values of the unknown radial functions.

We start the iteration for U(r) using Coulomb radial functions, then we recalculate the radial wavefunctions using U(r), and continue in this way until convergence is achieved. Unfortunately, such an iteration scheme does not always converge, but often oscillates with increasing amplitude at each step. A simple modification of the iteration, obtained by replacing the potential determined after each step by its average with the potential calculated at the previous step,

$$U^{(n+1)} \to \frac{1}{2}\,[U^{(n+1)} + U^{(n)}] \ ,$$

is sufficient to damp the oscillations and to achieve convergence. In practice, one lets this iteration scheme continue until the energy of each occupied orbital has converged to about one part in a thousand:

$$|\epsilon_k^{(n+1)} - \epsilon_k^{(n)}| < 0.001\,|\epsilon_k^{(n)}| \quad , \text{ for all } k \ .$$

At this point one is in a position to go on to the second step, solving the DHF equations with proper attention to exchange.

Step 2. Iterate using Fock's Generalized Green's Function

To set up an iteration scheme suitable for the complete DHF equations, we first rewrite Eqs.(67a,b) in the form

$$[\, H_a + V_{HF} \,]\, R_a(r) = \epsilon_a\, R_a(r) \quad , \tag{76}$$

where H_κ is the two-component radial hamiltonian:

$$H_a = \begin{bmatrix} (V_{nuc} + mc^2) & c\,(\frac{d}{dr} - \frac{\kappa_a}{r}) \\ -c\,(\frac{d}{dr} + \frac{\kappa_a}{r}) & (V_{nuc} - mc^2) \end{bmatrix} \ .$$

Equation (76) can in turn be written:

$$[H_a + U - \epsilon_a^{(0)}]R_a = [U - V_{HF}]\, R_a + (\epsilon_a - \epsilon_a^{(0)})\, R_a \ , \tag{77}$$

where $\epsilon_a^{(0)}$ is the eigenvalue of the local potential problem discussed above:

$$[H_a + U]\, R_a^{(0)} = \epsilon_a^{(0)}\, R_a^{(0)} \quad . \tag{78}$$

We write Eq.(77) in an the iterative form:

$$[H_a + U - \epsilon_a^{(0)}]\, R_a^{(n+1)} = [U - V_{HF}^{(n)}]\, R_a^{(n)} + (\epsilon_a^{(n+1)} - \epsilon_a^{(0)})\, R_a^{(n)} \ . \tag{79}$$

Equation (78) is first solved to give $R_a^{(0)}$ for each a, then Eq.(79) is iterated for n = 0, 1, ···, until satisfactory convergence is achieved.

The inhomogeneous eigenvalue equation (79) has a solution regular at r = 0 and r = ∞ only if its right-hand side is orthogonal to the solution of the corresponding homogeneous equation (78). This orthogonality constraint permits one to determine the value of $\epsilon_a^{(n+1)}$ from the solutions $R_a^{(n)}$ at the nth step of iteration:

$$\epsilon_a^{(n+1)} = \epsilon_a^{(0)} + \langle R_a^{(0)} |\ V_{HF}^{(n)} - U\ | R_a^{(n)} \rangle / \langle R_a^{(0)} | R_a^{(n)} \rangle \quad . \tag{80}$$

Using this value of $\epsilon_a^{(n+1)}$, we are to be able to find regular solutions to Eq.(78).

We determine solutions to Eq.(78) using the generalized Green's function technique originally applied to calculate nonrelativistic wavefunctions for sodium by Fock and Petrashen. To describe the technique, let us consider the two-component linear system of equations

$$(\mathcal{L} - e)y = -\mathfrak{R} \quad , \tag{81}$$

where e is an eigenvalue of the linear operator $\mathcal{L}$,

$$\mathcal{L}f = ef \quad , \tag{82}$$

belonging to eigenfunction f. The eigenfunction f(r) is assumed to be normalized,

$$\int_0^\infty f(r)\cdot f(r)\, dr = 1 \ .$$

Introduce the two functions y_0 and y_∞ as solutions to

$$(\mathcal{L} - e)y_0 = f \quad , \text{ regular at } r = 0 \ , \tag{83a}$$

$$(\mathcal{L} - e)y_\infty = f \quad , \text{ regular at } r = \infty \ . \tag{83b}$$

A radial Green's function may be constructed from y_0 and y_∞:

$$G(r,r') = \begin{cases} y_\infty(r)\ f(r') + f(r)\ y_0(r') & \text{for } r' < r \\ f(r)\ y_\infty(r') + y_0(r)\ f(r') & \text{for } r' > r \end{cases} \ . \tag{84}$$

This generalized Green's function satisfies the equation

$$(\mathcal{L} - e)\ G(r,r') = f(r)\ f(r') - \delta(r-r') \quad . \tag{85}$$

A particular solution to Eq.(81) may now be written as

$$y = \int_0^\infty G(r,r')\cdot\mathcal{R}(r')\, dr'$$

$$= y_\infty(r) \int_0^r f(r')\cdot\mathcal{R}(r')\, dr' + f(r) \int_0^r y_0(r')\cdot\mathcal{R}(r')\, dr'$$

$$+ f(r) \int_r^\infty y_\infty(r')\cdot\mathcal{R}(r')\, dr' + y_0(r) \int_r^\infty f(r')\cdot\mathcal{R}(r')\, dr' \quad . \tag{85}$$

This particular solution is orthogonal to the homogeneous eigenfunction, $f(r)$:

$$\int_0^\infty y(r)\cdot f(r)\, dr = 0 \ .$$

A normalized solution to the equation may be obtained by adding a multiple of f to the particular solution y:

$$y_{norm} = \lambda\, f + y \quad , \tag{86}$$

where

$$\lambda = \sqrt{1-\langle\, y \mid y\, \rangle} \quad . \tag{87}$$

In our application, $\mathcal{L} = H_a + U$ and $e = \epsilon_a^{(0)}$. Our procedure is to construct the generalized Green's function for each orbital a. Actually only the arrays, $y_0(r_i)$ and $y_\infty(r_i)$, are generated and stored for each orbital a. These arrays are used at each stage of the iteration to solve the inhomogeneous equations by quadratures using Eqs.(85-86). After the iteration has converged, we have solutions to the DHF equations and are in a position to study the correlation corrections and obtain really precise atomic wavefunctions. Numerical aspects of the correlation problem are considered in the next lecture.

IV. NUMERICAL ASPECTS OF THE ATOMIC MANY-BODY PROBLEM

To go beyond the independent-particle approximation and obtain energies and wavefunctions that include correlation corrections, one must be able to carry out excited state sums such as those in Eqs.(23) and (24). The direct approach to such sums is to construct the required solutions of the single-particle equations (including all discrete bound state solutions together with the positive energy continuum) and carry out the sums directly. An alternative method, often used in nonrelativistic calculations, is to convert the sums into integrals over solutions to inhomogeneous differential equations (single sums lead to the so-called Dalgarno-Lewis or Sternheimer equations and double sums lead to partial differential equations in two variables called pair-equations). We concentrate on the direct approach, since it appears to be the most powerful method to attack the relativistic problem.

To simplify the treatment of the sums, we approximate the real spectrum of the Dirac equation by a discrete spectrum obtained by placing the atom in a cavity of finite radius R. This radius is chosen so large that the low-lying states of interest are unaffected by the boundary (in practice, we choose R = 40 a.u. for neutral atoms). The effect of the cavity is to discretize the continuum; the spectrum of the Dirac equation in a cavity consists of discrete positive energy states and discrete negative energy states. In the relativistic case we impose the MIT bag model boundary condition [8],

$$Q_{n\kappa}(R) = P_{n\kappa}(R) \quad , \tag{88}$$

at the cavity boundary. We use the MIT bag model boundary conditions in order to avoid problems with the "Klein paradox" that arise when one attempts to confine a particle to a cavity using an infinite potential barrier at $r = R$ [9]. The low-lying positive energy states in the cavity spectrum agree very closely with the low-lying states in the original discrete spectrum; but as the principal quantum number increases, the boundary of the cavity modifies the spectrum more and more strongly. It is the discrete finite-cavity spectrum, not the original spectrum of the Dirac equation, that we approximate by a finite basis set. Because of the projection operators that appear in Eq.(6), only positive energy states contribute to the sums over excited states in expressions for wavefunctions and energies.

The infinite discrete spectrum is approximated by expanding the exact states in terms of a finite set of basis functions. Finite basis sets for the Dirac equations have received a great deal of attention during the past few years, and a number of successful attempts have been made to apply finite basis techniques to problems in relativistic quantum mechanics [10-12]. One way to understand our particular finite basis set approximation is to imagine that we are approximating solutions to the Dalgarno-Lewis differential equation in the one-dimensional case, or to the pair equations in the two-dimensional case. The solutions to these differential equations are smooth functions of r that vanish outside atomic dimensions. It is possible to approximate such smooth compact functions to high accuracy using piecewise polynomials. First, the radial grid from $r = 0$ to $r = R$ is divided up into segments. We then introduce a family of polynomials of a fixed degree on each segment and adjust the coefficients of the resulting piecewise polynomials to fit the functions of interest as well as possible. Approximating known functions by piecewise polynomials can be done in a completely systematic way using B-splines [13].

Our aim is to approximate solutions of the Dirac equation in a cavity of radius R with B-splines. Following the notation of deBoor [13], we designate the endpoints of the segments on the x axis by the "knot sequence" $\{x_i\}$, $i = 1, 2, \cdots$. The B-splines of order k are then defined in the interval spanned by these knots by the relations:

$$B_i^1(x) = \begin{cases} 1 \ , & x_i \leq x < x_{i+1} \\ 0 \ , & \text{otherwise} \end{cases} \ , \tag{89a}$$

$$B_i^k(x) = \frac{x - t_i}{t_{i+k-1} - t_i} B_i^{k-1}(x) + \frac{t_{i+k} - x}{t_{i+k} - t_{i+1}} B_{i+1}^{k-1}(x) \ . \tag{89b}$$

The B-splines of order k are piecewise polynomials of degree k-1. The function $B_i^k(x)$ vanishes outside the interval $[x_i, x_{i+k}]$. A software package written in FORTRAN is available [13] to generate all of the nonvanishing B-splines of order k associated with the knot sequence $\{x_i\}$ using the above recursion relations.

The set of B-splines of order k on a knot sequence $\{x_i\}$ forms a complete basis for piecewise polynomials of degree k-1 on the intervals defined by the knot sequence. We represent the solutions to the Dirac equation as linear combinations of these B-splines and we work with the B-spline representation of the functions, rather than with the functions themselves. The radial Dirac equation can be written

$$\begin{bmatrix} V(r) & c\,(d/dr - \kappa/r) \\ -c\,(d/dr + \kappa/r) & (V(r) - 2mc^2) \end{bmatrix} \begin{bmatrix} P_\kappa(r) \\ Q_\kappa(r) \end{bmatrix} = \epsilon \begin{bmatrix} P_\kappa(r) \\ Q_\kappa(r) \end{bmatrix} \ . \tag{90}$$

This form is slightly different from the one appearing in Eqs.(67a,b) in that we have replaced the energy ϵ by $\epsilon \rightarrow \epsilon - mc^2$ to facilitate comparisons with nonrelativistic calculations. The negative energy states are those with $\epsilon < -2mc^2$, whereas the positive energy states have $\epsilon < 0$ for low-lying states, and $\epsilon > 0$ for the higher states. One approach to approximating Eq.(90) is to expand the equations in terms of basis functions and then demand that the resulting equations be satisfied at a suitably chosen set of *collocation points*. This approach has been successfully applied to the Dirac equations by Bottcher and Strayer [14]. Another approach, which is employed here, is to use a *Galerkin method* [15]; the equations are expressed in terms of an action principle, and the action integral is expanded in terms of basis functions. The radial Dirac equation (90) can be derived from an action principle $\delta S = 0$, with

$$S = \frac{1}{2}\int_0^R [c\, P_\kappa(r)\, (\, d/dr - \kappa/r\,)\, Q_\kappa(r) - c\, Q_\kappa(r)\, (\, d/dr + \kappa/r\,)\, P_\kappa(r) + V(r)\, (\, Q_\kappa(r)^2 + P_\kappa(r)^2\,) - 2mc^2\, Q_\kappa(r)^2\,]\, dr - \frac{1}{2}\epsilon \int_0^R [\, P_\kappa(r)^2 + Q_\kappa(r)^2\,]\, dr\ . \tag{91}$$

From the point of view of the variational principle, the parameter ϵ is a Lagrange multiplier introduced to ensure that the normalization constraint,

$$\int_0^R [\, P_\kappa(r)^2 + Q_\kappa(r)^2\,]\, dr = 1 \quad , \tag{92}$$

is satisfied. The first variation of S is given by

$$\delta S = \int_0^R \Big[\delta P_\kappa(r)\, [\, c\, (\, d/dr - \kappa/r\,)\, Q_\kappa(r) + V(r)\, P_\kappa(r)\,] - \delta Q_\kappa(r)\, [\, c\, (\, d/dr + \kappa/r\,)\, P_\kappa(r) + (\, 2mc^2 - V(r)\,)\, Q_\kappa(r)\,] \Big]\, dr - \epsilon \int_0^R [\, \delta P_\kappa(r)\, P_\kappa(r) + \delta Q_\kappa(r)\, Q_\kappa(r)\,]\, dr + \frac{1}{2} c \Big[\, P_\kappa(r)\, \delta Q_\kappa(r) - Q_\kappa(r)\, \delta P_\kappa(r)\, \Big]_0^R\ . \tag{93}$$

The requirement $\delta S = 0$ for variations of $P_\kappa(r)$ and $Q_\kappa(r)$ satisfying the constraints $\delta P_\kappa(0) = \delta Q_\kappa(0) = 0$, and $\delta P_\kappa(R) = \delta Q_\kappa(R) = 0$, leads to the radial Dirac equations written down in Eq.(90). Specific boundary conditions can be imposed on the solutions by removing the boundary constraints from the variational functions $\delta P_\kappa(r)$ and $\delta Q_\kappa(r)$, and by adding appropriate boundary terms to the action integral S. In the present case, we add to S the boundary term:

$$S' = \frac{c}{4}[P_\kappa(R)^2 - Q_\kappa(R)^2] + \frac{c}{2} P_\kappa(0)^2 - \frac{c}{2} P_\kappa(0)Q_\kappa(0) \ , \ \text{for } \kappa < 0 \ , \tag{94a}$$

$$S' = \frac{c}{4}[P_\kappa(R)^2 - Q_\kappa(R)^2] + c^2 P_\kappa(0)^2 - \frac{c}{2} P_\kappa(0)Q_\kappa(0) \ , \ \text{for } \kappa > 0 \ . \tag{94b}$$

Combining the boundary term in Eq.(92) with the the terms obtained from the variation of Eqs.(94a,b) lead to boundary terms given by

$$\delta(S+S')_{\text{boundary}} = \frac{c}{2}[P_\kappa(R) - Q_\kappa(R)]\delta Q_\kappa(R) + \frac{c}{2}[P_\kappa(R) - Q_\kappa(R)]\delta P_\kappa(R)$$
$$-c\, P_\kappa(0)\delta Q_\kappa(0) + cP_\kappa(0)\delta P_\kappa(0) \ , \quad \kappa < 0 \ , \quad \text{or} \tag{95a}$$

$$\delta(S+S')_{\text{boundary}} = \frac{c}{2}[P_\kappa(R) - Q_\kappa(R)]\delta Q_\kappa(R) + \frac{c}{2}[P_\kappa(R) - Q_\kappa(R)]\delta P_\kappa(R)$$
$$- c\, P_\kappa(0)\delta Q_\kappa(0) + 2c^2 P_\kappa(0)\delta P_\kappa(0) , \quad \kappa > 0 \ . \tag{95b}$$

For variations of $P_\kappa(r)$ and $Q_\kappa(r)$, not subject to boundary constraints, the boundary terms in Eqs.(95a,b) vanish if

$$P_\kappa(R) = Q_\kappa(R) \ , \tag{96a}$$

which is just the MIT bag model boundary condition, and if

$$P_\kappa(0) = 0 \ . \tag{96b}$$

The choice of the boundary terms in Eqs.(95a,b) is to some extent arbitrary. The particular choice made here was found to eliminate spurious terms that can occur in the spectrum for $\kappa > 0$ when the Dirac equation is expanded in terms of finite basis sets [10].

We expand the radial functions $P_\kappa(r)$ and $Q_\kappa(r)$ in terms of B-splines of order k as:

$$P(r) = \sum_{i=1}^{n} p_i \; B_i(r) \quad ,$$
$$Q(r) = \sum_{i=1}^{n} q_i \; B_i(r) \quad . \tag{97}$$

The subscript "κ" has been omitted from the functions $P_\kappa(r)$ and $Q_\kappa(r)$ and the subscript "k" has been omitted on the functions $B_{i,k}(x)$ for notational simplicity. Substituting these expressions into the action functional leads to a quadratic form in p_i and q_i for $S + S'$. The variational principle reduces to the system of algebraic equations:

$$d(S+S')/dp_i = 0, \quad i = 1, n \; , \tag{98a}$$

and,

$$d(S+S')/dq_i = 0, \quad i = 1, n \; . \tag{98b}$$

These equations can be expressed in the form of a 2n × 2n symmetric generalized eigenvalue equation

$$A \; v = \epsilon \; B \; v, \tag{99}$$

where the 2n vector v is given by

$$v = (\; p_1, \; p_2, \cdots, \; p_n, \; q_1, \; q_2, \cdots, \; q_n \;) \; , \tag{100}$$

and where A and B are symmetric 2n × 2n matrices. Specifically:

$$A = \begin{bmatrix} (V) & c\,[\,(D) - (\kappa/r)\,] \\ -\,c\,[\,(D) + (\kappa/r)\,] & -\,2\,mc^2\,(C) + (V) \end{bmatrix} + A' \; , \tag{101}$$

$$B = \begin{bmatrix} (C) & 0 \\ 0 & (C) \end{bmatrix} , \tag{102}$$

where the 2n × 2n matrix A' is a contribution from the boundary terms

$$A'_{ij} = c\,\delta_{i,1}\,\delta_{j,1} - \frac{c}{2}\,\delta_{i,1}\,\delta_{j,n+1} - \frac{c}{2}\,\delta_{i,n+1}\,\delta_{j,1} + \frac{c}{2}\,\delta_{i,n}\,\delta_{j,n} - \frac{c}{2}\,\delta_{i,2n}\,\delta_{j,2n}\,, \quad \kappa < 0\,, \quad \text{or} \tag{103a}$$

$$A'_{ij} = 2c^2\,\delta_{i,1}\,\delta_{j,1} - \frac{c}{2}\,\delta_{i,1}\,\delta_{j,n+1} - \frac{c}{2}\,\delta_{i,n+1}\,\delta_{j,1} + \frac{c}{2}\,\delta_{i,n}\,\delta_{j,n} - \frac{c}{2}\,\delta_{i,2n}\,\delta_{j,2n}\,, \quad \kappa > 0\,. \tag{103b}$$

The $n \times n$ matrices (C), (D), (V), and (κ/r) are given by

$$(C)_{ij} = \int B_i(r)\;B_j(r)\;dr\;, \tag{104a}$$

$$(D)_{ij} = \int B_i(r)\;(dB_j(r)/dr)\;dr\;, \tag{104b}$$

$$(V)_{ij} = \int B_i(r)\;V(r)\;B_j(r)\;dr\;, \tag{104c}$$

$$(\kappa/r)_{ij} = \int B_i(r)\;(\kappa/r)\;B_j(r)\;dr\;. \tag{104d}$$

One important numerical point to be mentioned is that the matrices (C), (D), (V), and (κ/r) are diagonally-dominant banded matrices, so that the numerical solution of the generalized eigenvalue equation (99) can be carried out with high accuracy even when the dimensions of the basis set become very large. To solve the eigenvalue problem, we use subroutines from the FORTRAN EISPACK library [16]. If we let v_i^λ be the eigenvector associated with eigenvalue, ϵ_λ , then we find

$$\sum_{i,j} v_i^\lambda\;(B)_{ij}\;v_j^\sigma\; = \delta_{\lambda\sigma}\;. \tag{105}$$

Written in terms of the approximate eigenfunctions $P^\lambda(r)$ and $Q^\lambda(r)$, this equation expresses the orthonormality conditions

$$\int [\;P^\lambda(r)\;P^\sigma(r) + Q^\lambda(r)\;Q^\sigma(r)\;]\;dr\; = \delta_{\lambda\sigma}\;. \tag{106}$$

To summarize, the symmetric generalized eigenvalue equation (99) is solved to give 2n eigenvalues, ϵ_λ , and 2n orthonormal eigenfunctions:

Table II. Eigenvalues of the symmetric generalized eigenvalue equation using the B-spline approximation to the radial Dirac equation with $\kappa = -1$ in a Coulomb potential with $Z = 2$. Cavity radius $R = 40$ au. First interior knot = 0.002 au. We use $n = 40$ B-splines of order $k = 7$.

State	B-spline Negative Energy $+ mc^2$	B-spline Positive Energy $- mc^2$	Coulomb field Dirac Equation $n\ell$	$E_{n\ell} - mc^2$
1	-0.07847197	-2.00010651	1s	-2.00010651
2	-0.10914310	-0.50003330	2s	-0.50003329
3	-0.14295485	-0.22223407	3s	-0.22223406
4	-0.18003609	-0.12500646	4s	-0.12500541
5	-0.22119817	-0.07998339	5s	-0.08000290
6	-0.29438507	-0.05380099	6s	-0.05555728
7	-0.41086999	-0.02740878	..	...
8	-0.43435448	0.00654298	..	...
9	-0.64606974	0.05464323		
10	-1.01337089	0.10068188		
11	-1.63730978	0.29987895		
12	-2.71545051	0.40955814		
13	-4.61461875	1.24849717		
14	-8.02433776	3.20191043		
15	-14.25275433	7.42845945		
16	-25.79649563	16.20928403		
17	-47.43772681	34.02134879		
18	-88.35222309	69.61421924		
19	-166.13428773	139.99016771		
20	-173.52324422	173.42003277		
21	-314.40061326	277.92960523		
22	-596.70231465	545.83074309		
23	-1130.83631738	1059.86161826		
24	-2127.13856652	2028.07181496		
25	-3938.56928968	3800.19510168		
26	-7106.58537009	6913.11174828		
27	-12381.54723242	12110.64378928		
28	-20732.29633731	20352.19833424		
29	-33412.08426286	32877.22380726		
30	-52139.48448356	51383.70508368		
31	-79416.56837423	78342.16398555		
32	-119041.60191054	117500.66951358		
33	-177001.37025039	174761.83089322		
34	-263213.77492433	259891.66907592		
35	-395381.45868381	390289.40659954		
36	-608884.96376640	600640.06250182		
37	-987203.92683466	972520.98050020		
38	-1779801.13063116	1748823.57610166		
39	-4087410.99630964	3994928.81660704		
40	-23278423.46230817	22637209.28829002		

$$P^{\lambda}(r) = \sum_{i=1}^{n} v_i^{\lambda} B_i(r) \quad , \tag{107a}$$

$$Q^{\lambda}(r) = \sum_{i=n+1}^{2n} v_i^{\lambda} B_i(r) \quad . \tag{107b}$$

Let us consider, as an illustration, the case of a Coulomb field with charge Z = 2. We choose a cavity radius R = 40.0 a.u., and examine the approximate cavity spectrum of the Dirac equation for $\kappa = -1$. We express our wavefunctions as linear combinations of B-splines with n = 40 and k = 7. In Table II we list the 80 eigenvalues given by soving Eq.(99). As is seen from the table, 40 of these eigenvalues lie below $-2mc^2$, while the remaining 40 eigenvalues (the positive energy group) grow from small negative values to large positive values. The lowest few terms of the positive energy spectrum given in column 3, the eigenvalues of the low-lying bound states, are seen to be very close to the exact eigenvalues of Coulomb field Dirac equation in the absence of a cavity, given in column 5 of the table.

There are several tests of the quality of this spectrum that can be easily carried out. One test is to determine how well the Thomas-Reiche-Kuhn (TRK) sum rule is satisfied. This rule can be formulated in the relativistic case in terms of the two partial waves, $\kappa = \ell$, and $\kappa = -\ell - 1$, associated with orbital angular momentum ℓ :

$$\left[\frac{\ell}{2\ell+1} \sum_n \omega_{n0} \;|\langle \kappa = -1,\, n = 0 \;|\; r \;|\; \kappa = \ell,\, n \rangle|^2 \right.$$
$$\left. + \frac{\ell+1}{2\ell+1} \sum_n \omega_{n0} \;|\langle \kappa = -1,\, n = 0 \;|\; r \;|\; \kappa = -\ell-1,\, n \rangle|^2 \right] = 0 \; . \tag{108}$$

where $\omega_{n0} = \epsilon_{n,\kappa} - \epsilon_{0,-1}$. For the nonrelativistic case the corresponding TRK sum rule is:

$$\sum_n \omega_{n0} \;|\langle \ell = 0,\, n = 0 \;|\; r \;|\; \ell,\, n \rangle|^2 = \frac{\ell\,(\ell+1) + 1}{2} \; . \tag{109}$$

The sum over the positive energy states in the relativistic TRK sum rule gives a value close to the nonrelativistic expression, but this value is precisely cancelled by the negative energy sum. A test of the TRK sum rule using the spectra for higher values of angular momenta generated using B-splines with n = 40 and k = 7 in a Coulomb potential with Z = 2 is given in Table III. It is seen in the second column of the table that the cancellation between positive and negative terms is better than one part in 10^8 for all ℓ from 0 to 9. By increasing the number of splines to 50 and the order to k = 9, we can improve the cancellation to one part in 10^{10}. The deviation of the sum over positive energy states from the nonrelativistic limit shown in column 3, is a real relativistic effect, independent of the radius of the cavity.

Table III. Test of the relativistic TRK sum rule Eq.(108) for various ℓ using a B-spline representation of the Coulomb field Dirac equation confined to a cavity of radius R = 40 au. We use 40 B-splines with k = 7. The basis set is the same as that used in the previous table.

ℓ	Sum over positive and negative energy states	Sum over positive energy states only	Nonrelativistic limit $((\ell+1)\ell+1)/2$
0	6.6×10^{-9}	0.49974	0.5
1	6.8×10^{-9}	1.49973	1.5
2	6.3×10^{-9}	3.49733	3.5
3	6.3×10^{-9}	6.48821	6.5
4	6.3×10^{-9}	10.46637	10.5
5	6.4×10^{-9}	15.42440	15.5
6	6.8×10^{-9}	21.35369	21.5
7	6.6×10^{-9}	28.24458	28.5
8	7.0×10^{-9}	36.08654	36.5
9	7.7×10^{-9}	44.86831	45.5

Another test of the quality of the spectrum is the value of the dipole polarizability α_d, which is known exactly for an electron in a Coulomb field in the nonrelativistic limit, and analytically to order $(\alpha Z)^2$ in the relativistic case; for Z = 2, the n = 50, k = 9 spline routine gives

$$\alpha_d = 0.281187877\cdots a_0^3 \quad , \tag{110}$$

compared to Kaneko's analytic result[17],

$$\alpha_d = [\ 9/2 - 14/3\ (\alpha Z)^2 + \cdots\ ]\ Z^{-4}\ a_0^3$$
$$= 0.281187873\cdots\ a_0^3\ . \qquad (111)$$

Other tests of the spectrum include energy weighted sum rules such as the second energy-weighted sum rule [10]:

$$S_2 = \frac{2}{9}\sum_n \omega_{n0}^2\ |\ \langle\ \kappa = -1,\ n = 0|\ r\ |\ \kappa = 1,\ n\ \rangle\ |^2 +$$
$$+ \frac{4}{9}\sum_n \omega_{n0}^2\ |\ \langle\ \kappa = -1,\ n = 0|\ r\ |\ \kappa = -2,\ n\ \rangle\ |^2 = \alpha^{-2}\ . \qquad (112)$$

We obtain the value α^{-2} to an accuracy of one part in 10^8 using the $n = 40$, $k = 7$, $Z = 2$, Coulomb field spline spectrum, and to 1 part in 10^{10} using a spectrum with $n = 50$ and $k = 9$.

As a direct application of the B-spline basis to a problem arising in relativistic many-body perturbation theory, we have considered the calculation of the second-order corrections to the energy of helium starting from a Dirac Coulomb field approximation [18]. The expression for the second-order energy, which contains a relativistic pair function implicitly, is

$$E^{(2)} = \sum_\ell \Bigg[\frac{\ell}{(2\ell+1)^3} \sum_{n,m} \frac{R_\ell(n\kappa,\ m\kappa,\ 1s,\ 1s)^2}{(\ 2\ \epsilon_{1s} - \epsilon_{n\kappa} - \epsilon_{m\kappa}\ )}\Bigg|_{\kappa\ =\ \ell}$$
$$+ \frac{\ell+1}{(2\ell+1)^3} \sum_{n,m} \frac{R_\ell(n\kappa,\ m\kappa,\ 1s,\ 1s)^2}{(\ 2\ \epsilon_{1s} - \epsilon_{n\kappa} - \epsilon_{m\kappa}\ )}\Bigg|_{\kappa\ =\ -\ell-1} \Bigg]\ . \qquad (113)$$

The quantities $R_\ell(a,b,c,d)$ are Slater integrals, defined by

$$R_\ell(a,b,c,d) = \int dr\ dr'\ \frac{r_<^\ell}{r_>^{\ell+1}}\ [\ P_a(r)P_c(r) + Q_a(r)Q_c(r)\ ]\ \times$$
$$[\ P_b(r')P_d(r') + Q_b(r')Q_d(r')\ ]\ . \qquad (114)$$

In Eq.(113), the sums over n and m are restricted to the positive energy states. The nonrelativistic sums corresponding to Eq.(113) have been carried out to high accuracy [19] giving $E^{(2)}_{nr} = -0.15766643\cdots$. Using our B-spline basis, with n = 50 and k = 9 for helium, we find

$$E^{(2)} = -0.15768147 = -0.15766643 - 0.071\ (\alpha Z)^2 \ . \qquad (115)$$

The corresponding second-order Breit interaction (one Breit + one Coulomb) gives the result:

$$B^{(2)} = -0.563\ (\alpha Z)^2 \ . \qquad (116)$$

The resulting relativistic corrections to the second-order energy,

$$\Delta E^{(2)}_{rel} = -(0.071 + 0.563)\ (\alpha Z)^2 = -0.634\ (\alpha Z)^2 \ , \qquad (117)$$

are in excellent agreement with values obtained by Drake[20] who calculated the leading coefficients in a 1/Z expansion of the relativistic interaction using correlated nonrelativistic basis sets:

$$\Delta E^{(2)}_{rel} = -(0.070 + 0.565)\ (\alpha Z)^2 = -0.635\ (\alpha Z)^2 \ . \qquad (118)$$

It is seen that the B-splines provide a basis of high enough quality to permit the study of the small relativistic effects in an atom as light as helium.

Various applications of B-splines to calculate correlation corrections to alkali atoms and alkali-like ions in a relativistic framework have been carried out. These applications are at the cutting edge of research in atomic physics. You now have the basic tools to carry out such relativistic atomic structure calculations yourselves; all that remains before you become an expert in relativistic atomic physics is practice!

This work was supported in part by National Science Foundation grant No. PHY-86-08101 and in part by the Service de Physique Atomique at the Centre d'Etudes Nucleaires de Grenoble. A debt of gratitude is owed to C. Bottcher for an introduction to B-spline methods.

REFERENCES

1. G.E. Brown and D.G. Ravenhall, Proc. R. Soc. London, Sec.A **208**, 552 (1951).
2. J. Sucher, Phys. Rev. **A22**, 348 (1980).
3. M. Mittleman, Phys. Rev. **A24**, 1167 (1981).
4. S.A. Blundell, W.R. Johnson, W.Z. Liu, and J. Sapirstein, Phys. Rev. **A39**, 3768 (1989).
5. C.W. de Jager, H. de Vries, and C. de Vries, Atomic Data and Nuclear Data Tables **14**, 479 (1974).
6. R. Engfer, H. Schneuwly, J.L. Vuilleumier, H.K. Walter, and A. Zehnder, Atomic Data and Nuclear Data Tables **14**, 509 (1974).
7. *Handbook of Mathematical Functions* (Dover, New York, 1972) ed. M. Abromowitz and I.A. Stegun, p.915.
8. A. Chodos, R.L. Jaffe, K. Johnson, C.B. Thorn, and V.W. Weisskopf, Phys. Rev. **D9**, 3471 (1974).
9. J.J. Sakurai, *Advanced Quantum Mechanics* (Addison-Wesley, Reading Mass., 1967), p 120.
10. G.W.F. Drake and S.P. Goldman, Phys. Rev. **A23**, 2093 (1981).
11. I.P. Grant, Phys. Rev. **A25**, 1220 (1982).
12. W. Kutzelnigg, Int. J. Quantum Chem. **25**, 107 (1974).
13. Carl deBoor, *A Practical Guide to Splines* (Springer, New York, 1978).
14. C. Bottcher and M.R. Strayer, Ann. Phys. (N.Y.), **175**, 64 (1987).
15. C.W.A. Fletcher, *Computational Galerkin Methods* (Springer, New York, 1984).
16. B.S. Garbow, J.M. Boyle, J.J. Dongarra, and C.B. Moler, *Matrix Eigensystem Routines - EISPACK Guide Extension*, (Springer, Berlin, 1977), p 250.
17. S. Kaneko, J. Phys. **B10**, 3347 (1977).
18. W.R. Johnson and J. Sapirstein, Phys. Rev. Letts., **57**, 1126 (1986).
19. J.D. Baker, D.E. Freund, R.N. Hill, and J.D. Morgan, Phys. Rev. (in press).
20. G.W.F. Drake, Phys. Rev. **A19**, 1387, (1979), and private communication.

A CHAOS PRIMER *

M. Baranger

Center for Theoretical Physics
Laboratory for Nuclear Science
and Department of Physics
Massachusetts Institute of Technology
Cambridge, Massachusetts 02139 U.S.A.

It is certainly impossible to do justice to all of chaos in two lectures. The best I can hope to do is to highlight a few topics here and there, trying to emphasize those subjects which have more relevance in Atomic and Nuclear Physics. Fortunately, there is an excellent book which recounts many aspects of the history of chaos for the general public, and which you should all read. It is

James Gleick
Chaos: Making a New Science
(Viking, New York, 1987)

Its only shortcoming is that it does not give proportional emphasis to the enormous contribution made by the Russians to the field, which is understandable because the author is a journalist and he did all his work by personal interviews. Otherwise, the book gives a wonderful picture of how the new ideas arose simultaneously in many independent fields of research, including mathematics, physics, chemistry, biology, several branches of engineering, meteorology, economics, and of course the stock market. Now you can go to workshops on Chaos and Psychology, Chaos and Jungian Thought, *etc.* ... Gleick describes the struggles that the pioneers in these various fields went through to make sense out of the crazy stuff they were looking at, to overcome self-doubt, to overcome the doubts of their colleagues, until they eventually discovered each other and realized that they were all grappling with the same very, very big thing. And he explains how this thing came to receive the catchy name of Chaos.

Originally, the Greek work Chaos meant the infinite empty space which existed before all things. But in present usage, chaos means total disorder. As you will see, the technical Chaos I am going to speak about is nothing like total disorder. Because you cannot say anything about total disorder, by definition. While

* This work was supported in part by funds provided by the U. S. Department of Energy (D.O.E.) under contract #DE-AC02-76ER03069.

our Chaos, people are saying lots and lots of things about it. It is not disorder, but it is complication. It is pervasive and recurrent complication. You have all heard of fractals, of cellular automata, of scaling theory for critical phenomena: these are all examples of Chaos. Perhaps the best definition of Chaos is the following. You might well have been told in the past, by teachers or by parents, that every simple question has a simple answer. I know I have. This is the axiom upon which schools are built. Of course, we have all learned from mathematicians that some questions don't have any answers at all, and that some questions are meaningless, therefore they are not questions. But Chaos is what happens when a *simple question* has a *complicated answer*. That such a thing could be possible may run completely against one's esthetic sense, and yes it does happen, it happens all the time. Chaos is *not* what happens when you put together one million different objects and let them interact. *Chaos is a property of simple systems.* The first person to ever see Chaos and know what it was, technically, was the French mathematician Henri Poincaré in 1892; he was studying the three-body problem, the sun, the earth, and Jupiter; he was not studying the 1000-body problem. Put together four kinds of atoms, hydrogen, oxygen, nitrogen, and carbon; cook them together in a primordial soup; and you get life! Simple data; complicated, rich results. That is Chaos. The big discovery of the 60's and 70's is just that: the realization that Chaos is a property of simple systems. To see Chaos, you don't need to get a nucleus with 200 neutrons and protons and 10,000 energy levels per MeV; you don't need to look at the whole Milky Way with its trillions of stars and its hundreds of kinds of astronomical objects, from pulsars to globular clusters. You need a couple of variables, a couple of equations, and at least one non-linear term somewhere. The meteorologist Edward Lorenz of MIT, after becoming increasingly frustrated by the unpredictability of the weather, in spite of the availability of very sophisticated and very complex mathematical models, decided that he would reduce the whole thing to three innocent-looking, but non-linear, differential equations

$$\begin{aligned} \dot{X} &= \sigma(Y - X) \\ \dot{Y} &= rX - Y - XZ \\ \dot{Z} &= XY - bZ \quad . \end{aligned} \tag{1}$$

Of course, he was not able to convince most of his colleagues that these trivial-appearing equations had anything to do with the weather and all the complication that we know it to possess. But, lo and behold, when he ran his equations on the computer, their solutions were chaotic and he discovered the first "strange attractor." This was in 1963. And then he understood why the weather was unpredictable.

Simple questions, complicated answers. That is Chaos. Simple equations, complicated solutions. At this point, some of you are crying to say: "Look, this can't be. The solutions cannot contain more information than the equations. If the equations are simple, which means that they contain very little information, the solutions have got to be simple in some way." And you are right: in some way! But not in the way that you think they ought to be simple. Not in the way that the old mechanics textbooks have taught you to expect of simple systems. That's the marvelous thing about Chaos. It is simple, but in a hidden way. That is why Chaos is such a fun game to play. You are chasing after the hidden simplicities. They are there, but they are very subtle, and they open all new realms of mathematics. Most of you have certainly seen the famous color pictures of the Mandelbrot set by Peitgen and Richter. These amazing pictures have done the rounds of the museums of the world, and now you can also get them as a book. The definition of the Mandelbrot set is quite simple: you write

$$z_{n+1} = z_n^2 + c \ .$$

Both z and c are complex. You start with $z_0 = 0$ and you iterate. When the modulus of z_n becomes bigger than some fixed number R, you quit. Then, in the complex plane of c, you paint each point with a color which is in one-to-one correspondence with the number of iterations necessary to reach R. That's all! Simple process; astounding result. When you look at these pictures, you can tell intuitively that there must be something simple about them, but what is it? How would you express it? Such is Chaos. Chaos is not disorder. Chaos is not randomness. Chaos is simple, but its simplicity does not have a local character. You cannot isolate a little piece of the picture and say: look how simple that is. As a matter of fact, no matter how small a piece you isolate, it is always complicated. That is the typical property of a fractal. And the Mandelbrot set is still much more complex in detail than your elementary, garden-variety fractal. The simplicity that is there is a kind of global simplicity. It does not appear under successively finer divisions. It appears some other way. Exactly what way? That is the challenge!

But, at any rate, we know where the origin of this simplicity lies. It lies in the fact that the original equations were simple. And where is the origin of all the complexity? It comes from the initial conditions! When you solve such a system of equations, you always need to start with certain initial conditions. For instance, in the equation of the Mandelbrot set, the initial condition is the complex number c. What happens in all cases is that the solution is a completely unsmooth function of the initial conditions. This *sensitivity to initial conditions* is one of the main characteristics of Chaos. Lorenz called it "the butterfly effect": a butterfly flaps its wings in the fresh air of Colorado today, and the weather in Boston next week is completely changed. That was really a shock for people of the old school, who somehow, for no good reason, had always assumed that a tiny change in initial conditions would produce a tiny change in the results. Once again, the pictures of Peitgen and Richter give you a direct view of this sensitivity. We shall see another example later in these lectures, in connection with the main theorem concerning Chaos in Hamiltonian systems. There we shall get completely different results depending on whether the initial condition is a rational number or an irrational number sufficiently close to a rational number on the one hand, or whether it is some other irrational number on the other hand. Knowing how the rational and the irrational numbers are interlaced, you can tell right away that we shall have a fractal situation.

I am now ready to tell you which particular dynamical systems are capable of being chaotic and which are not. Generally speaking, they are all chaotic, with two kinds of exceptions. One class of exceptions is made up of all linear systems, which can always be solved exactly by Fourier transformation and therefore never show Chaos. The other class, about which I shall say more later, contains some exceptionally simple systems, which of course happen to be the only systems mentioned in the old textbooks, because they are the only ones to be analytically soluble. And that is how generations of students grew up with warped ideas about classical mechanics! There are two kinds of chaos, conservative chaos and dissipative chaos, and the people who work on one tend not to speak to the people who work on the other. Most physicists belong with the conservative bunch, and most others (engineers, biologists, heart doctors, *etc.*...) belong to the dissipative. Conservative chaos was discovered by Henri Poincaré in 1892; dissipative chaos was discovered by Edward Lorenz in 1963. The principle symbol of conservative chaos is the KAM theorem of which I shall speak later; the principle symbol of dissipative chaos is the strange attractor, which I shall discuss briefly. A conservative dynamical system is one in which the volume of phase space is conserved over time. The most important case is that of a Hamiltonian system, for which the conservation of the phase space volume is known as Liouville's theorem. A dissipative dynamical system is one in which the volume of phase space shrinks as time evolves, with the result that the system ends up in a special region of phase space, with a lower number of dimensions, and this region is called an attractor.

Another distinction that can be made is that between a "flow" and a "map." A flow is a continuously evolving dynamical system described by differential equations, for instance the Lorenz system mentioned earlier. A map is a system whose state is specified at discrete instants of time, and its evolution is described by a set of iterative formulae, such as the following one-dimensional example

$$x_{n+1} = rx_n(1 - x_n) \tag{2}$$

which is often used by biologists to describe how an animal population varies from year to year. A flow can be turned in many ways into a map of lower dimensionality by doing what is called "a section." For instance, take the Lorentz system and consider only points with $Z = 0$. Given one such point (X_n, Y_n), you can calculate the next place on the trajectory for which Z will vanish. Call it (X_{n+1}, Y_{n+1}), then calculate the next one (X_{n+1}, Y_{n+2}), *etc.*... . This defines a two-dimensional mapping. For another example, consider a non-linear, forced, damped oscillator described by the equation

$$m\ddot{x} + f(\dot{x}, x) = F_0 \sin 2\pi t/\tau \quad , \tag{3}$$

and plot the variables x and $\dot{x}$ at successive time intervals separated by one period, τ. This is called stroboscopic mapping. These sections are useful because, by reducing the number of dimensions, they make the problem more accessible to our imperfect intuition, which is rarely capable of going beyond three dimensions.

In order to classify these flows and these maps according to whether they are chaotic or not, I need to define phase space. It is the space spanned by the set of coordinates which are needed in order to define completely the present state of the system, and therefore in order to predict its evolution. For instance, for the Lorenz system (1), phase space is the three-dimensional space (X, Y, Z). For a flow governed by the equations

$$\begin{aligned} \dot{X} &= f(X, Y, Z, t) \\ \dot{Y} &= g(X, Y, Z, t) \\ \dot{Z} &= h(X, Y, Z, t) \end{aligned} \tag{4}$$

phase space would be (X, Y, Z, t), *i.e.* four-dimensional, since obviously I need to know t if I want to calculate what happens next. For the map (2), phase space is just (x), one-dimensional. For the oscillator (3), phase space is $(x, \dot{x}, t)$, three-dimensional. There is a theorem which says that, for a flow to show chaos, its phase space needs to be at least three-dimensional. A non-linear flow whose phase space is three-dimensional or more will generally have chaos, unless it can be reduced to independent systems with fewer than three dimensions. Examples are the Lorentz system, of course, or the oscillator (3). For Hamiltonian systems with a time-independent Hamiltonian, phase space always has an even number of dimensions, since it includes N positions and N momenta. Therefore, when $N = 1$, which is conventionally called a one-dimensional system, there is no chaos. But when $N = 2$, $2N = 4$, there is chaos, unless the system can be reduced to two one-dimensional systems, in which case it is called "integrable." A Hamiltonian system with $N = 1$ and a time-dependent Hamiltonian has a three-dimensional phase space; therefore, it can be chaotic. These are the simplest Hamiltonian systems to possess chaos, and therefore they are good ones to study. Let us switch now from flows to maps. There, the situation is very simple: they are all capable of being chaotic (as long as they are non-linear), even if their phase space is one-dimensional. For instance, the map (2), which is a particular quadratic map often called "the logistic map," is famous for its chaos.

I shall now highlight rapidly a few of the best-known things in dissipative chaos, and then I shall go at more length into conservative chaos, because this is a school for atomic and nuclear physicists, and in our work we deal with conservative systems much more than with dissipative ones. First, here are some good books:

1) R. H. Abraham and J. E. Marsden, *Foundations of Mechanics* (Benjamin/ Cummings, 1978).
2) R. H. Abraham and C. D. Shaw, *Dynamics: The Geometry of Behavior* (in four volumes) (Aerial Press, Santa Cruz, 1982–1986).

3) A. J. Lichtenberg and M. A. Liebermann, *Regular and Stochastic Motion* (Springer-Verlag, New York, 1982).
4) H. G. Schuster, *Deterministic Chaos* (Physik–Verlag, 1984).
5) P. Bergé, Y. Pomeau and C. Vidal, *Order Within Chaos* (Hermann-Wiley, Paris and New York, 1984).
6) J. M. T. Thompson and H. B. Stewart, *Non-Linear Dynamics and Chaos* (Wiley, New York, 1986).
7) M. Tabor, *Chaos and Integrability in Non-Linear Dynamics* (Wiley, New York, 1989).

Reference 3) is mostly about conservative chaos; 5) and 6) deal almost exclusively with dissipative chaos; 7) is a balanced mixture, also including some quantum chaos.

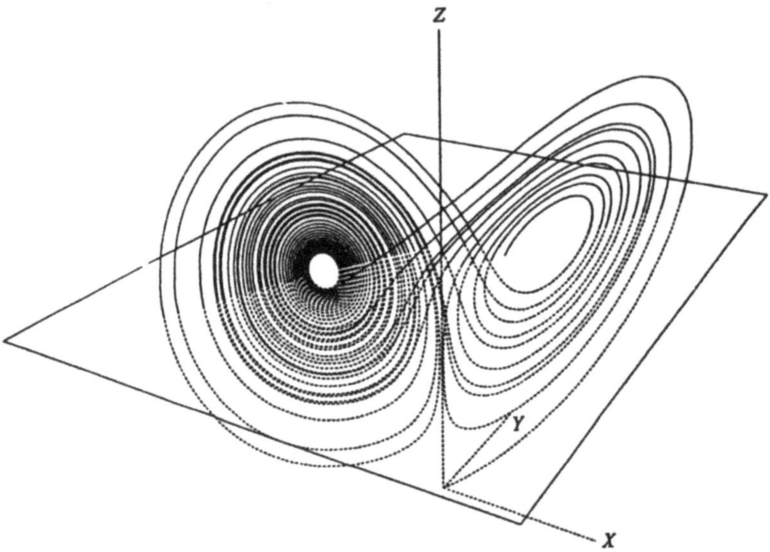

Fig. 1: The Lorentz attractor (from O. Lanford).

For a dissipative dynamical system, as I said earlier, the volume of phase space shrinks as time evolves so that, with arbitrary initial conditions, the system ends up in a special region of phase space, with a lower number of dimensions, and consisting of one or several "attractors." Each attractor has its "basin of attraction," which is the locus of the initial positions leading to it. The simplest attractor is a point. Next comes a cyclic, or periodic, attractor, consisting of a one-dimensional loop described periodically. Next comes a "quasi-periodic attractor," which is a two-dimensional torus with two incommensurate frequencies. After that you begin to see "strange attractors," also known as "chaotic attractors." They are characterized by aperiodicity, sensitivity to initial conditions, and a fractal number of dimensions. Figure 1 shows the famous Lorenz attractor. Its dimensionality is 2.06, so that it looks almost like two flat disks embedded in three dimensions. If the initial conditions are a point on the attractor, the system will describe the entire attractor as it evolves over infinite time. However, for two very close starting points both on the attractor, the two subsequent trajectories will eventually look completely different, although they both describe the entire attractor. I am going to try to give you a feeling for how this kind of object can arise out of the equations of motion. I have two examples: one is a map, which is two-dimensional, so you can see it well; the other is a three-dimensional flow.

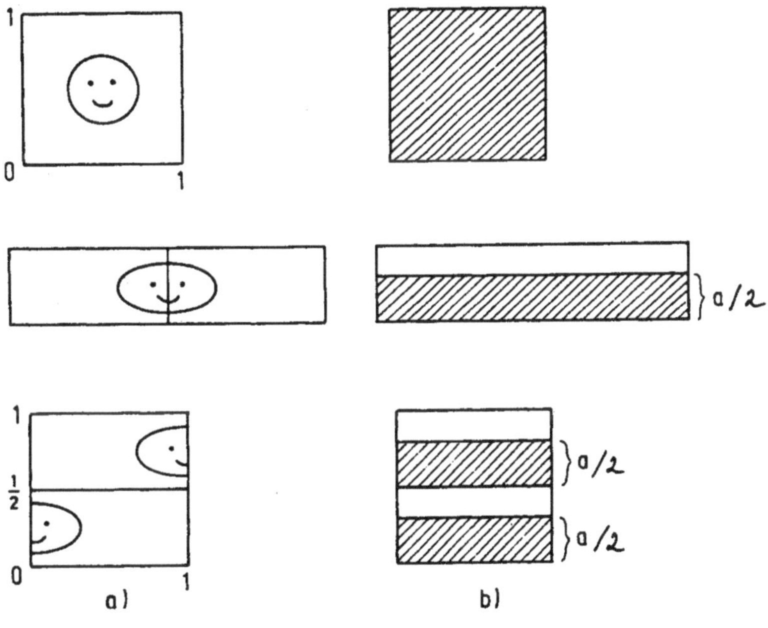

Fig. 2: a) baker's transformation; b) dissipative baker's transformation (from H. G. Schuster, Deterministic Chaos, p. 95).

Figure 2 explains the definition of the baker's transformation: you flatten the dough, then you cut it and you stack the two pieces. This is one iteration, and you keep repeating the process. When $a = 1$, this is a conservative system, the area is preserved. (One presumes that, in the pastry shop, the volume of the dough is conserved.) When $a < 1$, the system is dissipative, which is the case we are looking at now. Figure 3 shows the result of the second iteration for the particular value $a = 1/2$. Obviously the attractor will consist of a set of horizontal lines. Let's find out their y values. After one iteration, a given horizontal line is included if its y, in binary notation, contains a 0 in second position: X0XXXX, where X can be 0 or 1. After two iterations, a value of y is

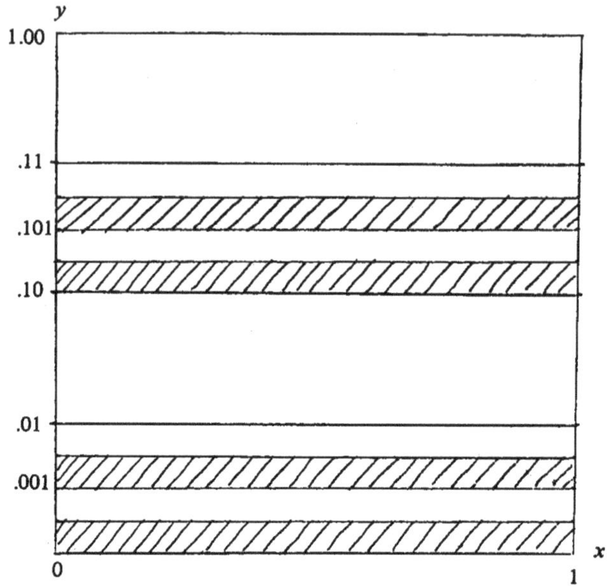

Fig. 3: (y in binary).

included if it contains a 0 in second and a 0 in fourth position: X0X0XXX. *Etc.*... in the limit, the attractor will contain all those horizontal lines whose ordinates in binary notation have a 0 in every even position: X0X0X0X0X0 *etc.*... . What is the number of dimensions of this object? We have 1 for the x-dimension, of course. But for the y-dimension, we need the standard definition of the Hausdorff dimension of a fractal. We ask how many line segments of size ϵ does it take to cover the set, then we let $\epsilon \to 0$ and the number should go like ϵ^{-d}. Here, if $\epsilon = 1/4$, we need two segments; if $\epsilon = 1/16$, we need four segments; *etc.*... . So, is goes like $\sqrt{1/\epsilon}$ and $d = 1/2$. Hence, this strange attractor has dimension 1.5. In general, for arbitrary a, it would be $1 + \log 2/\log(2/a)$. The sensitivity to initial conditions is also easily seen. It comes from the behavior of the x coordinate. Given the coordinate x_n in binary notation

$$x_n = .a\,b\,c\,d\,e\ldots \qquad (a, b, c, d \ldots \equiv 0 \text{ or } 1) \;,$$

x_{n+1} is obtained by simply lopping off the first digit:

$$x_{n+1} = .b\,c\,d\,e\ldots$$
$$x_{n+2} = .c\,d\,e\,\ldots$$

etc.... . Obviously, two points x and x' which start off very close, differing only in the nth and subsequent digits, will become completely different after $n-1$

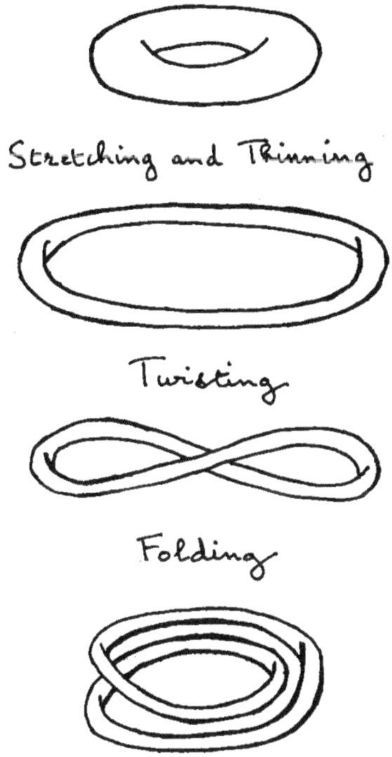

Fig. 4

iterations, and then y and y' will also go to completely different places, given the way in which y_{n+1} depends on x_n.

My second example is a qualitative one. It is a flow in three dimensions. I do not know the corresponding differential equations, but there are certainly some. Imagine a three-dimensional volume consisting of the inside of an inner tube (Fig. 4). The inner tube can be considered as generated by a small circle rotating around an axis in its own plane and not intersecting it. Imagine now the action of the dissipative flow as stretching and thinning this inner tube: the small circle (the cross section of the torus) gets smaller; the large circle (the locus of the center of the small circle) gets larger. The volume of the torus decreases, as it must in a dissipative flow. Imagine that further action of the flow has the effect of twisting the torus into a figure 8, then folding the 8 upon itself (see Fig. 4), so that the big circle has become a two-stranded closed spiral, which happens to be almost exactly the size of the original big circle. Imagine that the flow packs this object inside the original inner tube, considered as a box, with space to spare of course, since the total volume has decreased, and therefore the area of the small circle is less than half the original area. Imagine that, upon further action of the flow, this process repeats itself, and repeats itself, *ad infinitum*. In the limit, we get an infinitely-layered piece of pastry whose three-dimensional volume vanishes, but whose fractal dimension is somewhere between two and three. The sensitivity to initial conditions comes from the constant stretching of the big circle, whose effect is that, no matter how close to each other two points are originally, they will eventually end up in different layers of pastry.

The textbooks contain descriptions and calculations of the genesis of several famous strange attractors, such as Lorenz's or Rössler's, but the qualitative ideas are those that I just used. It takes two things to manufacture a strange attractor: stretching and folding. You stretch the torus, then you fold the figure 8. For the baker's transformation, you also start with stretching, then you have cutting and stacking, which is more radical than folding, but it has the same effect. These two components, stretching and folding, are very important to chaos; we shall meet them again as the primary agents of conservative chaos, in which case there is no strange attractor, but there is sensitivity to initial conditions. The stretching is responsible for this sensitivity. No matter how close two points start up, the stretching will eventually pull them apart. Due to this stretching, the average separation between two points varies like $e^{\lambda t}$. λ is called the Lyapunov exponent. But stretching is not enough. Stretching is a linearized approximation according to which the point in phase space would eventually end up at infinity. To keep the system bound, a non-linearity is needed: this is the folding (or the cutting and stacking). It is this folding which brings back the point to a finite distance, where it mixes with other points which have followed different routes and creates chaos.

As a final example of this, consider the one-dimensional logistic map (2) with $r = 4$, for which it is known to be chaotic.

$$x_{n+1} = 4x_n(1 - x_n) \quad .$$

This is a mapping of the interval (0,1) upon itself as Fig. 5 shows. But it can also be represented as in Fig. 6, in which the original interval has been labelled with x_n before being subjected to the transformation. On Fig. 6 you see the stretching, which is very uneven, but on the average it is a factor 2 (hence $e^{\lambda} = 2$). Then there is complete folding back over the interval (0,1).

For my second lecture, I shall discuss conservative chaos. The starting point is a Hamiltonian, with n coordinates and n momenta. The equations of motion are Hamilton's equations. I shall assume a time-independent Hamiltonian, hence energy is conserved. By our earlier considerations, phase space has $d = 2n$ dimensions, and the theorem quoted earlier says that chaos occurs only when $n \geq 2$. We could also take a time-dependent Hamiltonian, and then chaos might occur even when $n = 1$, but lack of time prevents me from showing you this case. In general, for time-independent H, the energy is the only constant of the motion. In the very special cases where there are n constants of the motion, then there is a canonical transformation which reduces the problem to n one-dimensional problems. These are the only cases where the problem has a standard analytical solution, and therefore they are the only problems mentioned in the old textbooks. Such problems are called "integrable." The old textbooks create the impression

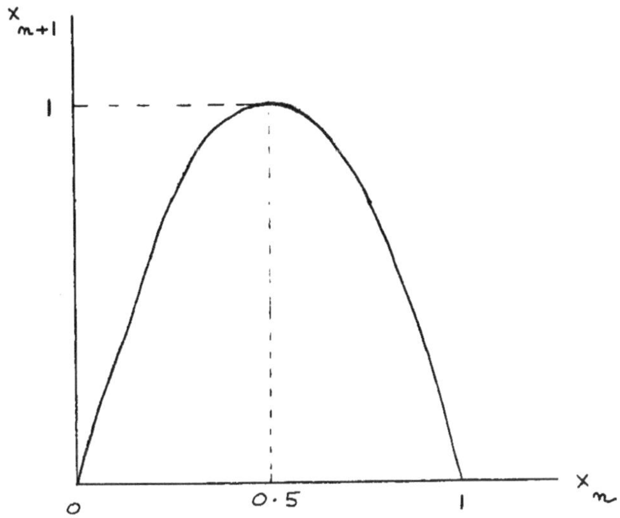

Fig. 5

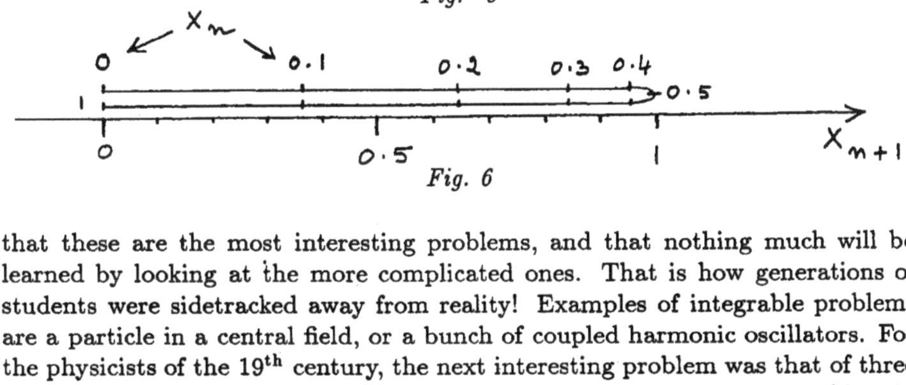

Fig. 6

that these are the most interesting problems, and that nothing much will be learned by looking at the more complicated ones. That is how generations of students were sidetracked away from reality! Examples of integrable problems are a particle in a central field, or a bunch of coupled harmonic oscillators. For the physicists of the 19th century, the next interesting problem was that of three bodies under the influence of their mutual gravitation. This is the problem in which Poincaré discovered chaos. Unfortunately, he did not have computers, or even calculators, and had to do everything analytically. So, after trying to draw some accurate pictures of what he was looking at, and proving a good number of fundamental theorems, he declared that this thing boggled his mind and seemed too complex for him to handle, or French words to that effect, and he left it to work on something else. For the next 60 years, very few people tried to pick up where he had left off; the subject was gathering dust in the museum. Then something happened: two Russians, Kolmogorov and Arnold, and an East German, Moser, proved a theorem, now known as the KAM theorem, which became the cornerstone of the work on conservative chaos. The theorem is very hard to prove, but I shall tell you what the result is, and even that takes a long time to explain.

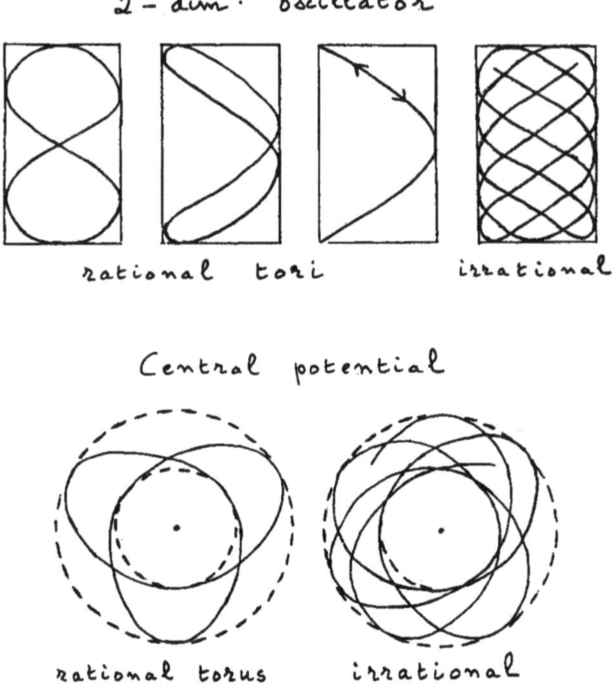

Fig. 7

Imagine that you have a non-integrable Hamiltonian H consisting of two parts

$$H = H_0 + \epsilon H_1$$

where H_0 is integrable and ϵ is a numerical coefficient. When H_0 is alone ($\epsilon = 0$) there are n constants of the motion

$$F_\alpha(q_i, p_i) = \text{constant} \qquad \alpha = 1, 2, \ldots n \ .$$

These equations define an n-dimensional surface in $2n$-dimensional phase space. One finds that it has the topology of a torus. Figures 7 shows this torus, projected on coordinate space, for two simple examples with $n = 2$: a two-frequency oscillator and a particle in a non-Keplerian central field. So, for integrable H, each trajectory resides on an n-dimensional torus, which itself resides on the $(2n - 1)$-dimensional energy shell. The motion on the torus is a composite of n one-dimensional motions. If the periods of these motions all have rational ratios, then the trajectory closes eventually and you have a periodic trajectory. If they do not, the trajectory never closes and covers the entire torus. In the first case, we call the torus a rational torus; in the second case, an irrational one (see

Fig. 7). In either case, we call this kind of motion, *i.e.* motion that remains on a torus, regular motion; it is characteristic of integrable systems. Note that a rational torus contains an infinite number of periodic trajectories; an irrational torus contains a single infinitely long trajectory.

Now, let ϵ differ from 0 and assume that all constants of the motion disappear, except the energy. Obviously, all trajectories stay on the energy shell, but what happens to the tori? The most straightforward answer would be that the tori disappear, but this is false. According to KAM, the correct answer is as follows: all rational tori disappear; some irrational tori remain. If fact, if ϵ is very small, most irrational tori remain. In the vicinity of every former rational torus, there is a finite region of phase space in which all tori, rational and irrational, have been destroyed. The size of this region decreases as ϵ decreases. More interestingly, the side of this region also decreases if the denominator of the frequency ratio (expressed as an irreducible fraction smaller than unity) increases. In other words, when a torus with a simple rational frequency ratio is destroyed, it takes along in its downfall a large region of phase space; but when a torus with a complicated rational frequency ratio is destroyed, it takes along a small region of phase space. Figure 8 tries to show schematically how this works. It happens in such a way that a finite fraction of irrational tori remain, and this fraction tends toward 1 as ϵ goes to 0. Obviously, a fractal is involved.

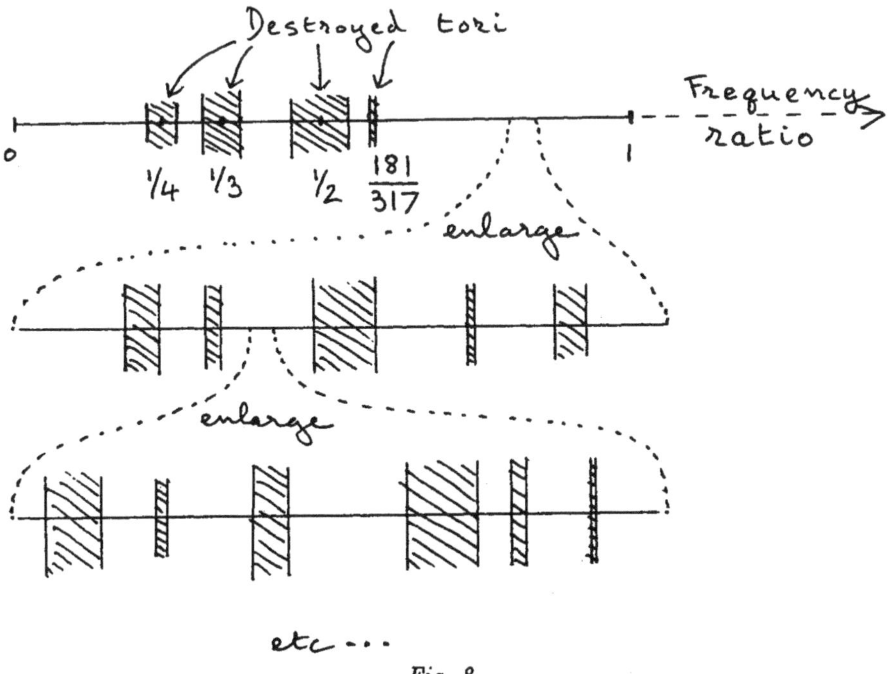

Fig. 8

The fate of the rational tori can be made more explicit. Although each and every rational torus disappears, a finite even number of its periodic trajectories remain. Half of these are stable and half are unstable. In the vicinity of each stable one, a new set of thinner, longer tori appears. Once again, this set is missing its rational tori, except for a few periodic trajectories, and so on *ad infinitum*. In the vicinity of each unstable periodic trajectory from the original rational torus, the motion is strange and complicated, but by no means random, and we call it chaotic or irregular. The upshot of all this is that, for finite ϵ, a finite fraction of phase space is occupied by regular motion, a finite fraction of phase space is occupied by chaotic motion, and the boundary between the two regions is a fractal. Figure 9 shows the possible behavior, as a function of ϵ and for fixed energy, of the fractional measure of the regular part of phase space.

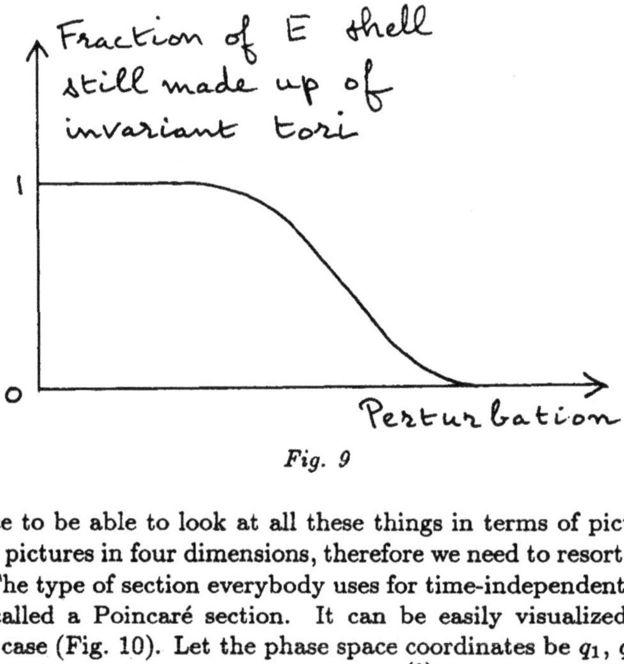

Fig. 9

It is nice to be able to look at all these things in terms of pictures, but we cannot draw pictures in four dimensions, therefore we need to resort to some kind of section. The type of section everybody uses for time-independent Hamiltonian systems is called a Poincaré section. It can be easily visualized in the two-dimensional case (Fig. 10). Let the phase space coordinates be q_1, q_2, p_1, p_2 and imagine that you fix one of them, say $q_2 = q_2^{(0)}$. You are left with a three-dimensional space q_1, p_1, p_2. Its intersection with the three-dimensional energy shell is a two-dimensional surface (Fig. 10). Project this surface on the plane of two of the variables, say q_1, p_1, and you have your Poincaré section. Of course, a trajectory is not usually localized in a section; it pierces it at a point. In bound systems, though, the trajectory will come back again and again and pierce the section an infinite number of times, and the pattern of these points identifies the kind of trajectory. The trace of a periodic trajectory consists of a finite number of points. The trajectory on an irrational torus eventually traces out the entire

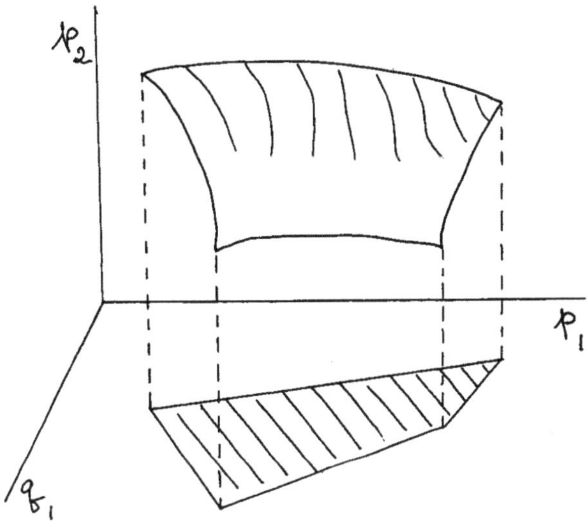

Fig. 10

section of the torus, a continuous closed curve. And a chaotic trajectory will eventually fill an entire area of the section.

Figure 11 shows schematically what happens in a Poincaré section according to the KAM theorem. In Fig. 11a we have an integrable Hamiltonian H_0. All trajectories stay on tori, and the sections of these tori appear as nested closed curves. I have labeled some of them by R, for "rational," and some by I, for "irrational." Let us now add a non-integrable ϵH_1 to H_0. The rational tori disappear, but an even number of periodic trajectories remain,half of them stable, the other half unstable. I have marked their traces for one of the rational tori of Fig. 11a. The trace of a stable periodic trajectory is called an elliptic point and I have represented it by a dot; that of an unstable one is called a hyperbolic point and I have shown it as a cross. The elliptic and hyperbolic points alternate as you go around the torus. In the general case you have only two periodic trajectories; the stable one goes through all the elliptic points and the unstable one through all the hyperbolic points. In the vicinities of these elliptic and hyperbolic points, the non-integrability generates the new structures shown in Fig. 11b. A new set of thinner, longer tori surrounds the stable trajectory, and therefore the elliptic points. On the Poincaré section, this looks like a chain of "islands of regularity." In between each pair of adjacent islands, the hyperbolic point becomes the center of a chaotic region to which I shall return shortly. Some of the original irrational tori are gone; these are the ones that were too close to a strongly perturbed rational one. But some remain, perhaps in somewhat distorted form. In addition, many of the rational-tori-island-chains that exist for not too large ϵ, cannot actually be seen on a numerical picture, because they are so thin as to be indistinguishable from unbroken tori, although you can

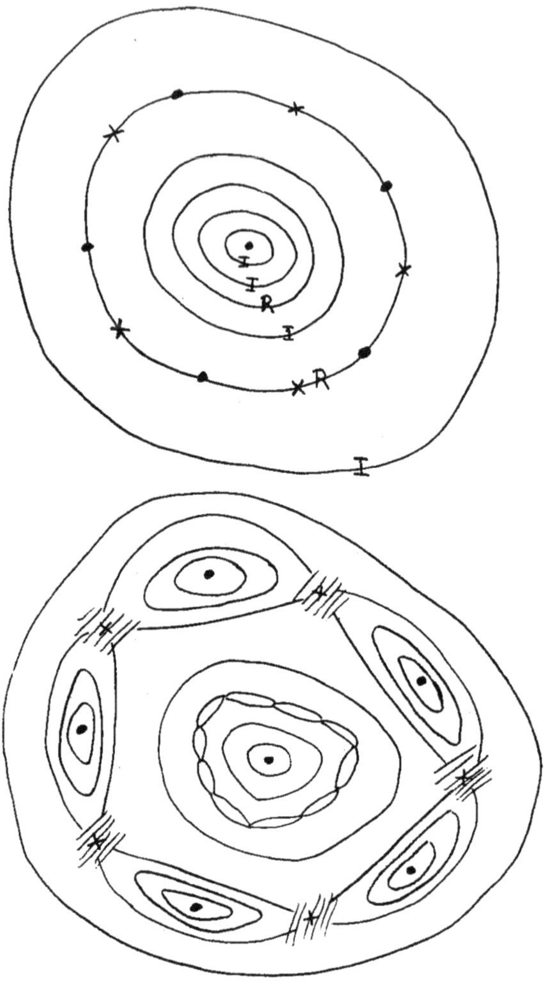

a

b

Fig. 11

always see them if you have sufficient computing power to enlarge the picture to the extent necessary. The boundary between an island chain and the neighboring unbroken tori is fuzzy and chaotic, but under small magnification it often appears rather sharp; it is called a "separatrix." Finally, it must be recalled that, within the islands of an island chain, the rational tori are broken and form second-order island chains, and so on *ad infinitum*, and this gives the picture its fractal structure. As ϵ gets larger and larger, it often happens that the regular regions get smaller and smaller; eventually the islands disappear and chaos reigns almost everywhere.

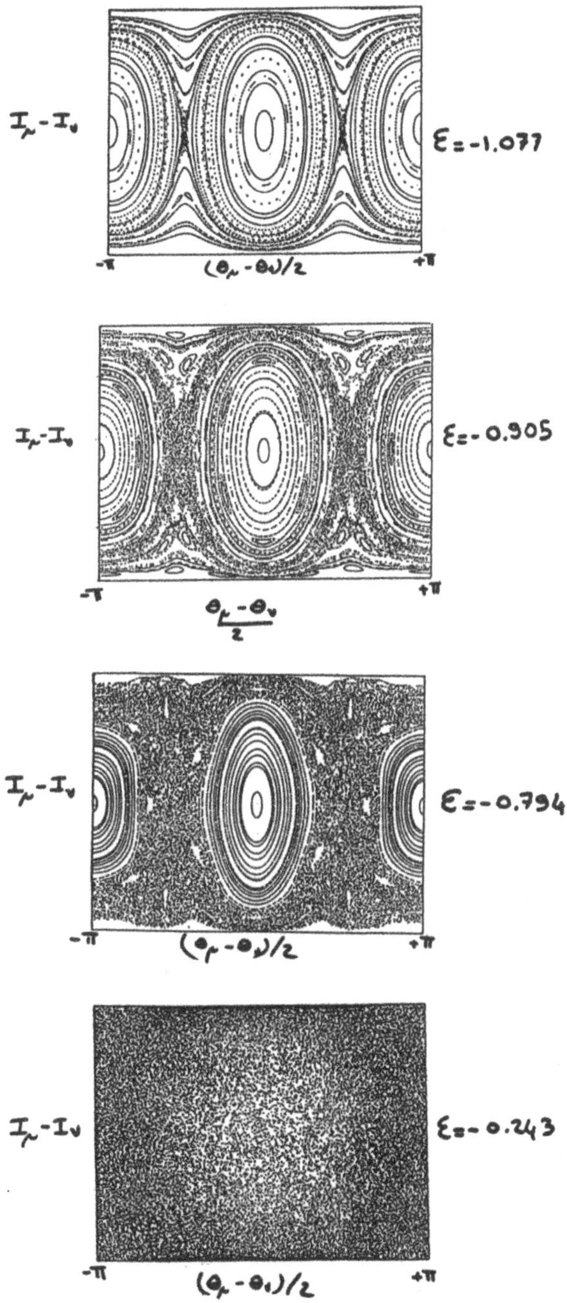

Fig. 12 (From D. Delande, 1988).

Figure 12 shows Poincaré sections for one of the most famous non-integrable real problems, the hydrogen atom in a magnetic field. The work is by D. Delande, from his doctor's thesis in Paris (1988). The parameter ϵ increases with the field. Its exact definition, and the labeling of the two axes, are unimportant for our present purpose. Note the tori, the island chains, the chaotic regions, and note how chaos increases as ϵ increases. In the last section ($\epsilon = -0.243$) chaos has taken over completely. These pictures, and many similar pictures that you will find all over this field, are generated simply by picking some initial conditions, integrating the equations of motion step by step, and marking each point where the trajectory pierces the Poincaré section.

Now, let us study more closely the chaotic region which surrounds a hyperbolic point. This is the part which fascinated Poincaré, and which he eventually gave up because of his lack of computer. First, we can linearize the Poincaré mapping in the vicinity of a fixed point, either elliptic or hyperbolic. We can define local coordinates $\vec{x}$ with origin at this fixed point,and point $\vec{x}$ will map into

$$\vec{x}' = M\vec{x}$$

where M is a 2×2 matrix. Because we are dealing with a conservative system, the mapping is area-preserving, and therefore M has unit determinant. Hence its two eigenvalues have product unity. Since M is real, this can happen in two ways:

1) The eigenvalues are $e^{i\alpha}, e^{-i\alpha}$; the point is an elliptic point.

2) The eigenvalues are $\pm e^{\beta}$, $\pm e^{-\beta}$ ($\beta > 0$); the point is hyperbolic.

We wish to consider the second case, and for definiteness we take the eigenvalues to be positive. The eigenvectors define two special directions at the fixed point O, along which the mapping consists of a simple stretching (eigenvalue e^{β}) or shrinking (eigenvalue $e^{-\beta}$). If now we continue these two special infinitesimal segments into the non-linear part of the plane, they become two special curves, known respectively as "the unstable manifold" U (eigenvalue e^{β} in the linear region) and "the stable manifold" S (eigenvalue $e^{-\beta}$ in the linear region). The latter is the locus of points which tend toward O under successive iterations. Such is the point A_0 in Fig. 13, which maps into A_1, A_2, $A_3, \ldots$ *etc.* On the other hand, U is the locus of points which tend toward O under iteration backwards in time, *i.e.* inverse mapping, such as the sequence B_0, B_{-1}, $B_{-2}, \ldots$ *etc.* on Fig. 13. Any other point, iterated either positively or negatively, will not tend toward O. For instance, point C_0 on Fig. 13 iterates into C_1, C_2, C_3 which brings it closer to O, but eventually the trajectory misses O, does a hyperbolic-shaped turnaround, and recedes from O very much like B_0, B_1, $B_2, \ldots$ do. The big question is: what do S and U look like far from the linear region around O?

The game has several important rules. One is that areas are conserved. Another is that trajectories such as S and U cannot intersect themselves, as this would mean that the mapping is not unique. A third rule is that trajectories have to keep going forever; they cannot end for no good reason. If the mapping

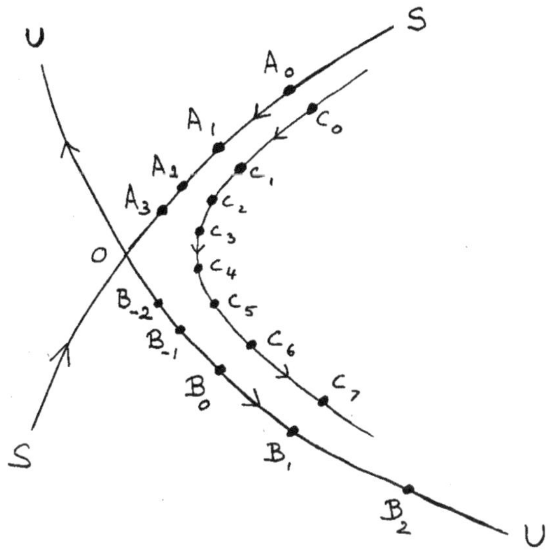

Fig. 13

is linear, then of course the trajectories are straight lines and go on to infinity. But we want to study the non-linear case. Moreover, if the system is bound and the energy is conserved, the trajectories are necessarily confined to a finite region of the map. There, S and U will have a hard time avoiding each other, especially since they must already avoid themselves. Two special cases come readily to mind. In one of them (Fig. 14a), S and U join and turn into each other. In the other (Fig. 14b), S becomes the unstable manifold U' of another hyperbolic point O', and U becomes its stable manifold S'. Both of these cases are characteristic of integrable systems. In the generic, non-integrable case, what happens is simply that S and U intersect at a point H_0, which Poincaré called a ***homoclinic point***. We shall see that, from the existence of this homoclinic point, a host of consequences come out, providing us with our first picture of Hamiltonian chaos. A good reference is Michael Tabor's book mentioned earlier, pp. 142 ff.

Consider two points D_0 and E_0 close to H_0. Let D_0 be on S and E_0 be on U. Let D_1 and E_1 be the iterates of D_0 and E_0, respectively (Fig. 15a). Let us look for the iterate H_1 of H_0. The problem is that, since H_0 is "ahead" of D_0 on S and also "ahead" of E_0 on U, its iterate H_1 should be ahead of both D_1 and E_1, by continuity of the mapping (see Fig. 15a). The only way this can happen is if one of the manifolds, say U, loops back as in Fig. 15b. Thus we have a second intersection between S and U, and H_1 is also a homoclinic point. We repeat the argument for H_1 and we find a third homoclinic point H_2 (Fig. 15c), *etc.* … . Moreover, since H_2 is closer to the hyperbolic point than H_1, and H_1 closer than H_0, the distance between H_2 and H_1 is shorter than that between H_1 and H_0. But area conservation says that the areas of the two loops must be the same. Therefore (see Fig. 15c) the second loop (from H_1 to H_2)

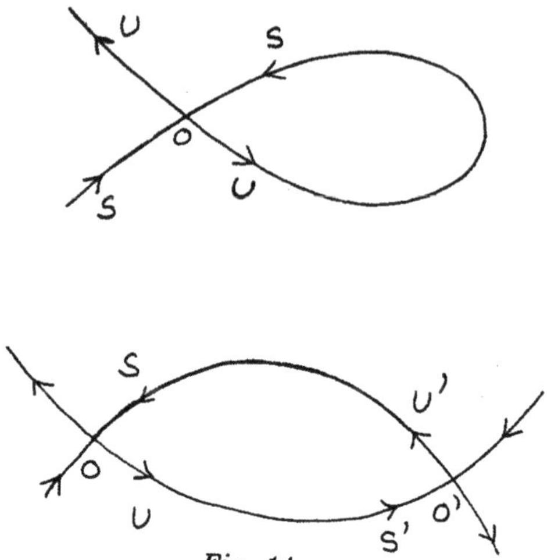

a

b

Fig. 14

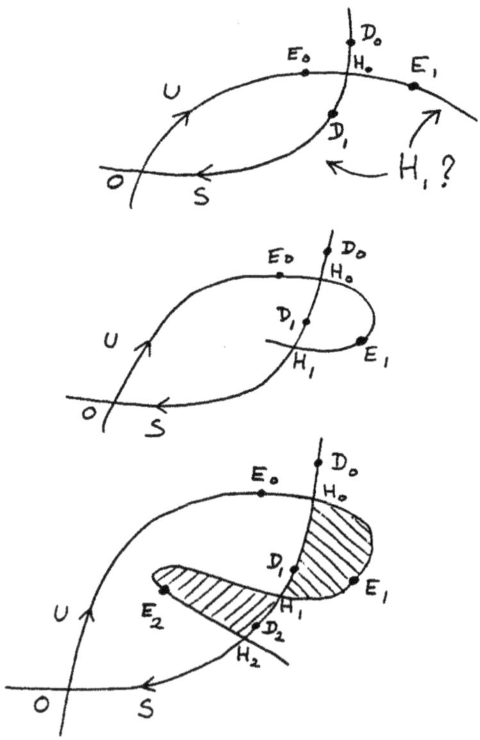

a

b

c

Fig. 15

must be longer than the first (from H_0 to H_1). As the iterations continue, the loops become longer and longer and they must curve and fold in order to fit in the space available. Then of course one could iterate backwards in time from H_0, and the same king of thing would happen again, with the roles of the two manifolds reversed. The net result is a beautifully complicated picture, often referred to as "the homoclinic tangle," of which Poincaré said "I am not even attempting to draw it." An imperfect attempt is shown in Fig. 16, taken from Ivar Ekeland, *Le Calcul, L'Imprévu* (Editions du Seuil, Paris, 1984). Every intersection in this picture is a homoclinic point, except for the original hyperbolic point O. One can easily convince oneself also that, within the confines of the homoclinic tangle, there are an infinity of unstable periodic trajectories, and therefore an infinity of hyperbolic points, out of which the whole story can repeat itself. In other words, the self-similarity typical of fractals is present once again.

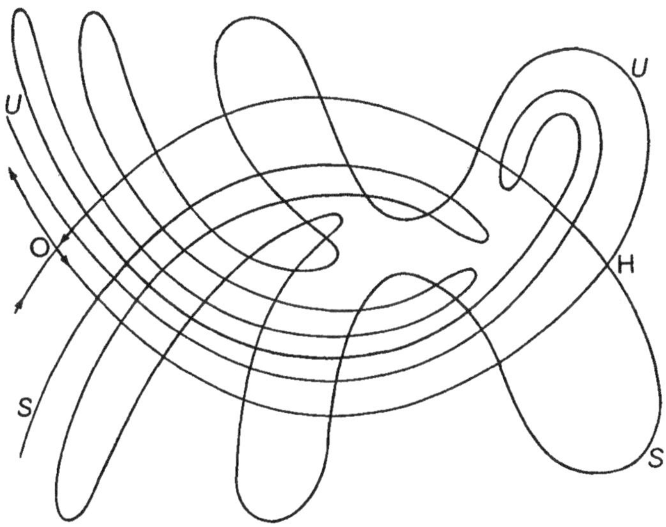

Fig. 16

What this has to do with chaos is now obvious. Choose a set of initial conditions surrounding a hyperbolic point (and remember that there are hyperbolic points everywhere, therefore choose any set of initial conditions). Through successive mappings, the set will narrow down in the S direction and stretch in the U direction, eventually turning into a thinner and thinner ribbon with constant area. Then this ribbon will proceed to fold, following the folds of the homoclinic tangle, and the folds will pile up against each other, as they begin to do in Fig. 16. We have all the elements of chaos generation by stretching and folding that we discussed earlier.

I have come to the end of my time. I hope I have said enough about classical chaos to enable you to play with these ideas in your work in atomic and nuclear physics. The next topic, if there were time, would be quantum chaos. Much work has been done on it also, but the issues are very different from those in classical chaos and I am leaving it to other speakers to raise them. The difference between the two should be evident if you consider that, in classical Hamiltonian chaos, phase space is divided into two distinct regions, regular and chaotic; the boundary between them is precisely defined and it is a fractal. When you make the transition to quantum mechanics, phase space becomes lumpy. Cells of volume h^3 are indivisible. Therefore the beautiful fractal boundary disappears and quantum chaos must be characterized in other ways. Actually, nuclear physicists were the first people to deal with quantum chaos, and that was long before classical chaos had become a fashionable discipline. But when the discoveries in classical chaos became generally known, they did much to spur the developments in quantum chaos. I have no time to talk about this.

SPLINE GALERKIN METHODS FOR ATOMIC STRUCTURE CALCULATIONS

Charlotte Froese Fischer

Department of Computer Science, Vanderbilt University
Nashville, Tennessee 37235 USA

ABSTRACT

Spline algorithms are described for both bound and continuum state problems in atomic structure theory. The Galerkin method leads to a generalized eigenvalue problem for which the eigenvalue is known in the case of a continuum orbital, but needs to be determined for bound state problems. Procedures for several simple problems arising in multiconfiguration Hartree-Fock methods will be described and analysed.

1. INTRODUCTION

The multiconfiguration Hartree-Fock (MCHF) method[1)] has been an effective method for the prediction of atomic properties. It is a general method that can be applied both to one- and two-electron systems outside closed shells and to complex open-shell systems, though calculations for the latter are considerably more extensive. An atomic structure software package based on this theory (MCHF_ASP) has been developed[2)] for the prediction of a number of properties including energy levels, transition rates, autoionization rates, and lifetimes.

The MCHF method is based on the configuration model. The total wave function, $\Psi(\gamma LS)$, is an atomic state function (ASF) describing a many-electron system labeled γ (usually the dominant configuration in the expansion) and total angular and spin momenta LS, and is approximated by

$$\psi(\gamma LS) = \sum_{i=1}^{m} c_i \Phi(\gamma_i LS) \tag{1}$$

where $\Phi(\gamma LS)$ is a configuration state function (CSF) with a specified coupling scheme. The latter is defined in terms of one-electron orbitals for which the radial functions, $P_{nl}(r)$, are to be determined. In the MCHF procedure, the radial functions are solutions of integro-differential equations with many of the properties of the Coulomb equation,

$$\mathcal{L}P_{nl}(r) \equiv \left(\frac{d^2}{dr^2} + \frac{2Z}{r} - \frac{l(l+1)}{r^2} + \epsilon\right) P_{nl}(r) = 0, \tag{2}$$

where $P_{nl}(0) = 0$. For bound state orbitals, the second boundary condition,

$$P_{nl}(r) \rightarrow 0 \text{ as } r \rightarrow \infty \tag{3}$$

makes ϵ a discrete eigenvalue problem with $\epsilon < 0$, whereas for the continuum problem, the asymptotic boundary condition,

$$P_{nl}(r) \rightarrow \sqrt{\frac{2}{\pi\kappa}} \sin(\kappa r + \delta), \tag{4}$$

is satisfied for all $\epsilon > 0$, with $\kappa = \sqrt{\epsilon}$. The mixing coefficients, c_i of Eq. (1) are components of an eigenvector of a configuration interaction matrix.

With the availability of supercomputers, new research studies become feasible:

1. **More complex systems**
   Few studies have been performed for excited states of transition metals. The near degeneracy of $3d^n$, $3d^{n-1}4s$, and $3d^{n-2}4s^2$ configurations greatly complicates prediction of reliable energy levels, and large matrix eigenvalue problems need to be solved.

2. **Better Accuracy for Small Systems**
   Few electron systems are test cases for different theories. One such test case is the helium ground state. Methods which explicitly introduce the inter-electron distance, r_{12}, outperform all others[3)], but cannot readily be generalized to other systems. A group at Chalmers University[4)] have developed programs for solving the two-electron "pair-equations", using finite differences and extrapolation. Some excellent energies have been obtained[5)] [6)] . In the configuration model,

$$\psi(1s^2\,{}^1S) = \sum_{l=0}^{l=\infty} \sum_{n=l+1}^{n=\infty} c_{nl} \Phi(nl^2\,{}^1S).$$

Recently, the MCHF method was used to compute the wave function in this form, but the finite difference schemes used by MCHF_ASP started to limit the numerical accuracy at $l = 6$.

3. **Continuum Calculations**
The study of continuum processes require calculations for a whole range of energies. The MCHF method has been extended to the study of photoionization, but in the presence of resonances, some numerical difficulties remain.

Theoretical calculations for complex systems need to be able to solve large, sparse interaction matrices. The Davidson Algorithm[7] frequently used in quantum chemistry where matrices of dimension 10^6 have been encountered, has now been installed in MCHF_ASP and its performance evaluated[8].

For the latter two research areas, spline algorithms are being designed to improve the accuracy and stability of the MCHF calculations. At the same time, the algorithms should be vector in nature so as to perform efficiently on a wide range of supercomputers. These methods are the subject of this paper.

In the next sections we will describe the spline basis for approximation, the Galerkin method for differential equations, inverse iteration as a method for computing eigenvectors, followed by the application of spline Galerkin methods, first to the solution of the Hartree-Fock equation for the Helium ground state and then to some continuum problems. In each case, an analysis of accuracy will be presented as the step-size parameter h and the order K of the spline is varied.

2. SPLINE APPROXIMATION

A spline approximation of order k is a piecewise polynomial approximation of degree $k-1$ in C^{k-2}. An interval of approximation, say $[0, b]$, is divided into a series of subintervals by a sequence of "knots" or grid points. Thus discontinuities occur only in the $k - 1^{st}$ derivative and only at the knots. A B-spline basis for this approximation can be obtained by applying de Boor's well-known spline algorithms[9]. With n subintervals, a complete set of $n + k - 1$ basis functions can be obtained by allowing the two end grid-points to be duplicated k times. The B-spline basis has the property that each basis element is non-zero over k adjacent intervals.

A spline approximation has several advantages over finite difference methods.

2.1 Grid Selection

Many problems in atomic physics require the solution of systems of coupled, non-linear, integro-differential equations[10)]. Numerical methods for solving these equations must be able to solve the radial equation accurately and reliably.

The radial equation in atomic physics has a singularity at the origin so solutions may change rapidly in that region. With splines, an unequally spaced grid introduces no additional complexity. In fact, the grid points can be selected to suit the problem. Because of the way in which the solution to the radial equation scales with respect to Z near the origin, the knots should be equally spaced with respect to the variable Zr. For large r, in the case of a continuum equation for an electron with energy κ^2 the knots should be equally spaced with respect to κr. A variable grid can easily be defined *a priori* to meet these constraints.

2.2 Higher-Order Methods

The cubic-spline is the spline of lowest degree with a continuous second derivative. Thus it is in the class of C^2 functions. The basis with minimal support (i.e. minimal region over which the function is non-zero), is the B-spline basis. In the case of cubic splines, the basis elements are positive, piecewise cubic polynomials over four intervals. In general, a B-spline of order k, consists of positive, piecewise polynomials of degree $k-1$ in the class C^{k-2}, over k adjacent intervals. Clearly, $k+1$ knots are required to define k intervals. If the range of approximation is [0, b], and the first interval is [0, r_1], then the knot sequence

$$(\underbrace{0, \ldots, 0}_{k \text{ zeros}}, r_1)$$

defines the left most B-spline whose support terminates at r_1. It is convenient to designate this basis element as $B_0(r)$. Similarly, the knots at b must be duplicated k times. Then the knot sequence,

$$0 = r_0 < r_1 < r_2 < \cdots < r_{n-1} < r_n = b,$$

will define a basis $\{B_i(r), i = 0, \ldots, n+k-2\}$ of B-splines over the range [0, b]. At any interior point of an interval, k splines will be non-zero. A stable evaluation of these splines is given by the subroutine BSPLVB, part of the PPPACK software prepared by de Boor[9)]. The values of the splines as well as their derivatives can be calculated using the subroutine BSPLVD. In all these routines, k is a parameter so that the order of the method can easily be varied.

3. SPLINE GALERKIN METHOD

Let us illustrate the spline Galerkin approach for the Coulomb equation. In this approach, the solution of the differential equation is approximated by a linear combination of the spline basis elements, i.e.

$$P_{nl}(r) \approx \tilde{P}_{nl}(r) = \sum_{i=0}^{N} c_i B_i(r). \tag{5}$$

where $N = n+k-2$. The boundary conditions determine some of these cefficients. The only basis element different from zero at the origin is $B_0(r)$. Hence $c_0 = 0$. For bound orbitals, the boundary condition at $r = b$ leads to $c_N = 0$, but for the continuum orbital, there is no such condition. The Galerkin method[11] requires that the residual of the differential equation , $\mathcal{L}\tilde{P}_{nl}(r)$ be orthogonal to each of the basis functions, i.e.

$$< B_j|\mathcal{L}\tilde{P}_{nl}(r) >= 0, \quad j = 1, \ldots, N' \tag{6}$$

where

$$< f|g >= \int_0^b f(r)g(r)dr.$$

In the above equation, $N' = N$, in the continuum case, and $N' = N - 1$ for the bound case. The requirement of Eq. (6) leads to the Galerkin system of equations

$$\sum_j [h_{ij} + \tilde{\epsilon} b_{ij}]c_j = 0, \quad i = 1, \ldots, N' \tag{7}$$

where

$$\begin{aligned} h_{ij} &= < B_i|\frac{d^2}{dr^2} + \frac{2Z}{r} - \frac{l(l+1)}{r^2}|B_j > \\ b_{ij} &= < B_i|B_j > . \end{aligned}$$

and $\tilde{\epsilon} \approx \epsilon$.

Eq. (7) is a generalized eigenvalue problem for which an eigenvector defines the expansion of the approximate solution associated with the corresponding eigenvalue. For the bound state problem, the matrices are symmetric, but for the continuum problem the matrices are "almost symmetric" in that only the last row and column do not satisfy the symmetry conditions.

The matrix elements of Eq. (7) are defined as inner products which, in general, require integration over [0, b]. Because of the finite support of the B-splines, this integration reduces to integration over at most k adjacent intervals that form the intersection of the intervals of support for the two splines, $B_i(r)$ and $B_j(r)$. In each interval, the integrand is continuous and can be evaluated at any point. In this case, guassian quadrature provides excellent accuracy in that, with k evaluations of the integrand in an interval of length h, the error would be $O(h^{2k+1})$. It would also depend on $f^{2k}(\eta)$, where $f(x)$ is the integrand and η is some point in the interval[12]. Thus the b_{ij} matrix elements, for example, would be computed exactly, except for rounding errors, since the integrand would be a polynomial of degree $2k-2$. Unless stated otherwise, this procedure was used in all cases. Clearly, if the intersection of support of the two splines is null, the inner product will be zero. It is easy to see that

$$b_{ij} = 0 \quad \text{if } |i-j| \geq k.$$

For the simple radial equation, this leads to banded systems of equations.

4. INVERSE ITERATION

In EISPACK[13) 14)], whenever a selected number of eigenvectors are required, the eigenvalues are determined and then eigenvectors computed using inverse iteration. Eq. (7) may be written in matrix form as

$$\mathbf{Ac} = 0.$$

Because the given ϵ is not an exact eigenvalue of the Galerkin system of equations, the desired vector $\mathbf{c}$ is the one that corresponds to the smallest eigenvalue of $\mathbf{A}$. Note that the definition of $\mathbf{A}$ depends on ϵ.

The algorithm that converges to a relative error of 10^{-12} in 3-4 iterations is the following.

1. Perform an LU decomposition of the matrix $\mathbf{A}$ such that $\mathbf{A} = \mathbf{L}\ \mathbf{U}$.
2. Set $c_i^{(0)} = 1,\ i = 1, \ldots, N'$.
3. Until the maximum, absolute change in $c_i < 10^{-12}$
   - 3.1 Solve $\mathbf{Ly} = \mathbf{c^{(m)}}$ for $\mathbf{y}$
   - 3.2 Solve $\mathbf{Ux} = \mathbf{y}$ for $\mathbf{x}$
   - 3.3 Set $\mathbf{c^{(m+1)}} = \mathbf{x}/||\mathbf{x}||_\infty$ and determine maximum absolute change, $\left|c_i^{(m+1)} - c_i^{(m)}\right|,\ i = 1, \ldots, N'$.

The LINPACK[15] routines DGBFA and DGBSL were used in the banded matrix case, DGEFA and DGESL for the dense matrix that arises in the hydrogen scattering problem.

5. HARTREE-FOCK EQUATION FOR HELIUM

The Hartree-Fock equation for the radial function, $P_{1s}(r)$, of the $1s$ orbital defining the $1s^2$ ground state of a two-electron system, with nuclear charge Z, is the solution of the boundary value problem[1],

$$\left(\frac{d^2}{dr^2} + \frac{2}{r}(Z - Y^0(1s, 1s; r)) - \epsilon\right) P_{1s}(r) = 0, \tag{8}$$
$$P_{1s}(r) = 0 \text{ and } P_{1s}(r) \to 0 \text{ as } r \to \infty.$$

The function, $Y^0(1s, 1s; r)$, represents the screening of the nucleus by the other electron and is defined as

$$Y^0(1s, 1s; r) = \int_0^r P_{1s}^2(s)ds + \int_r^\infty \frac{r}{s} P_{1s}^2(s)ds \tag{9}$$

5.1 The Self-Consistent Field Method

Note that the presence of the Y^0 term makes the problem a non-linear, integro-differential equation of eigenvalue type. This problem is usually solved iteratively by the self-consistent field (SCF) method as follows[16] :

- Let an initial estimate of $P_{1s}(r)$ to be a hydrogenic function
- until converged
  - Compute Y^0 using the current estimate of $P_{1s}(r)$.
  - Solve the linear differential equation for a new estimate of $P_{1s}(r)$

The Galerkin method with the SCF iteration leads to the symmetric, generalized eigenvalue problem, where now

$$h_{ij} = \int_0^\infty B_i(r) \left(\frac{d^2}{dr^2} + \frac{2}{r}\left(Z - Y^0(1s, 1s; r)\right)\right) B_j(r)dr.$$

Only one eigenvector is required, and since the estimates of $\mathbf{c}$ improve as the SCF iterations proceed, inverse iteration was used along with the Rayleigh-quotient for the estimate, $\tilde{\epsilon} = \mathbf{c^tHc/c^tBt}$.

To solve the differential equation numerically by the spline method, the range $[0, \infty)$ was divided into a set of subintervals, over $[0, r_{max}]$, with $\tilde{P}_{1s}(r) = 0$

for $r > r_{\max}$. Earlier numerical procedures used a grid, which required special series expansions near the origin . With splines, a linear grid near the origin was combined with an exponential grid at larger radii.

The grid was defined in terms of a step parameter, $h = 2^{-m}$ as follows:

$$\begin{aligned} Z * r_i &= 0 \text{ for } i = 1, ..., k \\ &= (i-k)h \text{ for } i = k+1, k+2, ..., k+2^m \\ &= r_{i-1} * (1+h) \text{ for } i = k+2^m+1, ..., N+k \\ &= r_{N+k} \text{ for } i = N+2k-1 \end{aligned}$$

Because of the iterative nature of the SCF method, additional features could be added to the algorithm such as the automatic adjustment of $r_{\max} = r_{MX}$, $MX \leq N + k$. As long as the initial range was sufficiently large, MX could be reduced to eliminate the negligible "tail" of the expansion of Eq. (5) which, in turn, reduces the size of the matrix that needs to be considered in the eigenvalue problem. The criterion used for range reduction was $|c_{N+k-1}| \leq 10^{-13}$. Similarly, if the last two coefficients in the expansion were too large, namely their sum of absolute values was $> 10^{-10}$, the value of MX was increased. In this manner, the range of the solution was allowed to adapt to the solution and the requested level of accuracy, eliminating the possibility of an erroneous or inefficient choice of $r_{\max}$.

5.2 Solving For Functions Y^k

The presence of the $Y^0(1s, 1s; r)$ function in Eq. (8) makes the problem non-linear. This function is a special case of a more general integral entering into two-electron Slater integrals, namely

$$Y^k(nl, n'l'; r) = \int_0^r (\frac{s}{r})^k P_{nl}(s) P_{n'l'}(s) ds + \int_r^\infty (\frac{r}{s})^{k+1} P_{nl}(s) P_{n'l'}(s) ds \qquad (10)$$

Several approaches can be taken to determining Y^k. One would be to evaluate the integrals for those values of r needed for the Gaussian quadrature calculation of the matrix elements h_{ij}. Such integrations were done in the continuum calculations but lead to lengthy procedures. More typically, the Y^k functions are determined from differential equations. When finite differences were used, it was customary to integrate first order differential equations outward for the first term in the of Eq. (10) and inward for the Y^k function. But this pair of equations can be combined into a single second-order differential equation,

$$\frac{d^2}{dr^2}Y^k(r) = \frac{k(k+1)}{r^2}Y^k(r) - \frac{2k+1}{r}P_{nl}(r)P_{n'l'}(r), \tag{11}$$

with boundary conditions

$$\begin{aligned} Y^k(0) &= 0, \\ \frac{d}{dr}Y^k(r) &= \frac{-k}{r}Y^k(r) \text{ as } r \to \infty, \end{aligned}$$

and solved by the spline Galerkin method, in the same way as the equation for $P_{1s}(r)$. This leads to a banded linear system of equations for the expansion of $Y^k(r)$. The last equation in this system arises from applying the boundary condition at $r = r_{MX}$.

5.3 Accuracy Studies

The procedures outlined above depend on a number of parameters. The parameter $r_{\max}$ depends, in general, on the eigenfunction and the accuracy to which the function is to be represented. For a higher eigenstate, the wave functions extend further to infinity, and $r_{\max}$ needs to be larger. For the ground state, with the cut-off criterion of 10^{-12}, $r_{\max}$ was found to be in the neighborhood of 22 au.

There are two other fundamental parameters of the calculation. One is the step-size parameter h and the other is the order k. Ideally, one would like to find that combination of parameters for which a given accuracy can be obtained in the least amount of CPU time. Such a criterion ignores the role of memory which could be a critical resource but, as in the studies mentioned earlier, the evaluation of an algorithm is usually based on arithmetic operations or CPU time. In order to gain an understanding of the interplay of step-size and order, timing studies have been performed for all combinations of $h = 2^{-m}$, $m = 0$ to 5 and $k = 3$ to 15.

For the helium ground state, the total energy of the system is

$$E = 2I(1s) + F^0(1s, 1s),$$

where

$$I(1s) = -(1/2)\int_0^\infty P_{1s}(r)\left(\frac{d^2}{dr^2} + \frac{2}{r}\left(Z - Y^0\right)\right)P_{1s}(r)dr.$$

The first term in this expression is part of the kinetic energy operator, say T; the rest together with the Slater integral, F^0, represent the potential energy, V. For

Table 1: Hartree-Fock results for $h = 1/8$ and different values of k, using the spline Galerkin method. Included is the total Energy (in au.), the potential energy V, the kinetic energy T, and $2 + V/T$ which is a measure of the error. The notation n.nn (−mm) is used to represent n.nn×10^{-mm}.

k	Energy	V	T	$2+V/T$
3	−2.8616663245256	−5.7232844124722	2.8616180879466	−1.69(−05)
4	−2.8616799825485	−5.7233599248684	2.8616799423199	−1.41(−08)
5	−2.8616799955797	−5.7233599912031	2.8616799956234	1.53(−11)
6	−2.8616799956110	−5.7233599912233	2.8616799956124	4.74(−13)
7	−2.8616799956113	−5.7233599912235	2.8616799956122	3.18(−13)
8	−2.8616799956113	−5.7233599912235	2.8616799956122	3.02(−13)
9	−2.8616799956114	−5.7233599912236	2.8616799956122	2.63(−13)
10	−2.8616799956114	−5.7233599912235	2.8616799956122	2.87(−13)
11	−2.8616799956114	−5.7233599912236	2.8616799956122	2.79(−13)
12	−2.8616799956114	−5.7233599912236	2.8616799956122	2.78(−13)
13	−2.8616799956114	−5.7233599912236	2.8616799956122	2.72(−13)
14	−2.8616799956114	−5.7233599912236	2.8616799956122	2.81(−13)
15	−2.8616799956115	−5.7233599912236	2.8616799956122	2.56(−13)

an exact variational calculation and a complete basis, the virial theorem states that

$$V/T = -2.$$

Because in this case the exact energy is not known, the value of 2 + V/T was used as a measure of the error in the calculation. Table 1 reports the results for h=1/8. More details can be found elsewhere[17] .

6. CONTINUUM CALCULATIONS

The simple equation,

$$y'' + \kappa^2 y = 0,$$

shows that the boundary condition $y(0) = 0$ and the amplitude condition uniquely determine the solution, $y(r) = \sin \kappa r$.

In our tests we have assumed without loss of generality that $\kappa = 1$ and that a solution was required over the interval [0,10]. For any other value of κ, the simple transformation $r' = \kappa r$ would transform the equation to this form. Clearly, for this problem a uniform grid is appropriate.

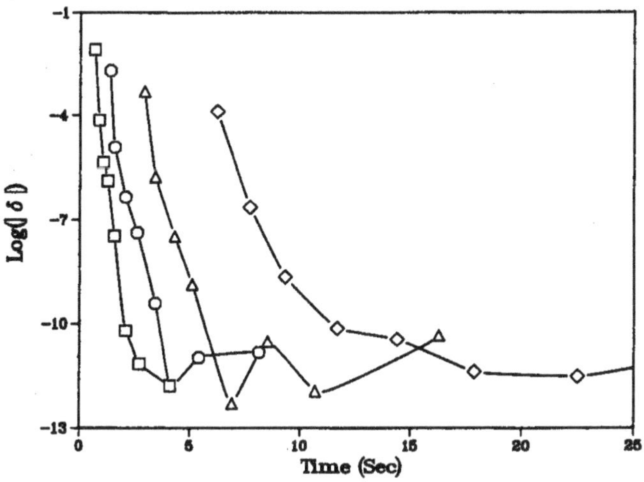

Figure 1: Plot showing the log of the phase shift in the solution of $y'' + y = 0$, over the interval [0, 10], as a function of the execution time for different values of parameters h and k. For an exact solution, $\delta = 0$. Going from left to right, the curves are for $h = 1/2$, 1/8, 1/16, and 1/32, respectively. Each curve starts with $k = 4$.

The spline Galerkin method cannot be expected to produce an exact solution. The boundary condition at the origin selects the $\sin \kappa r$ behavior, but at the other end of the range,

$$\begin{aligned} Y(r) &\approx C_s \sin \kappa r + C_c \cos \kappa r \\ &= C \sin(\kappa r + \delta) \end{aligned}$$

where

$$\begin{aligned} \tan \delta &= C_c / C_s \\ C^2 &= C_s^2 + C_c^2 \end{aligned}$$

Thus the "phase shift" δ is a measure of the error in the solution. The amplitude C^2 and phase shift were determined by fitting the approximate form to the last two, interior knots of the region.

Figure 1. shows how the phase shift error, for different values of h and k, depends on the execution time on a SUN 3/160 with a floating point board. The four curves, starting from the left, are for $h = 1/4$, 1/8, 1/16, and 1/32, respectively. The first data point on each curve, starting from the top, is for

$k = 4$, and increases by one going to the next data point on the curve. The optimum choice of parameters would be one for which the required accuracy is met in the least amount of time, assuming an efficient implementation. Generally the larger values of h combined with a higher-order method are favored, though this combination appears to be more subject to rounding errors. Large values of h are also desirable in that they reduce the size of the matrix and hence the memory requirements. For each k we have analyzed the observed phase shift and found that

$$\delta \approx C_k h^{k-2},$$

provided the rounding region has not been reached.

The hydrogen scattering problem[18)] combines many features of the helium problem with those of the $\sin(\kappa r)$ problem, except that the exchange function leads to a *dense* matrix $\mathbf{H}$ but the problem is a linear one. An analysis of the errors show the same types of behavior as the two examples discussed here.

7. CONCLUSION

Spline Galerkin algorithms have been applied and analysed in a number of test cases that include both bound and continuum problems. In each case, the higher-order methods were found to have the better performance characteristics. Attempts have been made to apply these procedures to the study of photoionization in the presence of resonances, using the MCHF method, but the iterative, SCF nature of the solution process introduces instabilities near resonance. With splines, direct methods become feasible requiring the solution of large, sparse systems of equations.

ACKNOWLEDGEMENTS

This research was supported, in part, by a grant from the US Department of Energy, Office of Basic Energy Sciences and, in part, by a grant from the National Science Foundation.

References

1. C. Froese Fischer, *The Hartree-Fock Method for Atoms* (John Wiley & Sons, New York, 1977).
2. C. Froese Fischer, , Comput. Phys. Commun. (in preparation).
3. Y. Accad, C.L. Pekeris, and B. Schiff, Phys. Rev. A **4** 516–536 (1971).
4. A.-M. Mårtenson, J. Phys. B: At. Mol. Phys. **12** 3995 (1979).
5. I. Lindgren, Phys. Rev. A **31** 1273 – 86 (1985)
6. S. Salomonson and P. Öster, (unpublished)
7. E. R. Davidson, J. Comput. Phys. **17** 87 – 94 (1975).
8. V. Umar and C. Froese Fischer, , Int'l J. Supercomputer Appl. (submitted).
9. C. de Boor *A Practical Guide to Splines*, Applied Math. Series 27, (Springer Verlag, New York, 1978)
10. C. Froese Fischer, , Comput. Phys. Reports, **3** 273 – 326 (1986).
11. W.A. Fletcher, *Computational Galerkin Methods* (Springer, New York, 1984)
12. S. D. Conte and C. de Boor, *Elementary Numerical Analysis* (McGraw-Hill Book Co., New York, 1980).
13. B.T. Smith, J.M. Boyle, B.S. Garbow, Y. Ikebe, V.C. Klema, C.B. Moler, *Matrix Eigensystem Routines — EISPACK Guide*, Lecture Notes in Computer Science, Vol. **6** (Springer-Verlag, New York, 1977).
14. B.S. Garbow, J.M. Boyle, J.J. Dongarra, and C.B. Moler, *Matrix Eigensystem Routines — EISPACK Guide Extensions*, Lecture Notes in Computer Science, **51** (Springer-Verlag, New York, 1977).
15. J.J. Dongarra, J.R. Bunch, C.B. Moler, and G.W. Stewart, *LINPACK User's Guide*, Soc. Ind. Appl. Math., (Philadelphia, PA 1977).
16. C. Froese Fischer, and W.-Q. Guow, J. Comput. Phys. (submitted).
17. C. Froese Fischer, and M. Idrees, Computers in Physics, **3** 53 – 58 (1989).

CALCULATING COHERENT PAIR PRODUCTION WITH MONTE CARLO METHODS

C. Bottcher and M. R. Strayer

Physics Division, Oak Ridge National Laboratory

Oak Ridge, TN 37831

ABSTRACT

We discuss calculations of the coherent electromagnetic pair production in ultra-relativistic hadron collisions. This type of production, in lowest order, is obtained from tree diagrams which contain two virtual photons. We discuss simple Monte Carlo methods for evaluating these classes of diagrams without recourse to involved algebraic reduction schemes.

1. INTRODUCTION

In this lecture we shall address the question of pair production from classical time-dependent electromagnetic fields. These types of fields are present in the peripheral ultrarelativistic collisions of heavy nuclei [1,2]. Because of the possible coherence in the field, it is speculated that relatively large fluxes of exotic particles may be produced. Such particles include μ and τ pairs,[3,4,5], magnetic monopoles

[6], W-pairs, b-quark pairs, and possibly the Higgs [7,8]. Most of these studies address colliding beams of heavy nuclei at energies per nucleon in the range $3-10^3$ GeV, corresponding to suggested experiments at the AGS, CERN, the proposed RHIC facility at Brookhaven, and possibly the SSC.

Out of these speculations arise a series of questions:

- In view of the interest in dilepton production and J/ψ suppression as a diagnostic of the formation of the quark-gluon plasma, it is important to know whether electromagnetic pair production exhibits features which clearly distinguish it from hard hadronic processes.

- Pairs of W bosons have not yet been produced in laboratory collisions. The Z^4 scaling implicit in the coherent production of these particles may amplify cross section yields sufficiently to be of interest experimentally. Simple calculations suggest that the processes for producing W-pairs are also candidates for producing the Higgs. Thus it may be possible to use heavy-ion machines to search for the Higgs.

- Many of the properties of vector mesons have emerged from the Drell-Yan measurements of electron pair production [9,10]. However, there are anomalies in this analysis which have recently been resolved with a more careful consideration of the hadronic current matrix elements [11,12].

Concerning the coherent production of μ and τ pairs perturbatively, other calculations of cross sections have been performed using the Weizsäcker-Williams approximation [4,8]. However, this approximation may not always work, since the constraint that the photons must be *on-shell* is not always realizable; an example is the case where the virtual photons carry an explicit transverse momentum [12]. Furthermore, the coherent production of pairs from nuclei is a collective process which is very different than that arising from the collision of point charges. Nucleons in nuclei move at high velocities and carry a large momentum. Form factors control the coupling of these nucleons to photons, and the strength of this coupling regulates the magnitude of the pair cross section. Thus, nuclear and nucleon electromagnetic form factors must be included in these calculations.

Pair production in heavy-ion collisions is nonperturbative at high energies. It is plausible that the introduction of nucleon form factors may regularize the perturbation theory order by order. Thus as noted in Ref. [11,12], Feynman perturbation theory becomes an efficient tool to address details of the hadron-photon couplings and their effect on the lepton cross sections. Calculations in this framework for electron–positron pairs demonstrate a substantial yield at large pair invariant masses and transverse momenta, where equivalent photon approximations are invalid.

In comparing computation to experiment, it is important to understand the experimental constraints placed on the data. The Monte Carlo algorithms discussed in this talk have the advantage that complicated constraints on the reaction phase space can be dealt with in a straightforward manner. We shall apply these methods to calculate cross sections for a variety of complex experiments as follows: anomalous $e^{\pm}$ production in 17 GeV $\pi^{-}p$ collisions, and μ and τ lepton production in ultrarelativistic collisions of heavy nuclei.

2. SEMICLASSICAL THEORY

Formally, we employ a semiclassical treatment of pair production wherein the leptons are coupled to the classical electromagnetic four-potentials of the colliding nuclei. The source currents are the Lorentz-boosted charge distributions of the colliding ions. In the limit of structureless point charge nuclei, these potentials are the retarded Liénard-Wiechert interactions. However, it is clear from general arguments that the electromagnetic coupling of μ and τ pairs to protons in nuclei depends on the nucleon currents and form factors in each ion. In our present calculations, we are primarily concerned with peripheral heavy-ion collisions, and we construct classical four-vector potentials employing phenomenological nuclear currents and form factors which fit elastic electron scattering data from nuclei.

The formulation of electromagnetic pair production used here is based on the Lagrangian density for the coupling of the classical electromagnetic four-vector potential to the lepton fields given by

$$\mathcal{L}_{int}(x) = -\overline{\Psi}(x)\gamma_{\mu}\Psi(x)A^{\mu}(x) \; . \tag{1}$$

This Lagrangian conserves lepton number [2], and we treat the production of electrons, muons, and tauons as occurring independently. The equations of motion for lepton state vectors is obtained from (1) using variational methods. We use a semiclassical action corresponding to a time-dependent *many-lepton* state $\Phi(t)$,

$$\mathcal{S} = \int d^4x < \Phi(t) \mid : \mathcal{L}_0(x) + \mathcal{L}_{int}(x) : \mid \Phi(t) > \quad , \tag{2}$$

where the normal ordering is with respect to a reference state, and where $\mathcal{L}_0$ is the usual noninteracting fermion Lagrangian,

$$\mathcal{L}_0(x) = \overline{\Psi}(x)[\gamma_\mu i\partial^\mu - m]\Psi(x) \quad . \tag{3}$$

In (2), the dynamical coordinates which are varied to make the action stationary are the parameters labeling the state vector $\Phi(t)$. We additionally assume that the initial state vector corresponds to a single Slater determinant, $\mid O >$,

$$\lim_{t\to-\infty} \mid \Phi(t) > \to \mid O > \quad . \tag{4}$$

Since we shall only consider pair production out of the vacuum, we choose $\mid O >$ as the vacuum state, which we also identify as the reference state for the ordering in (2). By construction, we have a well-defined initial Hamiltonian with a corresponding complete and orthonormal set of single-particle states, $\chi_q^{(+)}$ and $\chi_q^{(-)}$, satisfying

$$H_0 \mid \chi_q^{(\pm)} > = E_q^{(\pm)} \mid \chi_q^{(\pm)} > \tag{5}$$

where

$$E_q^{(\pm)} = \pm(m + \vec{q}^{\,2})^{1/2} \tag{6}$$

and

$$H_0 = \vec{\alpha} \cdot \vec{p} + m\alpha_0. \tag{7}$$

Since the above model results in rules that are different than the usual Feynman diagrams for two-photon processes, we find it expedient to norm our states as follows,

$$\sum_q |\chi_q^{(+)}><\chi_q^{(+)}| + |\chi_q^{(-)}><\chi_q^{(-)}| = 1 ,$$
$$<\chi_q^{(i)} | \chi_p^{(j)}> = \delta_{q,p}\delta_{i,j} . \tag{8}$$

With this choice of reference state, the states $|\chi^{(+)}>$ and $|\chi^{(-)}>$ are single-particle and single- antiparticle states, respectively. The dynamics governing the time evolution of the states in (2) is unitary, and given by

$$|\Phi(t)> = K(t, -\infty) | O >, \tag{9}$$

where $KK^\dagger = K^\dagger K = 1$. There are several important consequences of these assumptions. Equations (4) and (9) guarantee that the state Φ is at all times a representation of a single Slater determinant, and equations of motion can be cast into the form

$$H(x)K(t,t') = i\partial_t K(t,t'), \tag{10}$$

where

$$H(x) = H_0(x) + H_{int}(x),$$
$$H_{int}(x) = \gamma_0 \not{A}(x) \tag{11}$$
$$= -\vec{\alpha} \cdot \vec{A}(x) + A_0(x) .$$

With the above-noted assumptions, all orders of processes can be obtained from the solutions to (10). In particular, those solutions which are perturbative in H_{int} can be expressed as the series

$$K(t,\infty) = K_0(t,-\infty) + (-i)\int_{-\infty}^{t} d\tau K_0(t,\tau)H_{int}(\tau)K_0(\tau,-\infty) + (-i)^2\int_{-\infty}^{t} d\tau \int_{-\infty}^{\tau} d\tau' K_0(t,\tau)H_{int}(\tau)K_0(\tau,\tau')H_{int}(\tau')K_0(\tau',-\infty) + \cdots, \tag{12}$$

where in (12), the lowest order term is simply

$$K_0(t,t') = \exp[-iH_0(t-t')]. \tag{13}$$

In our model the electromagnetic coupling to the nuclear currents is mediated by the classical four-vector potentials A_μ. The source currents for these fields come from the motion of the nucleons, both collectively, and as motion within each nucleus. For definiteness, we consider a collision between two nuclei a and b, viewed in the equal velocity frame. The nucleus a moves with a velocity $+\beta$ parallel to the z-axis. The nuclei have charges Z_a and Z_b, respectively, and their trajectories are separated by an impact parameter, $\vec{b}$. Hereafter, in all equations, we employ natural units with $\hbar = c = m = e = 1$, m being the lepton mass. In presenting calculational results for physical observables, we restore the appropriate dimensional scaling. This is consistent with and facilitates comparisons with formulae in the literature [11]. In our units, each nucleus has a kinetic energy per nucleon, $E = M_N(\gamma - 1)$, given in terms of the Lorentz factor, $\gamma = (1-\beta^2)^{-1/2}$, and the nucleon mass, M_N. The electromagnetic potential is the sum of the two terms coming from nuclei a and b, respectively

$$A^\mu(x) = A^\mu_a(x) + A^\mu_b(x)$$

with corresponding currents

$$\mathcal{J}^\mu(x) = \mathcal{J}^\mu_a(x) + \mathcal{J}^\mu_b(x).$$

The four-potentials are obtained from Maxwell's equations

$$\partial_\mu \partial^\mu A^\nu(x) - \partial^\nu(\partial_\mu A^\mu) = \mathcal{J}^\nu(x). \tag{14}$$

The development of these equations is straightforward; we work in the rest frame of each respective nucleus, solve for the four potentials in momentum space, and then Lorentz boost the result to the frame in which the collision is taking place. We shall specialize to spin-zero nuclei, and use time-reversal invariance; in the rest frame of a nucleus of charge Z

$$\begin{aligned} \mathcal{J}^0(Q) &= 8\pi^2 \delta(Q_0) Z f^Z(Q^2) G_E(Q^2) \\ \mathcal{J}^\mu(Q) &= 0 \qquad \mu \neq 0\,, \end{aligned} \tag{15}$$

where $f^Z(Q^2)$ is the momentum distribution of a proton in the nucleus, and $G_E(Q^2)$ is the electric distribution function of the proton. Details of this development are given in Ref. [13]. From (14) and (15), the four-vector potential associated with nucleus a is a function of the four-momentum Q, and in the center-of-momentum frame of the heavy ions, is given by

$$A^0_a(Q) = -8\pi^2 Z_a \gamma^2 \frac{\delta(Q_0 - \beta Q_z)}{Q^2} G_E(Q^2) f^{Z_a}(Q^2)\ \exp\left(i\vec{Q}_\perp \cdot \vec{b}/2\right),$$

$$A_a^x(Q) = 0 \ , \tag{16}$$

$$A_a^y(Q) = 0 \ ,$$

$$A_a^z(Q) = \beta A_a^0(Q) \ .$$

The potentials from nucleus b can be easily obtained from (16) by the substitutions

$$\vec{b} \to -\vec{b}, \ \ \beta \to -\beta.$$

3. CROSS SECTIONS FOR PAIR PRODUCTION

If we focus on the second-order processes expressed by the diagram in Fig. 1, the relevant S-matrix element is given by

$$< k \mid S_{ab} \mid q > = (-i)^2 \int_{-\infty}^{\infty} d\tau \int_{-\infty}^{\tau} d\tau' < k \mid K_0(0,\infty) K_0(\infty,\tau)$$

$$\times \not{A}_a(\tau) K_0(\tau,\tau') \not{A}_b(\tau') K_0(\tau',-\infty) K_0(-\infty,0) \mid q >, \tag{17}$$

where

$$\mid q > = \mid \chi_q^{(-)} > \ , \mid k > = \mid \chi_k^{(+)} > \ , \tag{18}$$

and S_{ba} is obtained from S_{ab} by reversing the signs of the velocity β and the impact parameter $\vec{b}$.

The pair production cross section is given, adequately for the purposes in hand [11], by the expression

$$\sigma = \int d^2 b \sum_{k>O} \sum_{q<O} \mid < k \mid S_{ab} + S_{ba} \mid q > \mid^2 . \tag{19}$$

The amplitude for the diagram in Fig. 1 was given in (17) and can be understood as follows:

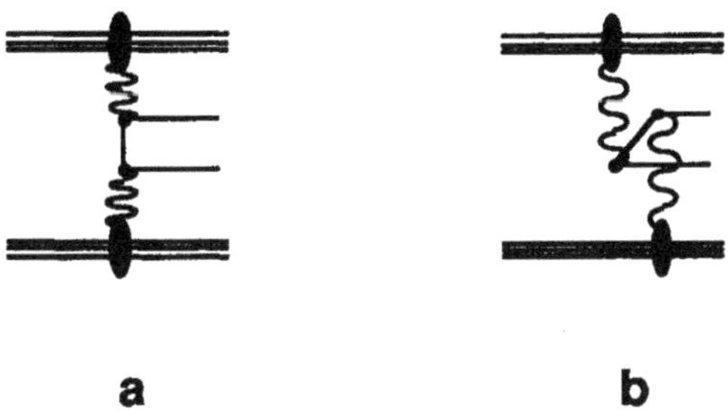

Fig. 1. Direct (a) and crossed (b) Feynman diagrams for pair production in a heavy-ion collision.

In the center-of- momentum frame of the colliding nuclei, the classical electromagnetic fields of the nuclei provide a source of "off-shell" photons with four momenta given as Q_a, and Q_b, respectively

$$Q_a = (\omega_a, \vec{Q}_{a\perp}, \frac{\omega_a}{\beta}) ,$$
$$Q_b = (\omega_b, \vec{Q}_{b\perp}, -\frac{\omega_b}{\beta}) . \tag{20}$$

The lepton and antilepton carry four-momentum k and q, respectively, and the total momentum of the pair

$$P = k + q . \tag{21}$$

The frequencies of the virtual photons are explicitly given by

$$\omega_a = \frac{E_k^{(+)} - E_q^{(-)} + \beta(k_z - q_z)}{2}, \tag{22}$$

$$\omega_b = \frac{E_k^{(+)} - E_q^{(-)} - \beta(k_z - q_z)}{2}.$$

The photon and lepton part of the diagram follow the usual Feynman rules so that

$$P = Q_a + Q_b\,. \tag{23}$$

Thus with a modest amount of analysis, the amplitude (17) can be written as

$$< k \mid S_{ab} \mid q >= \frac{i}{\beta}\int \frac{d^2\tau_\perp}{(2\pi)^2}\exp(-i\vec{\tau}_\perp\cdot\vec{b})F_a(Q_a^2)F_b(Q_b^2)\mathcal{T}_{kq}\left(\frac{\vec{P}_\perp - \vec{\tau}_\perp}{2} : \beta\right)\,, \tag{24}$$

where $\tau_\perp$ is the transverse momentum transfer

$$\vec{\tau}_\perp = \vec{Q}_{a\perp} - \vec{Q}_{b\perp}. \tag{25}$$

The transverse momentum is not fixed by the kinematics, but depends on the distribution of momenta carried in the classical fields. The form factor $F(Q^2)$ expresses the momentum dependence of the vertex on the nucleon and nucleus distributions and is obtained from (18) for either nucleus a or b as,

$$F(Q) = 4\pi Z f^Z(Q)G_E(Q^2). \tag{26}$$

The reduced Feynman amplitude $\mathcal{T}$ in (17) relates the coupling of the photon lines to the outgoing fermion lines in terms of a projection matrix for the intermediate fermion state

$$\mathcal{T}_{kq}(\vec{p_\perp} : \beta) = < u_{\sigma_k}^{(+)} \mid (1 - \beta\alpha_z)\frac{H_0(\vec{p}) + \omega}{H_0(\vec{p})^2 - \omega^2}(1 + \beta\alpha_z) \mid u_{\sigma_q}^{(-)} >, \tag{27}$$

where

$$H_0(\vec{p}) = H_\perp(\vec{p}_\perp) + p_z\alpha_z,$$

$$H_\perp(\vec{p}_\perp) = \vec{\alpha}_\perp \cdot \vec{p}_\perp + \gamma_0. \tag{28}$$

The four-momentum p is that carried by the intermediate lepton state in Fig. 1

$$p = (\omega, \vec{p}_\perp, p_z)\ , \tag{29}$$

whose components (ω, p_z) depend on β, as in (31) below. The momentum in intermediate states is composed of parts transverse and parallel to the motion of the heavy ions, $\vec{p} = \vec{p}_\perp + p_z\vec{e}_z$, and the frequencies ω_a and ω_b are related to the lepton energy ω through

$$\omega_a = E_k^{(+)} - \omega, \quad \omega_b = \omega - E_q^{(-)}. \tag{30}$$

The frequencies and intermediate longitudinal momentum are thus determined as functions of β

$$\begin{aligned} p_z(\beta) &= \frac{-E_k^{(+)} + E_q^{(-)} + \beta(k_z + q_z)}{2\beta} \\ \omega(\beta) &= \frac{E_k^{(+)} + E_q^{(-)} - \beta(k_z - q_z)}{2}. \end{aligned} \tag{31}$$

Utilizing the above analysis, the expression for the cross section (19) becomes,

$$\sigma = \frac{1}{4\beta^2} \sum_{\sigma_k, \sigma_q} \int \frac{dk_z dq_z d^2P_\perp}{(2\pi)^4} \int \frac{d^2Q_{a\perp} d^2Q_{b\perp}}{(2\pi)^4}$$

$$\times \mid F_a(Q_a^2) F_b(Q_b^2) \mid^2 \mid \mathcal{T}_{kq}(\frac{\vec{P}_\perp - \vec{r}_\perp}{2} : \beta) + \mathcal{T}_{kq}(\frac{\vec{P}_\perp + \vec{r}_\perp}{2} : -\beta) \mid^2 . \tag{32}$$

The expression (32) will be evaluated, without further approximation, using the Monte Carlo technique described in the next section.

4. MONTE CARLO INTEGRATION

The evaluation of (32) entails the integration of a positive function over eight variables. We can transform all of the integration into new variables which span the interval $(0,1)$, as follows. The rule, which greatly simplifies the numerics, is to scale all longitudinal momenta by the Lorentz factor associated with the collisions. Thus we map the coordinates (k_z, q_z) onto a vector, $\vec{\kappa}$, so that

$$\int_{-\infty}^{\infty} dk_z \int_{-\infty}^{\infty} dq_z = \int_0^{\infty} \kappa d\kappa \int_0^{2\pi} d\phi_\kappa \,. \tag{33}$$

The magnitude of the vector κ is scaled by the Lorentz factor γ associated with the collision. The integration over $\vec{P}_\perp$ is given as

$$\int d^2 P_\perp = \int_0^{P_u} P_\perp dP_\perp \int_0^{2\pi} d\phi_\perp \,. \tag{34}$$

The integration on the magnitude of $P_\perp$ is scaled with the mass of the lepton and is carried out with a simple quadrature formula, thus effectively mapping the integration onto the interval $(0, P_u)$. The integration on the transverse momenta of the two photons is given by the following transformation which is applied equivalently to both a and b.

$$\int d^2 Q_\perp = C \int_0^{\pi/2} d\theta \tan\theta \int_0^{2\pi} d\phi \,, \tag{35}$$

where the constant C is the frequency carried by the field associated with either nucleus a or b, given in (22),

$$C = \omega/\gamma \,.$$

Thus we can transform (32) in terms of an eight-fold integration over a unit hypercube having the form

$$\sigma = F_0 \int_0^1 dx_1 dx_2 \cdots dx_8 f(x_1, x_2, \cdots, x_8) \ , \tag{36}$$

where F_0 is chosen so that $f \leq 1$, and where the coordinates $\{x_i\}$ are simply related to the set $\{\kappa, \phi_\kappa, P_\perp, \phi_\perp, \theta_a, \phi_a, \theta_b, \phi_b\}$. At this point there are several different ways to proceed. We shall discuss two approaches that are equally successful.

Method 1

In the present case we have found it necessary to start with an estimate F_0, which is periodically improved by noting the largest value of f for varying numbers of Monte Carlo points. The algorithm proceeds by making *throws* in a nine-dimensional space, $\vec{y} = \{y_1, \ldots, y_9\}$, where $0 < y_j < 1$. We then calculate

$$x_9 = f(y_1, y_2 \cdots, y_8), \tag{37}$$

so that x_j lies in the range $0 < x_j < 1$. The *throw* is said to be *successful* if $x_9 > y_9$. After T throws, let the number of successes be S. The cross section is given by

$$\sigma = F_0 \frac{S}{T} \ , \tag{38}$$

with a proportional error in σ / F_0 of

$$\Delta = \left(\frac{1}{S} + \frac{1}{T}\right)^{\frac{1}{2}} . \tag{39}$$

Clearly, a good estimate of F_0 is desirable. We are able to compensate for the lack of *a priori* knowledge of F_0 by analysing each throw for a set of F_0 in parallel. At the end of a run, the estimates of σ thereby obtained from (38) were interpolated to the best value of F_0 inferred from the run itself. The improved value of F_0 was used in subsequent runs. In Fig. 2 we display the error Δ as a function of the number of Monte Carlo points for a test run, $\Delta \simeq 30 - 100/T^{\frac{1}{3}}$.

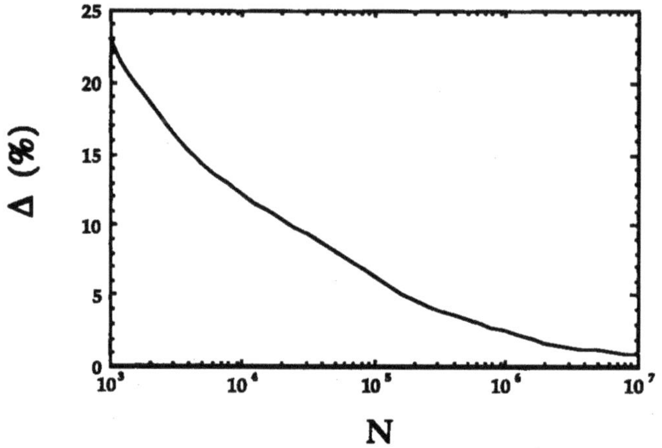

Fig. 2. Percentage error in Monte Carlo integration, Δ, vs. the number of Monte Carlo points, N.

We observe a rapid and stable convergence of the sampling algorithm as described above. Typical success rates were $S/T \simeq 10^{-3} - 10^{-4}$.

Method 2

In this approach, an eight-fold integration on the unit hypercube is approximated by a *sampling* of the integrand T times and then taking

$$\sigma = \frac{F_0}{T} \sum_{i=1}^{T} f(x_1(i), x_2(i), \cdots, x_8(i)) \, , \tag{40}$$

where the set $\{x_1(i), x_2(i), \cdots, x_8(i)\}$ denotes the ith sampling of the coordinates. If these variables are uniformly distributed on the interval $(0, 1)$, then (40) is just the

Stiltjes definition of integration. Here, a single sampling of the coordinates requires eight independent pseudo-random numbers. Each evaluation of the function f can be thought of as a statistical estimate of σ/f_0 with a corresponding variance. Thus, after T throws, the corrsponding error works out to be

$$\Delta_T^2 = \frac{1}{T}\sum_{i=1}^{T}(f(x_1(i), f_2(i), \cdots, x_8(i)) - \sigma/F_0)^2 \ , \tag{41}$$

so that both the cross section and its variance can be computed at the same time. In both methods the error goes as $1/\sqrt{T}$. This rate of convergence may seem slow compared to the more sophisticated prescriptions utilizing Gauss points; for example, using K^{th} order Gauss–Laguerre integration, the error goes as $\Delta_G = 1/T^{2K-1}$. However, in eight dimensions, the overall error goes as $\Delta_G^{1/8}$ so that Monte Carlo wins. Generally, any calculation is a balance between accuracy and economy; we summarize some of the features of Monte Carlo calculations which we find advantageous.

- Both of the above Monte Carlo algorithms are basically very simple to use without a great deal of specialized knowledge, and both methods vectorize easily on supercomputers. Thus modern calculations can be effected with a large number of integration points.

- Very complicated constraints can be placed on the summations, therby allowing experimental "cuts" on phase space to be directly calculated. This is explored in depth in the section on $\pi^- p$ collisions.

- The integration can be "binned" in terms of other variables during the evaluation. Thus a variety of distributions can be directly computed without the need for calculating difficult Jacobians.

- We are basically studying two-photon diagrams which entail eight independent integrations. Diagrams of much higher order can be evaluated with this approach. Although not discussed in these lectures, we have also examined the radiative corrections to pair production, which reduces to a ten dimensonal problem.

In the following two sections we shall discuss two very different application of these techniques.

5. μ, τ PAIR PRODUCTION

In this section we study the effects of the nuclear form factors on the μ and τ pair production. In particular, we consider the symmetric collision of two gold nuclei at RHIC energies. From (18), the charge current distribution in the gold nucleus involves the product of the momentum distribution of a proton in the nucleus, f^Z, times the electric distribution function of the proton, G_E. For small values of the momentum transfer, f^Z and G_E are related to the Fourier transforms of the gold and proton charge distributions, respectively. However, for large values of Q^2, these are difficult to determine experimentally [14)]. Theoretical and empirical models, as for example, perturbative QCD [15)], vector meson dominance [16)], and dimensional scaling [17)] give predictions at large Q^2 which seem to agree with the measured values of the magnetic form factor [14]. Experimental tests of these predictions for the electric form factor are presently unavailable.

Accordingly, we have chosen empirical form factors for both the proton and the Au nucleus. For Au, we employ a Fermi function with two parameters, which are adjusted to fit low-energy elastic electron scattering data [18]. This is shown in Fig. 3 as the solid curve.

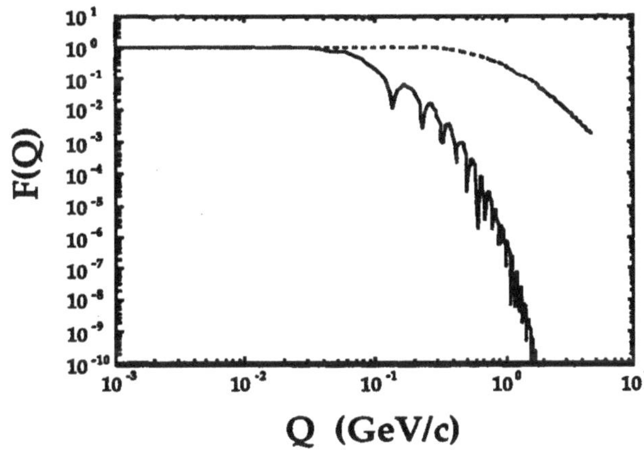

Fig. 3. The proton (dashed curve) and Au (solid curve) form factors as a function of the magnitude of the photon four momentum in units of GeV/c.

The proton form factor is assumed to have a dipole form [13] given by

$$G_E(Q^2) = \left(1 + \frac{Q^2}{\Lambda_p^2}\right)^{-2} , \quad \Lambda_p^2 = 0.71\text{GeV}^2/\text{c}^2 . \tag{42}$$

This function is depicted as the dashed curve in Fig. 3. These curves illustrate, in a simple way, that the electromagnetic vertex is strongly suppressed for values of Q^2 greater than about one GeV^2/c^2.

In Fig. 4 we present lepton pair cross sections calculated for colliding beams of

Au nuclei at a bombarding energy per nucleon of 100 GeV. Here, we give the total cross section in units of σ_0

$$\sigma_0 = \frac{(Z\alpha)^4\hbar^2}{M_\ell^2c^2} \tag{43}$$

as a function of the intrinsic mass of the lepton, M_ℓ, treated as a continuous variable. Thus, there are only three physical masses in the figure, 5.11×10^{-4}, 1.06×10^{-2}, and 1.784 GeV/c^2, corresponding to the electron, muon, and tauon respectively. The dotted curve represents results obtained with point charge nuclei, the dashed curve results obtained with the proton form factor, and the solid curve results obtained with form factors for the proton and for the Au nucleus included. We note that for masses less than about 10 MeV/c^2, neither the proton nor the Au form factor substantially affects the cross section. Thus, the total electron pair production is not strongly affected by the details of the nuclear currents. However, for values of the mass near about 2 GeV/c^2, the results with the proton form factor are more than an order of magnitude smaller than the corresponding point charge results, and the results which include the Au form factor are reduced by more than a factor of 10^3. Hence, tauon pair production is extremely sensitive to these form factors. The muon pair cross sections are suppressed by about a factor of four compared to the results with point nuclear currents.

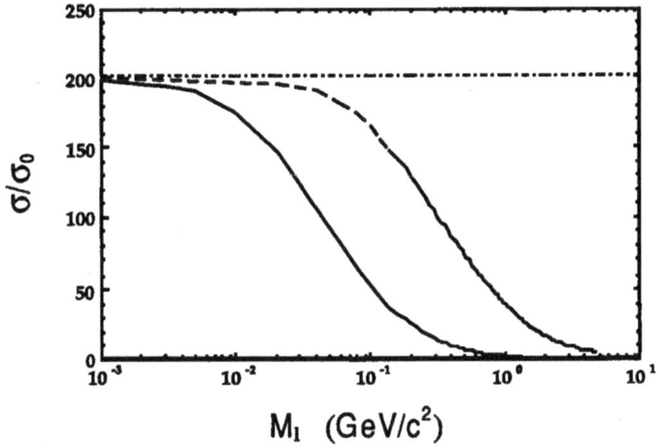

Fig. 4. Dependence of pair production cross sections on the intrinsic lepton mass M_ℓ for collisions of $Au + Au$ at 100 GeV per nucleon. The ratio of the cross section to the reduced cross section σ_0 is plotted vs. M_ℓ. Full line: result including Au and proton form factors; dashed line: result including only the proton form factor; dotted line: point source result.

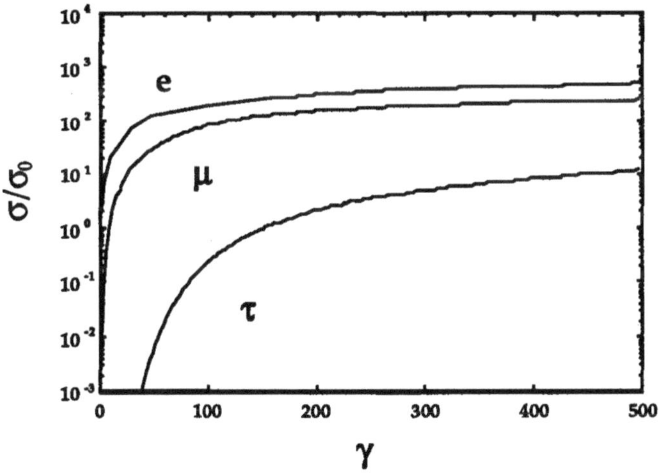

Fig. 5. The pair cross sections for colliding beams of Au nuclei as a function of the Lorentz γ of the beam. The reduced cross section σ_0 is defined in (43). Full line: electrons, dashed line: muons, long dashed line: tauons.

A comparison of the electron, muon, and tauon total cross sections as a function of the Lorentz γ of the beams is given in Fig. 5 for collisions of Au nuclei. These results include both nuclear form factors as described previously. Note that the results for the electrons are equivalent to the same results using point nuclear currents. Values of γ about 3 and 10 approximately correspond to experiments at the AGS and at CERN. Here we see that for all practical purposes, the form factors completely suppress the tauon cross sections and reduce the muon cross sections by more than a factor of 100 at $\gamma = 10$. Thus it is only at the highest bombarding energies, such as will be attained at RHIC, that heavy leptons are produced with large multiplicities. These results are summarized in Table 1.

Table 1

Results for the coherent muon and tauon pair production for 100 GeV per nucleon colliding beams of Au nuclei.

	μ (mb)	τ (mb)
Point Source	770	2.7
Form Factor	200	2.8×10^{-3}

The role of the nuclear form factors in heavy lepton production can be understood in greater depth by examining their effects on the pair distributions in rapidity, transverse momentum, and invariant mass. These are shown in Figs. 6, 7, and 8, for muons.

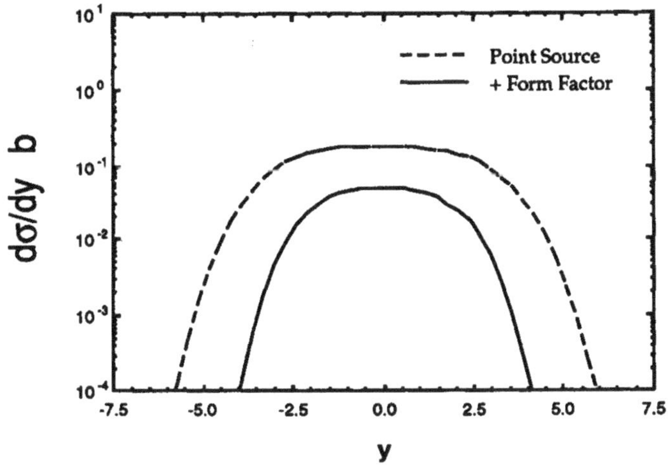

Fig. 6. Rapidity distribution of dimuons produced from colliding beams of Au nuclei at an energy per nucleon of 100 GeV. The solid curve is the calculated result including the proton and Au form factors, and the dashed curve is the result from point charge nuclei.

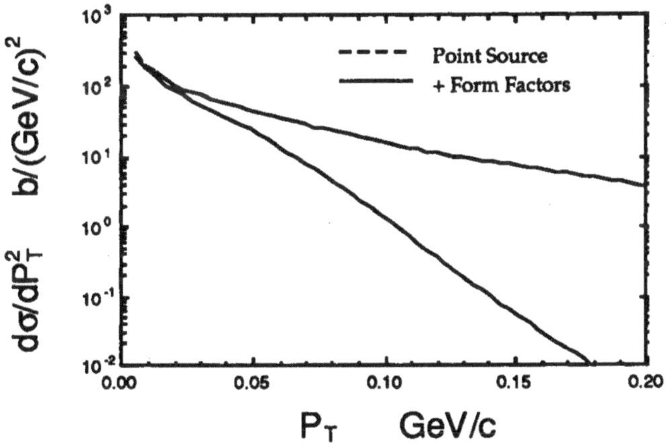

Fig. 7. Transverse momentum distribution of dimuons produced from colliding beams of Au nuclei at an energy per nucleon of 100 GeV. The solid curve is the calculated result including the proton and Au form factors, and the dashed curve is the result from point charge nuclei.

Again, we are considering colliding beams of Au nuclei at 100 GeV per nucleon.

In Fig. 6 we show the μ pair cross section as a function of the rapidity. The solid curve gives the results which include the nuclear form factors, and the dashed curve gives the results for point nuclear currents. The rapidity of the Au beams is approximately $Y = \pm 5.29$. Again, we are considering colliding beams of Au nuclei at 100 GeV per nucleon. In Fig. 6 we show the μ pair cross section as a function of the rapidity. The solid curve gives the results which include the nuclear form factors, and the dashed curve gives the results for point nuclear currents. The rapidity of the Au beams is approximately $Y = \pm 5.29$. Thus we observe that the form factors reduce the cross section by about a factor of ten and result in a distribution which is sharply confined to the region of central rapidity. In Fig. 7 the distribution in transverse momentum, P_T, is compared for the two cases. We note that the two curves are approximately the same for values of P_T near zero. The Compton momentum of the μ is about 0.1 GeV/c; at this value of P_T the nuclear form factors have reduced the cross section by an order of magnitude, and hence the resulting distribution is concentrated at small values of P_T. The distribution as a function of the invariant mass is shown in Fig. 8. Again we are comparing the results of calculations with form factors and with point currents. In this case the threshold for the mass distribution is about 0.2 GeV/c^2, and we see a strong reduction in the distribution due to the form factors. For τ pair production, these effects are even more pronounced.

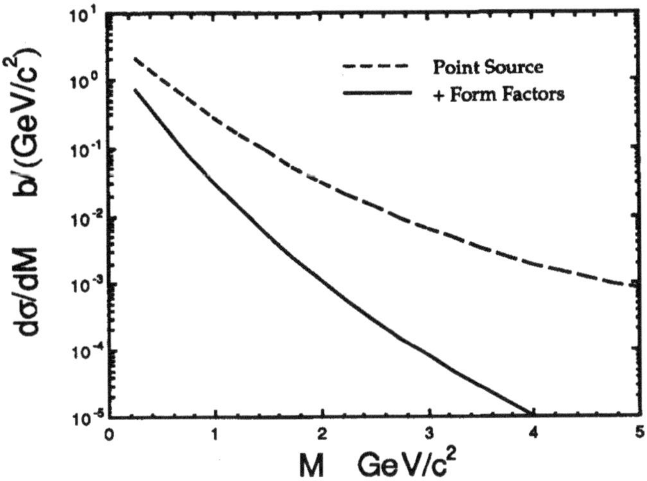

Fig. 8. Invariant mass distribution of dimuons produced from colliding beams of Au nuclei at an energy per nucleon of 100 GeV. The solid curve is the calculated result including the proton and Au form factors, and the dashed curve is the result from point charge nuclei.

6. $\pi^- P$ COLLISIONS

The emission of electron–positron pairs has been observed [9,10] in π^- p collisions with corresponding pion momenta of 16 and 17 Gev/c. The experiments detected pairs over a limited range of invariant pair mass, given by $0.2 \leq M \leq 1.2$ GeV/c^2. A large peak near $M = 0.8$ GeV/c^2 was observed. This peak is well understood as resulting from the direct electromagnetic decay of the ρ and ω mesons, $\rho \rightarrow e^+ e^-$, and $\omega \rightarrow e^+ e^-$. A continuum of pairs is also seen at lower invariant masses. These pairs were deemed anomalous, as conventional hadronic mechanisms failed to explain the data. We demonstrate here that an *ab initio* calculation of the two-photon mechanism produces results that are in quantitative agreement with the continuum data.

The experimental results of Refs. [9] and [10] are shown in Figs. 9, 10, and 11. The qualitative features of the cross sections are clear. The cross section decreases with increasing pair mass, decreases with increasing Feynman x (defined as the total longitudinal momentum of the pair divided by the maximum value allowed kinematically), and decreases with increasing transverse momentum.

If we calculate the two-photon process in the limit where the relative motion of the pion and proton is treated classically [19], the total cross section for electron–positron pair production is given by (19). For a 17 GeV/c pion, we can use a straight-line trajectory. The scattering amplitude has the two terms corresponding to crossed and uncrossed photons as shown in Fig. 1 and as discussed in Section 2. Here $A_a^\mu(q)$ and $A_b^\mu(q)$, which appear as prescribed external fields, are given by the photon propagator times the matrix element of the proton or pion current, as given in (15) and (16). However, the resulting expression for the proton current is incomplete. The most general form of the electromagnetic current of the proton is [13]

$$J^\mu(p'\sigma', p\sigma) = e\,\overline{u}^{(\sigma')}(p')\,[\gamma^\mu F_1(p'-p)]\,u^{(\sigma)}(p)$$
$$-\frac{e}{2M_p}\,\overline{u}^{(\sigma')}(p')\,[\sigma^{\mu\nu}(p'_\nu + p_\nu)\,\kappa\, F_2(p'-p)]\,u^{(\sigma)}(p)\,. \qquad (44)$$

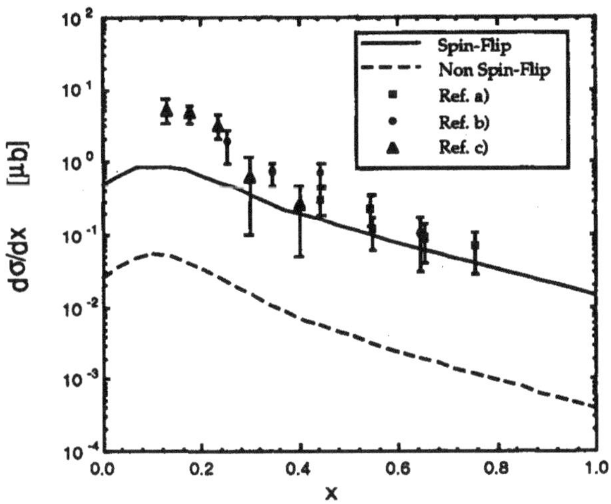

Fig. 9. The differential cross section for electron–positron pair production *vs.* Feynman x. The dashed curve is the result omitting the spin-flip terms. The data are from Ref. [9].

By inserting explicit expressions for the proton spinors in (44) and multiplying by the photon propagator, the vector potential from the proton in (16) becomes

$$A^{\mu}_{a}(\mathbf{b}; q) = a^{\mu}(\mathbf{b} : q) + \delta a^{\mu}(\mathbf{b} : q) ,$$

where the term which corresponds to non-spin-flip events is

$$a^{o}(\mathbf{b} : q) = 2\pi\, e\, \delta(q_o - \beta q_z)\, \gamma \frac{e^{-i\vec{q}\cdot\vec{b}/2}}{q_z^2 + \gamma^2 q_\perp^2} G_E(q^2) ,$$

$$a^{x}(\mathbf{b} : q) = 0 , \quad a^{y}(\mathbf{b} : q) = 0 , \quad a^{z}(\mathbf{b} : q) = \beta\, a_o^{(+)}(\mathbf{b} : q) . \tag{45}$$

The electric form factor of the proton is given in (42) [13]. The second term in

$A_a^\mu(q)$, which involves both $F_1(q)$ and $F_2(q)$, corresponds to events in which the proton spin flips. This term yields

$$\delta a^x(\mathbf{b}:q) = 2\pi \frac{e}{2M_p} \delta(q_o - \beta\, q_z) \frac{e^{-i\vec{q}\cdot\vec{b}/2}}{q_z^2 + \gamma^2 q_\perp^2} q_z\, G_M(q^2) \,,$$

$$\delta a^y(\mathbf{b}:q) = 2\pi \frac{e}{2M_p} \delta(q_o - \beta\, q_z) \frac{e^{-i\vec{q}\cdot\vec{b}/2}}{q_z^2 + \gamma^2 q_\perp^2}\, i\, q_z\, G_M(q^2) \,,$$

$$\delta a^z(\mathbf{b}:q) = -2\pi \frac{e}{2M_p} \delta(q_o - \beta q_z)\, \gamma \frac{e^{-i\vec{q}\cdot\vec{b}/2}}{q_z^2 + \gamma^2 q_\perp^2} (q_x + i q_y)\, G_M(q^2) \,,$$

$$\delta a^o(\mathbf{b}:q) = \beta\, \delta a_o^{(+)}(\mathbf{b}:q) \,, \tag{46}$$

where the magnetic form factor of the proton, $G_M(q^2)$, is normalized so that $G_M(0) = 1+\kappa = 2M_p\mu_p/e$, where μ_p is the magnetic moment of the proton. We use the same dipole form for the magnetic form factor, $G_M(q^2) = (1+\kappa)(1+q^2/\Lambda_p^2)^{-2}$, as for $G_E(q^2)$. For the spin-zero pion the electromagnetic current is given by

$$J^\mu(p',p) = e\,(p'^\mu + p^\mu)\, G_E\left((p'-p)^2\right) \;. \tag{47}$$

which produces a similar result for $A_b^\mu(q)$ as is given in (16); in the center-of-velocity frame, $A_b^\mu(q)$ is given by (45) with $\beta \to -\beta$ and $\mathbf{b} \to -\mathbf{b}$. We utilize a monopole form for the electric form factor of the pion, $G_E(q^2) = (1 + q^2/\Lambda_\pi^2)^{-1}$, where $\Lambda_\pi^2 = 0.59\ \ \mathrm{GeV}^2/\mathrm{c}^2$ [13]. The expression derived here for the total cross section is Lorentz invariant and is most easily evaluated in the center-of-velocity frame. Finally, the crossed photon term, which contributes to (19), is added coherently to the direct term.

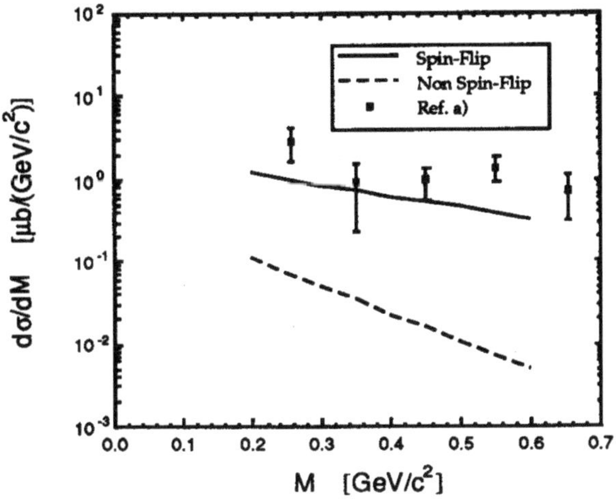

Fig. 10. The differential cross section for electron–positron pair production *vs.* the pair mass. The dashed curve is the result omitting the spin-flip terms. The data are from Ref. [9].

In summary, we take the classical limit on the relative motion of the proton and pion of standard QED perturbation theory for the two–photon process. The arguments of Ref. [19] indicate that the wavelength of the relative pion-nucleon motion is sufficiently large and the electromagnetic field sufficiently smooth that we are well into the classical region. In this region, the matrix element of the current, which couples the proton or the pion to the electromagnetic field multiplied by the photon propagator, takes the form of prescribed electromagnetic fields that have been boosted from the rest frame to a frame with velocity $-\beta$, as given in (45) for the non-spin-flip term, and (46) for the spin-flip term.

The eight-dimensional integration required to calculate the total cross section in

(32) is done utilizing the Monte Carlo technique described earlier. We impose on the Monte Carlo integration the experimental cut on the pair mass, $0.2 \leq M \leq 0.8$ GeV/c^2. Points which correspond to invariant masses outside this region are counted in the overall normalization of the calculation, but do not contribute to the data of present interest, and thus the integrand need not be calculated for these points. The total cross section for electromagnetic pair production is about 55 μb, of which only 0.5 μb lies within the cut. The differential cross sections follow simply by binning the points as a function of the relevant independent variable.

The results of a calculation which used about 25×10^6 Monte Carlo points are shown in Figs. 9-11. The contributions which arise from the non-spin-flip current, (45), are presented as the dashed curves. These results are significantly below the data. The cross sections for production from the currents given in (45) are large [11] for small electron and positron momenta. However, in the kinematical region imposed by the cut on the invariant mass, this process produces a very small cross section. In addition, the direct and crossed terms interfere destructively to reduce the predicted cross sections by two orders of magnitude.

The contributions for electron–positron pair production, including the magnetic spin-flip term, (46), are also shown in Figs. 9-11 as the solid curve. We see that the results are in reasonable agreement with the data. We stress that the calculation is made *ab initio* with no adjustable parameters. Although the overall agreement is good, the cross sections in Fig. 9 lie a little below the data for low Feynman x. This may be due to our use of the lower cut on the invariant mass at exactly

0.2 GeV/c^2. If we decrease this value slightly, the cross sections at low x and low $p_\perp$ are enhanced. The difficulty might also lie in our neglect of the final-state distortions of the electron and positron. The cross sections for the slower-moving leptons might well be somewhat higher if distorted waves were used.

The spin-flip term, (46), is of order q/M_p, as compared to the leading term, (45). The reduction due to this factor is more than compensated for by two features of the spin-flip current: 1) the current has transverse components, and 2) the 0 and z components are proportional to the transverse momentum of the photon. Thus the spin-flip current produces pairs at the large values of transverse momentum emphasized by the experimental cut on the invariant mass.

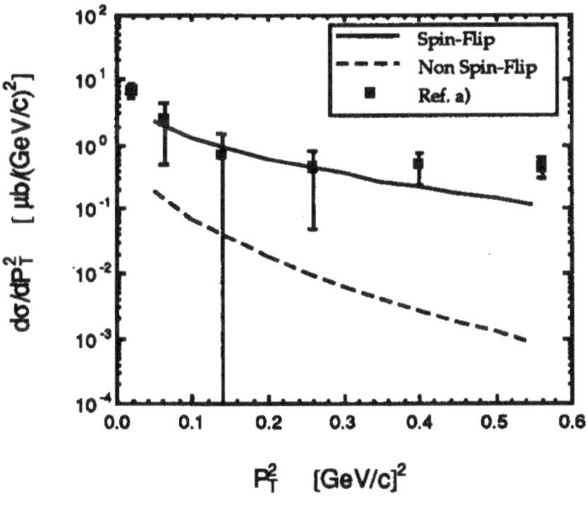

Fig. 11. The differential cross section for electron–positron pair production *vs.* the transverse momentum of the pair. The dashed curve is the result omitting the spin-flip terms. The data are from Ref. [9].

We conclude from the general quality of the agreement of the calculation with the data that the lepton pairs observed in Refs. [9] and [10] are produced, to a large extent, via the two-photon mechanism specifically by the magnetic part of the proton current. The calculation itself allows some internal consistency checks on this view of the physics. The presence of the proton and pion form factors reduces the predicted cross sections by less than a percent. This indicates that the cross section arises from a region where the scale for the momentum or energy of the virtual photon is given by several hundred MeV or less. At these momentum transfers, the dominant response of the proton or the pion is the coherent elastic response, although the excitation of the nucleon to a delta, or the pion to a rho might also contribute a non-negligible amount in this kinematic region. We note that the coincidence data given in Ref. [10)] are consistent with this low multiplicity interpretation of the data. This relatively low momentum transfer from the hadrons does not justify a quark level calculation of the two-photon process, as was done in Ref. [10)]. We further point out that the independent quark-level processes are suppressed by a factor of $Z_1^2 Z_2^2$, with Z_i the charge of the quark, and thus can only become the dominant term in kinematic regions where the elastic form factor suppresses the coherent elastic term. In our calculation we integrate over impact parameters starting at zero. For impact parameters less than one Fermi, the pion and proton will undergo strong interactions which would invalidate the pure electromagnetic calculation done here. However, we find that non-negligible numbers of lepton pairs are emitted out to impact parameters of greater than one hundred Fermis. Thus, neglecting the contribution which arises from impact parameters below one Fermi would have little effect on our results.

Electromagnetic lepton pair production in particle collisions is important for several reasons. It is well understood that the two-photon process can produce large numbers [5,11] of soft pairs. Here we have found that it can also produce pairs with significant transverse momentum, which could provide a background for the lepton pairs emitted in heavy-ion collisions. If the lepton pairs are to serve as a probe of the quark-gluon plasma, this background must first be understood. Also, the Lorentz-contracted fields, represented by the vector potentials of (45) and (46), become very large for the ultrarelativistic heavy ions that will be produced at RHIC and perhaps at the SSC. In this case, the perturbative approach utilized here will certainly be inadequate. Lepton production would then offer the opportunity to study QED in a region where it is not perturbative. The present results are needed as a baseline for this study.

7. CONCLUSIONS

We conclude by reviewing the questions raised in the Introduction concerning electromagnetic mechanisms for pair production. The present work supplies detailed information on the backgrounds necessary to interpret analysis of heavy-ion dimuon data. We have shown that muon production in heavy-ion collisions is largely coherent over distances of the dimension of the nucleus. In the case of tauons, the form factors break this coherence. The same result should be true for even more massive particles such as W-pairs. Thus it is expected that the incoherent production will dominate the two-photon mechanism for particles as massive as the tauon or W. In summary:

- We have calculated heavy lepton pair production in the two-photon approximation. We have employed Feynman perturbation theory to go beyond equivalent photon formulations, and in particular, to incorporate correctly the transverse momentum dependence.

- We have shown that form factors must be included in calculations of muon and tauon pair cross sections. We employed the best fits to the elastic electron nucleus scattering data that presently exist. This determines the form factor for Q^2 less than about one GeV^2/c^2. There are no form factor data for heavy nuclei at larger values, and the sensitivity of the cross sections to the form factor in this region is not known. We note that other models, which fit the electron scattering data less well, differ in their cross section predictions by as much as a factor of ten. Thus, these effects need to be explored in greater depth.

- There is a need for measurements which will test theories of two-photon pair production and its scaling with charge, mass, and energy. If these issues are settled, the two-photon pair production could be a method of probing the form factors of heavy nuclei at large values of momentum transfer.

Our future work will address incoherent mechanisms, the effects of inelastic structure functions, and central collisions.

8. ACKNOWLEDGEMENTS

This research was sponsored by the Division of Chemical Sciences, Office of Basic Energy Sciences, and by the Division of Nuclear Physics of the U.S. Department of Energy under contract No. DE-AC05-84OR21400 with Martin Marietta Energy Systems, Inc.

REFERENCES

1. H. Gould, in "Proceedings of the Atomic Theory Workshop on Relativistic and QED Effects in Heavy Atoms", AIP Conf. Proc. 136, p. 66, (AIP, New York, 1985).

2. C. Bottcher and M.R. Strayer, in "Physics of Strong Fields", edited by W. Greiner, (Plenum, New York, 1987), Vol.153, page 629.

3. C. Bottcher and M.R. Strayer, "Proceedings of the Second Workshop on Experiments and Detectors for the RHIC", Berkeley, California, May 25-29, 1987, page 279, (LBL-24604).

4. G. Bauer and C.A. Bertulani, Phys. Rev. C35, 836 (1987).

5. C. Bottcher and M.R. Strayer, Nucl. Inst. and Meth. B31, 122 (1988).

6. E. Teller, Nucl. Inst. and Meth. B24, 1 (1987).

7. Elena Papageorgiu, Phys. Rev. D40, 92 (1989); Nucl. Phys. A498, 593c (1989); M. Drees, J. Ellis and D. Zeppenfeld, Phys. Lett. 223b, 454 (1989).

8. M. Grabiak, B. Müller, W. Greiner, G. Soff, and P. Koch, J. Phys. G15, L25 (1989); G. Soff, J. Rau, M. Grabiak, B. Müller and W. Greiner, *Production of super-symmetric particles and Higgs bosons in ultrarelativistic heavy-ion collisions*, Preprint,1989, (GSI-89-55); J. Rau, M. B. Müller and W. G. Soff, *Production of super-symmetric particles in ultrarelativistic heavy-ion collisions*, Preprint, submitted to J. Phys. G, 1989 (GSI-89-76).

9. R. Stroynowski, D. Blockus, W. Dunwoodie, D. W. G. S. Leith, M. Marshall, C. L. Woody, B. Barnett, C. Y. Chien, T. Fiefuth, M. Gilchriese, D. Hutchinson, W. B. Johnson, P. Kunz, T. Lasinski, L. Madansky, W. T. Meyer, A. Pevsner, B. Ratcliff, P. Schacht, J. Scheid, S. Shapiro and S. Williams, Phys. Lett. **97B**, 315 (1980); D. Blockus, W. Dunwoodie, D. W. G. S. Leith, M. Marshall, R. Stroynowski, C. L. Woody, B. Barnett, C. Y. Chien, T. Fieguth, M. Gilchriese, D. Hutchinson, W. B. Johnson, P. Kunz, T. Lasinski, L. Mandansky, W. T. Meyer, A. Pevsner, B. Ratcliff, P. Schacht, J. Sheid, S. Shapiro and S. Williams, Nucl. Phys. **B201**, 205 (1982).

10. J. Stekas, G. Abshire, M. R. Adams, C. Brown, L. Cormell, E. Crandall, G. J. Donaldson, J. Goldberger, H. A. Gordon, P. D. Grannis, B. T. Meadows, G. R. Morris and P. Rehak, Phys. Rev. Lett. **47**, 1686 (1981); M. R. Adams, G. Abshire, C. Brown, E. S. Crandall, J. Goldberger, P. D. Grannis, B. T. Meadows, J. Stekas, G. J. Donaldson, H. A. Gordon, G. R. Morris, P. Rehak and L. Cormell, Phys. Rev. D. **27**, 1977 (1983).

11. C. Bottcher and M.R. Strayer, Phys. Rev. D39, 1330 (1989).

12. C. Bottcher, M.R. Strayer, C. J. Albert and D.J. Ernst, to appear in Phys. Lett. B, 1990.

13. F.E. Close, "An Introduction to Quarks and Partons", (Academic Press, New York, 1979).

14. R.C Walker, B.W. Filippone, J. Jourdan, R. Milner, R. Mc Keown, D. Potterveld, R. Arnold, D. Benton, P. Bosted, G. de Chambrier, L. Clogher, A. Lung, S.E. Rock, Z.M. Szalata, A. Para, F. Dietrich, K. Van Bibber, J. Button-Shafer, B. Debebe, R.S. Hicks, S. Dasu, P. de Barbaro, A. Bodek, H. Harada, M.W. Krasny, K. Lang, E.M. Riordan, J. Alster, R. Gearhart, L.W. Whitlow, OAP-697, March, 1989, submitted to Phys. Lett . B.

15. G.P. Lepage and S.J. Brodsky, Phys. Rev. Lett. 43, 545 (1979).

16. F. Iachello, A.D. Jackson, and A. Landé, Phys. Lett. 43B, 191 (1973).

17. S.J. Brodsky and G. Farrar, Phys. Rev. D11, 1309 (1975).

18. Nuclear Data Tables, 14 (1974).

19. A. S. Umar, M. R. Strayer, D. J. Ernst, and K. R. Sandhya Devi, Phy. Rev. C **30**, 1934 (1984).

PATH INTEGRALS AS A CALCULATIONAL FRAMEWORK FOR MANY-BODY PHYSICS*

John W. Negele

Center for Theoretical Physics
Laboratory for Nuclear Science
and Department of Physics
Massachusetts Institute of Technology
Cambridge, Massachusetts 02139 U.S.A.

ABSTRACT

These lectures explain the use of path integrals as a calculational framework in many-body physics. The stationary-phase approximation to an exact path integral is shown to give rise to computationally tractable self-consistent mean field solutions applicable to a wide range of systems, and the special case of tunneling decay is described in detail. In cases for which the stationary-phase approximation is inapplicable, stochastic evolution yields results which are exact to within controllable sampling errors. Examples are presented for one-dimensional Fermion systems, and a new approach to three-dimensional systems is described.

INTRODUCTION

One of the great contributions of Feynman to theoretical physics was the formulation of quantum mechanics in terms of path integrals. Whereas many of the advantages of path integrals are well-known, such as the physical picture they provide in terms of summation over time histories and their elegance and utility in quantizing field theories, their role as a calculation framework for many-body physics may not be fully appreciated. Thus, in this school devoted to computational nuclear and atomic physics, I would like to explain and illustrate the utility of path integrals in calculating the properties of complex many-particle systems.

There are basically two general approaches to the calculation of path integrals, each leading in the end to its own applications and computational techniques. The first is the stationary phase-approximation, which leads to a quantum mean

* This work was supported in part by funds provided by the U. S. Department of Energy (D.O.E.) under contract #DE-AC02-76ER03069.

field theory plus fluctuation corrections. This approach considerably extends the notion of mean field theory, and gives rise to important new applications such as tunneling.

The second approach, which is required when the fluctuation corrections are not small, is to use stochastic techniques to evaluate the path integral exactly, to within controllable sampling errors. Of the various strategies for calculating path integrals stochastically, I will concentrate on the path integral Monte Carlo method, which is particularly physical and well-suited to the many-body problems of interest. For Boson systems in any dimension and Fermions in one spatial dimension, the method is immensely powerful and can provide practical quantitative solutions. Special problems associated with antisymmetry plague Fermion calculations in more than one dimension, and I will describe a recent attempt to circumvent these problems.

The basic principles and specific examples treated in these lectures have been published in a recent textbook and in a number of articles published in readily accessible journals and books. Thus, both from the viewpoint of conservation and copyright infringement, it makes no sense to repeat essentially the same presentation in these Proceedings. Hence, I will briefly summarize the essential ideas here and refer to the relevant published work for details.

PATH INTEGRALS

The foundations of path integrals relevant to these lectures are described in chapters 1, 2 and 7 of Ref. 1). The basic idea is to rewrite the evolution operator for a many-body system $e^{-iT\hat{H}}$ as the product of infinitesimal evolution operators $e^{-i\epsilon\hat{H}}$ which propagate the system from one time slice to the next. By inserting unity expressed as an integral over a complete or overcomplete set of functions at each time step and evaluating the resulting matrix elements of $e^{-i\epsilon\hat{H}}$ in the limit of small ϵ, one obtains an exact integral for the evolution operator. The essential point is that as $\epsilon \to 0$, the non-commutativity of the operators in $\hat{H}$ becomes irrelevant, so that the time slicing eliminates all the complications of non-commuting operators in quantum mechanics by integration over complete sets of functions at each time slice. Thus, the path integral effectively reduces quantum mechanics to quadrature, and allows us to apply to quantum mechanics any powerful techniques we may have for evaluating integrals. It is essential to appreciate the great generality of the approach and the flexibility it provides in formulating approximations and computational schemes. Any complete or overcomplete set of functions is fair game, giving rise to path integrals in coordinate space, momentum space, phase space, coherent states, and the space of Slater determinants. In addition, there is nothing special about the time. An expression of the form $e^{-\beta\hat{H}}$ which arises in statistical mechanics or in trying to filter the ground state out of a trial wave function may be treated analogously by slicing β into infinitesimal steps. Since such a variable β is formally equivalent to imaginary time, the parameter appearing in the path integral is often referred to as the time or imaginary time, but has no direct connection with the physical time.

MEAN FIELD THEORY

Application of the stationary-phase approximation to an evolution operator expressed as an integral over all possible paths singles out a single path or a set of paths which dominate the result. In the applications arising in many-body physics, the result has the form and interpretation of mean field theory, and extends the familiar static or time-dependent Hartree–Fock approximations to a much broader and general class of quantum mean field theory applicable to large amplitude collective motion, tunneling, and S-matrix elements.

The basic ideas are described in Chapters 2 and 7 of Ref. 1) and Section V of Ref. 2). An instructive starting point is the quantization of states of a particle moving in one dimension in a potential having a single minimum. Applications of this stationary-phase approximation to the Feynman path integral leads to a geometric series summing all multiples of the periodic classical motion of a particle of energy E, from which the poles in the Fourier transform of the trace of the evolution operator yield the Bohr–Sommerfeld quantization condition. Application of this approach to nuclear collective motion is described in Refs. 2) and 3).

Generalization to a double well yields an important new feature. Since the stationary point in the integral over the time occurs for purely imaginary time, a new class of stationary trajectories arises in the classically forbidden region. For these solutions it is replaced by a real variable τ and since the Euler–Lagrange equation involves a second time derivative, the minus sign arising from the two factors of i may be associated with the potential, yielding a picturesque description in terms of evolution in an inverted well. This is the basic idea behind the "bubbles," "bounces," and "instantons" arising when similar arguments are applied to statistical mechanics and field theory, and yields a powerful technique for treating tunneling in systems with many degrees of freedom.

At this school, it is particularly appropriate to emphasize the application of this approach to tunneling to spontaneous fission. The argument is presented in general terms in Chapter 7 of Ref. 1) and in greater detail in Ref. 4). One key issue is how to write a path integral appropriate to a many-Fermion system, and the two most convenient options are to work in the space of Slater determinants or to introduce an auxiliary field via the Hubbard–Stratinovich transformation. In either case, the theory which emerges has the form of self-consistent Hartree–Fock eigenvalue equations, but in four space-time dimensions instead of three space dimensions. Omitting the exchange term for simplicity, the equations have the form

$$\left[\frac{\partial}{\partial\tau} - \frac{\nabla^2}{2m} \int dx' \sum_k \phi_k(x', -\tau)\phi_k(x', \tau)v(x - x')\right] \phi_j(x, \tau) = \lambda_j \phi_j(x, \tau) \quad . \quad (1)$$

These equations, which govern the evolution of a nucleus from the metastable parent state through the fission barrier to a final state comprised of two or more fission fragments, give rise to the richest class of instanton or bounce solutions ever to be solved in field theory or many-body theory. A first impression of the structure of the solutions is obtained by solving a simple model in one dimension, and the results, shown in Refs. 1) and 3), show how the self-consistent solution evolves from the parent to daughter configurations.

As essential part of the physics of many Fermions arises from the rearrangement of nodal surfaces, and the essence of this aspect of the problem is displayed in the simple pedagogical model presented in Ref. 5). This model has the virtue of allowing comparison between the exact solution, the mean field result of Eq. (1), and the conventional constrained Hartree–Fock approximation. Solution of this model shows that the mean field result yields an excellent approximation to the exact answer while the constrained Hartree–Fock approximation fails badly. The difference arises because the solution to Eq. (1) yields the minimum action path through the barrier which incorporates much larger symmetry breaking than the constrained Hartree–Fock theory.

Of particular interest to this school is the computational problem of solving Eq. (1). It is well-known that obtaining good Hartree–Fock solutions in three dimensions for heavy nuclei with all possible symmetries broken is already a major task, and Eq. (1) requires that we do same in four dimensions. Furthermore, Hartree's method of straightforward iteration of the equations doesn't even work in principle, so one is forced to take a hard look at the structure of the equations and develop a whole new technology for their solution. This was done in Ref. 6), where we looked quite generally at the stability of a large class of possible iteration procedures, and developed a generalized filter method which is both stable and numerically efficient for the fission problem. One implements the method by starting at the top of the fission barrier with an infinitesimal solution to the imaginary-time RPA equations, which have at least one mode in the fission direction. By gradually increasing the period and apply the iterative filter, the amplitude of this solution can gradually be increased until it satisfies the exact equation for fission. The result of such a calculation for 32 particles in three dimensions is shown in Ref. 7). The most salient feature of this result is that, as in the case of the pedagogical model, the self-consistent mean field solution defines a minimum action path through the barrier displaying far greater symmetry breaking than the conventional constrained solution.

STOCHASTIC EVALUATION OF PATH INTEGRALS

When the stationary-phase approximation is inadequate, we still have recourse to evaluating the full path integral numerically. Conventional quadrature methods become intractable in large dimensions. For example, Simpson's rule in d-dimensions yields an error of order $N^{-4/d}$ where N is the number of evenly spaced points at which the integrand is sampled. Hence, it is highly preferable to use stochastic techniques, for which the error is of order $N^{-1/2}$, independent of dimension.

An elementary introduction to stochastic methods is presented in Chapter 8 of Ref. 1). The main idea is to divide the integrand into a product of a probability distribution $P(x)$ and a remainder, and use the central limit theorem to write the integral as an average over points distributed according to $P(\vec{x})$ plus a Gaussian distributed error

$$\int dx^d\, f(\vec{x})P(\vec{x}) = \frac{1}{N} \sum_{\substack{i=1 \\ \vec{x}_i \in P(\vec{x})}}^{N} f(\vec{x}_i) \pm \frac{1}{\sqrt{N}} \left[\langle f^2 \rangle_P - \langle f \rangle_P^2 \right] \quad . \tag{2}$$

Although this approach appears completely general, there are two essential conditions for its validity. First, it is essential to sample $P(\vec{x})$ without bias over its entire range. Second, $f(\vec{x})$ must be dominantly positive. Otherwise, if we separate f into its positive and negative parts, although each part can be calculated to within $1/\sqrt{N}$, the difference may be so small that the result cannot be made statistically significant with any practical value of N.

There is a great deal of freedom in formulating stochastic methods, and, as elaborated in Ref. 1), this freedom may be exploited to improve the accuracy of a specific physical calculation. In general, we will be interested in calculating the ground state of a quantum system, and thus use $e^{-\beta H}$ to filter out the ground state. One is free to pick the quantity to be evaluated, and, for example, a specific matrix element with an approximate wave function may be far superior to a thermodynamic trace. One is also free to pick the functional integral representation, and for many applications, a Feynman path integral in coordinate space offers advantages over other alternatives such as integration over an auxiliary field. Another choice is whether to sample the global action, thereby updating an entire time history of configurations, or whether to iteratively apply the filter $e^{-\epsilon\beta}$ to an ensemble of configurations, and eventually obtain steady-state configurations sampling the exact ground state. This latter approach is known as the path integral Monte Carlo method and has the structure of an initial-value diffusion problem. It may be improved significantly by multiplying and dividing by an approximate trial wave function at each time slice. In this modified diffusion problem, the trial function guides the random walk, and the stochastic sampling effectively only has to generate the difference between the exact and trial function rather than generating all the structure of the exact wave function.

By now, there are a host of applications to nuclear systems, so I will only single out a few for specific discussion here. One illuminating case of interest in nuclear physics is the one-dimensional nuclear model of Ref. 8). This example, which exploits the simplicity of dealing with with Fermions in one dimension, defines a two-body interaction with the qualitative features of the nucleon-nucleon interaction. The parameters governing the short-range repulsion and long-range attraction are defined such that dimensionless ratios characterizing the binding energy and kinetic energy of nuclear matter, the core radius, and maximum attraction are the same in one and three dimensions. One is then able to calculate the nuclear matter saturation curve and ground state wave functions, energies, and density distributions exactly. This model thus provides a test of the Skyrme phenomenology, by allowing us to define a Skyrme-like effective interaction fit to the binding energies and rms spatial extent of finite nuclei, and calculate all the same observables in the mean field approximation. The results are quite striking, in that even in one dimension, where the mean field should be at its worst, this approximation is extremely successful. *A fortiori*, it lends strong support to the accuracy of the analogous approximation in three dimensions. Furthermore, this success suggests that it would be fruitful to use this model as a laboratory to study corrections to mean field theory and to attempt to obtain a quantitative understanding of the effect of correlations and occupation probability corrections on the density fluctuations. This is particularly timely, since there has been significant interest and controversy concerning the interpretation of differences between quantum density fluctuations observed by elastic electron scattering from nuclei and those calculated in mean field theory from various effective interactions.

The calculation of spontaneous fission in this one-dimensional model is discussed in Ref. 9). Since the trajectories corresponding to fissions configurations are exponentially suppressed relative to trajectories describing the metastable parent state or decay products, the key point is to formulate the problem so that both the choice of the observable to be calculated and the collective moves introduced in the stochastic algorithm emphasize the role of instanton-like trajectories. Although it has not yet been done, it would be interesting to use this calculation as another check on the mean field theory of fission by comparing the solutions based on Eq. (1) with exact stochastic calculations with the same model.

A second application is to a one-dimensional confining quark model described in Refs. 10) and 11). The physics focus of this work is to compare an effective hadronic theory with an underlying quark model. In this context, exact Monte Carlo solutions to both the hadronic theory and the quark model eliminate uncertainties arising from uncontrolled many-body approximations and thus facilitate direct study of the validity of the effective hadronic theory itself. The starting point of this comparison is to use the path integral Monte Carlo method to calculate the exact phase shifts for hadron-hadron scattering in the flux tube quark model. These phase shifts are then fit with a phenomenological two-body force, and this interaction defines the effective hadronic Hamiltonian.

The results obtained from comparing exact solutions of the quark model and effective hadronic theory are highly suggestive of the phenomena observed in real nuclei. First, the binding energy per particle in the hadronic model is an excellent approximation to the exact result up to twice nuclear matter density, mirroring the success of phenomenological potential models for the bulk properties of nuclei. Secondly, the quark momentum distribution in hadronic matter can be understood using a combination of the quark momentum distribution in an isolated hadron and the many-hadron ground state wave functions calculated with the effective hadron-hadron interaction. As in the case of the experimental EMC effect, the quark momentum distribution in hadronic matter differs from that in an isolated hadron at roughly the 10% level, and the simple convolution approximation is inadequate to resolve the discrepancy. When, however, quark exchange is properly taken into account by using an antisymmetrized product of the hadronic wave function for hadrons in hadronic matter times the ground state distribution of quarks in a hadron, one obtains excellent agreement with the exact quark momentum distribution. This result again illustrates the utility of exact stochastic solutions to make crisp tests unencumbered by uncontrolled many-body approximations.

The calculations described above were limited to one spatial dimension because of the problems associated with Fermion antisymmetry in more than one dimension. The basic problem arises from the fact that the filter $e^{-\beta H}$ filters out the lowest energy eigenstate irrespective of symmetry. In the systems of interest in nuclear and atomic physics, the lowest energy state is the totally symmetric Bosonic state rather than the totally symmetric Fermionic state, so at some stage in the calculation we must project onto an antisymmetric state. If the projection were exact, there would be no problem, but if it is only performed stochastically then the noise grows like $e^{\beta(E_{\text{Fermion}}-E_{\text{Boson}})}$ whereas the stochastic projection only has accuracy $1/\sqrt{N}$.

In one dimension, one can overcome this antisymmetry problem by working in the ordered subspace $x_1 < x_2 < x_3 \ldots < x_n$ and solving for the positive, essentially Bosonic wave function in this sector. In higher dimensions, there is no way by antisymmetry alone to define regions surrounded by nodal boundaries in which the wave function has a specific sign, so one either has to invent a new approach or be overcôme by stochastic noise.

Reference 12) describes a method we have recently explored to try to overcome the Fermion sign problem by defining a generalized Markov process in which one solves for two distinct positive functions, $\psi^+(x)$ and $\psi^-(x)$, such that the antisymmetric ground state is given by $\psi^+(x) - \psi^-(x)$. The basic idea is to define a ficticious repulsive interaction which acts between Feynman paths corresponding to regions in which the wave function has opposite signs, and then systematically correct for violations of the Schrödinger equation produced by the finite range of this interaction. For small numbers of Fermions, this method is extremely successful, as demonstrated in Ref. 12) for an example in two dimensions for which the standard path integral Monte Carlo method fails completely. For large systems, even this extension appears inadequate, and so more powerful techniques are required if the full power of stochastic methods for evaluating path integrals are to be brought to bear in realistic problems in nuclear and atomic physics.

In summary, path integrals provide an extremely useful tool in confronting the complex many-body problems arising in nuclear and atomic physics. I hope these lectures and the cited references provide a helpful introduction to their use.

REFERENCES

1. Negele, J. W. and Orland, H., *Quantum Many-Particle Systems* (Addison-Wesley, Reading, MA, 1987).

2. Negele, J. W., *Rev. Mod. Phys.* **54** (1982) 931.

3. Levit, S., Negele, J. W. and Paltiel, Z., *Phys. Rev.* **C21** (1980) 1603.

4. Levit, S. Negele, J. W. and Paltiel, Z., *Phys. Rev.* **C22** (1980) 1979.

5. Arve, P., Bertsch, G., Negele, J. W. and Puddu, G., *Phys. Rev.* **C36** (1987) 2018.
6. Puddu, G. and Negele, J. W., *Phys. Rev.* **B35** (1987) 1007.
7. Negele, J. W., in *Trends in Theoretical Physics*, P. J. Ellis and Y. C. Tang, eds. (Addison-Wesley, Reading, Ma, 1989).
8. Alexandrou, C., Myczkowski, J. and Negele, J. W., *Phys. Rev.* **C39** (1989) 1076.
9. Alexandrou, C. and Negele, J. W., *Phys. Rev.* **C37** (1988) 1513.
10. Horowitz, C., Moniz, E. J. and Negele, J. W., *Phys. Rev.* **D31** (1985) 1689.
11. Lissia, M. and Negele, J. W., *Phys. Rev.* **D39** (1989) 1413.
12. Brommer, K. D. and Negele, J. W., *Ann. Phys.* (NY) **192** (1989) 119.

HOT RELATIVISTIC NUCLEAR MATTER

Brian D. Serot
Physics Department and Nuclear Theory Center
Indiana University, Bloomington, IN 47405

ABSTRACT

The nuclear matter equation of state is studied in the framework of quantum hadrodynamics (QHD) using the mean-field approximation to the Walecka model (QHD–I).

1. INTRODUCTION

The accurate description of hot, dense matter is an important problem in theoretical physics. For example, shortly after the Big Bang, the universe was composed entirely of hot, dense matter. In the present universe, the nuclear equation of state as a function of temperature, density, and the ratios of protons to neutrons and nucleons to hyperons is needed to study stellar collapse through a possible supernova phase into a neutron star. On a smaller scale, we can explore the same nuclear dynamics through energetic collisions of heavy ions, where at least part of the hot matter can be described in terms of its hadronic constituents.

In this lecture, I discuss some techniques for computing the properties of hot, dense nuclear matter. The traditional approach to this problem relies on the Schrödinger equation for nonrelativistic nucleons interacting through static, two-body potentials. Although the Schrödinger equation has been useful in this program for more than fifty years, newly completed experimental facilities and existing initiatives will force us to go beyond the Schrödinger equation to compare calculations with the data of the future. A more complete treatment of hadronic systems should include relativistic motion of the nucleons, dynamical mesons and baryon resonances, modifications of the nucleon structure in the nucleus, and the dynamics of the quantum vacuum, while maintaining general properties of quantum mechanics, covariance, gauge invariance, and causality. These physical effects will be relevant regardless of the degrees of freedom used to describe the system; they must be studied simultaneously and consistently if we are to draw definite conclusions about nuclear dynamics at high temperatures and high densities.

Our basic goal is therefore to formulate a consistent microscopic treatment of strongly interacting, relativistic, quantum statistical mechanical systems. This will allow for investigation of a wide variety of phenomena. For example, we can compute both static thermodynamic properties (like energy, pressure, and entropy) and dynamical characteristics (such as viscosity, transport coefficients, and collective modes and their damping). In addition, we can study the production and absorption of particles under extreme conditions and map out the nuclear matter phase diagram. Ultimately, we would like to extend the techniques to deal with nonequilibrium systems, so that we can describe the development of two isolated nuclei into a single system in equilibrium, which may be relevant for heavy-ion collisions.

Since the physical phenomena of interest are relativistic and involve particle production and absorption, the only existing consistent framework for their description is relativistic quantum field theory. I will use a relativistic quantum field theory based on mesons and baryons, which is known as quantum hadrodynamics (QHD).[1] QHD is consistent in the sense that the dynamical assumptions (such as the relevant degrees of freedom, the form of the lagrangian, and the normalization conditions) are made at the outset, and one then attempts to extract concrete results from the implied formalism. In principle, the assumptions permit the formulation of systematic, "conserving" approximations[2,3] that maintain the important general properties mentioned above. Calculations can then be compared to data to see if the framework is related to the real world and to decide where QHD succeeds and where it fails.

There are several reasons for considering only hadronic degrees of freedom. First, these variables are the most efficient at low densities and temperatures and for describing particle absorption and production, as hadrons are the particles observed experimentally. Second, hadronic calculations can be calibrated by comparing to observed hadron-hadron scattering and empirical nuclear properties; we can then extrapolate to extreme conditions and test the predictions of the models. Most importantly, we must understand the limitations of QHD to deduce true signals of QCD behavior in nuclear systems.

If QHD is defined using renormalizable lagrangians, the number of parameters in any model is finite.[4,5] Calculations can be carried beyond the "tree" level without introducing additional parameters determined solely by short-distance phenomena; thus, the sensitivity of calculated results to short-distance input is minimized. The dynamical assumption underlying renormalizability is that the short-distance behavior and the quantum vacuum can be described in terms of hadronic degrees of freedom only. This assumption must ultimately break down, since hadrons are actually composed of quarks and gluons. For QHD to be useful, nuclear observables of interest must not be dominated by contributions from short distances (where QHD is inappropriate). This possibility can be tested (and its limitations uncovered) by performing detailed calculations in a consistent relativistic framework.

To provide a concrete example of these ideas, I will illustrate a finite-temperature calculation in a simple context: the mean-field theory of the Walecka model (QHD–I).[6,1] This model contains some basic elements of hadronic theories of nuclei, namely, baryons coupled strongly to neutral scalar and vector fields. We must remember, however, to keep this calculation in context. It is just a simple approximation to a simple model. In fact, much of the previous work in QHD has been performed in either the mean-field theory (MFT) or at the one-loop level ("relativistic Hartree approximation" or RHA), which includes the shift in the baryon vacuum energy.[6–8] The original motivation for these studies was that the MFT should become increasingly valid as the density increases, and that the MFT and RHA could be good *nonperturbative* starting points for calculations at normal nuclear density. These MFT and RHA calculations are successful in describing a variety of nuclear systems and phenomena.[1,9,10] For example, the bulk properties of spherical nuclei can be described accurately with parameters similar to those obtained by fitting meson-exchange potentials to nucleon-nucleon scattering observables. These results have stimulated many successful phenomenological calculations that use Dirac nucleons and that are modeled after QHD.

Nevertheless, detailed calculations are needed to justify the successes of the MFT and RHA, or to relegate them to pure phenomenology. One advantage of the Schrödinger formalism is that systematic approximation schemes have been developed and used to produce reliable conclusions. Such schemes have yet to be developed for QHD. This is an important topic for future work.

2. THE WALECKA MODEL

Quantum hadrodynamics is a general framework for the relativistic nuclear many-body problem.[1] The detailed dynamics must be specified by choosing a particular renormalizable lagrangian density. To illustrate the nuclear matter formalism as simply as possible, I will consider the Walecka model[6] (sometimes called QHD–I), which contains baryons (ψ) and neutral scalar (ϕ) and vector (V^μ) mesons.

The lagrangian density in the Walecka model is given by[1]

$$\mathcal{L} = \overline{\psi}\big[\gamma_\mu(i\partial^\mu - g_v V^\mu) - (M - g_s\phi)\big]\psi + \tfrac{1}{2}(\partial_\mu\phi\partial^\mu\phi - m_s^2\phi^2) \\ - \tfrac{1}{4}F_{\mu\nu}F^{\mu\nu} + \tfrac{1}{2}m_v^2 V_\mu V^\mu + \delta\mathcal{L}\,, \tag{1}$$

where $F^{\mu\nu} = \partial^\mu V^\nu - \partial^\nu V^\mu$ and $\delta\mathcal{L}$ contains counterterms. The parameters M, g_s, g_v, m_s, and m_v are phenomenological constants that may be determined (in principle) from experimental measurements. The counterterms are for renormalization purposes.

The motivation for this model has evolved considerably since it was introduced. At present, there is significant empirical evidence that when the nucleon-nucleon (NN) interaction is described in a Lorentz-covariant fashion, it contains strong Lorentz scalar and four-vector pieces.[11–13] These must be reproduced in any relativistic theory of nuclear structure, and the simplest way to do this is through the exchange of scalar and vector mesons. These two components are the most important for describing bulk nuclear properties, which is my main concern here. The other Lorentz components of the NN interaction, in particular the terms arising from pion exchange, average essentially to zero in spin-saturated nuclear matter and may be incorporated as refinements to the present model.[1] The important point is that even in more refined models, the dynamical features generated by scalar and vector mesons remain; thus it is important to first understand the consequences of these degrees of freedom for relativistic descriptions of nuclear systems. Note that eq. (1) resembles massive QED with an additional scalar interaction, so the relativistic quantum field theory generated by this Lorentz invariant lagrangian is *renormalizable*.[14]

The field equations for the model follow from the Euler–Lagrange equations and can be written as

$$(\partial_\mu\partial^\mu + m_s^2)\phi = g_s\overline{\psi}\psi\,, \tag{2}$$

$$\partial_\nu F^{\nu\mu} + m_v^2 V^\mu = g_v \bar{\psi}\gamma^\mu\psi \,, \tag{3}$$

$$[\gamma^\mu(i\partial_\mu - g_v V_\mu) - (M - g_s\phi)]\psi = 0 \,. \tag{4}$$

Equation (2) is simply the Klein–Gordon equation with a scalar source. Equation (3) looks like massive QED with the conserved baryon current

$$B^\mu \equiv (\rho_B, \underset{\sim}{B}) = \bar{\psi}\gamma^\mu\psi \,, \qquad \partial_\mu B^\mu = 0 \tag{5}$$

rather than the (conserved) electromagnetic current as the source. Finally, eq. (4) is the Dirac equation with scalar and vector fields introduced in a minimal fashion. These field equations also imply that the canonical energy-momentum tensor

$$\begin{aligned} T^{\mu\nu} &= i\bar{\psi}\gamma^\mu\partial^\nu\psi - \tfrac{1}{2}[\partial_\sigma\phi\partial^\sigma\phi - m_s^2\phi^2]g^{\mu\nu} + \partial^\mu\phi\partial^\nu\phi \\ &\quad + \tfrac{1}{2}[\partial_\sigma V_\lambda\partial^\sigma V^\lambda - m_v^2 V_\sigma V^\sigma]g^{\mu\nu} - \partial^\mu V_\lambda\partial^\nu V^\lambda \end{aligned} \tag{6}$$

is conserved ($\partial_\mu T^{\mu\nu} = \partial_\nu T^{\mu\nu} = 0$). Note that to write eq. (6), I discarded several total divergences (which are irrelevant for evaluating observables in infinite matter) and used $\partial_\mu V^\mu = 0$, which follows from the vector meson field equation (3) and the conservation of the baryon current (5).

I emphasize that eqs. (2) to (4) are *nonlinear quantum field equations*, and their exact solutions are very complicated. In particular, they describe mesons and baryons *that are not point particles*, but rather objects with intrinsic structure due to the implied (virtual) mesonic and baryon-antibaryon loops. It is here that the dynamical input of renormalizability is apparent, since we are assuming that (at least the long-range part of) this intrinsic structure can be described using hadronic degrees of freedom. The validity of this input and its limitations have yet to be tested consistently within the framework of QHD.

We also expect the coupling constants in eqs. (2) to (4) to be large, and thus perturbative solutions are not useful. We have therefore made little progress by writing down these equations without a suitable method for solving them. Fortunately, there is an approximate solution that should become increasingly valid as the nuclear density increases. Consider a system of B baryons in a box of volume V at zero temperature. For simplicity, take the total momentum to be zero, so that we are in the system's rest frame; it is easy to generalize the results by allowing for uniform motion.[15] As the baryon density B/V increases, so do the source terms on the right-hand sides of eqs. (2) and (3). When the sources are large, the meson field operators can be replaced by their expectation values, which are classical fields:

$$\phi \equiv \langle\phi\rangle\,, \qquad \delta^{\mu 0} V_0 \equiv \langle V^\mu \rangle \qquad \text{(classical)}\,. \tag{7}$$

For our static, uniform system, ϕ and V_0 are *constants* that are independent of space and time. Rotational invariance implies that the expectation value of the three-vector piece of V^μ vanishes.

It is important to emphasize the strategy involved in the preceding "mean-field" approximation. First, the resulting mean-field theory (MFT) should give the correct solution to the field equations in the high-density limit. More importantly, however, the MFT serves as a starting point for consistently calculating corrections within the framework of QHD, using Feynman diagrams, path-integral methods, and so forth.

When the meson fields in eq. (4) are approximated by the classical fields of eq. (7), the Dirac equation is linear,

$$[i\gamma_\mu \partial^\mu - g_v \gamma^0 V_0 - (M - g_s\phi)]\psi = 0\,, \tag{8}$$

and can be solved exactly. [It is this *linearization* of the full field equation (4) that allows the baryons to be interpreted now as point particles.] The resulting baryon solutions have a mass that is shifted by the scalar field:

$$M^\star \equiv M - g_s\phi\,, \tag{9}$$

and an energy spectrum that is shifted by the vector field:

$$\epsilon^{(\pm)}(k) = g_v V_0 \pm (\underset{\sim}{k}^2 + M^{\star 2})^{1/2} \equiv g_v V_0 \pm E^\star(k)\,, \tag{10}$$

but otherwise look precisely like free-particle Dirac solutions.[16] As expected, there are solutions with both positive and negative square roots characteristic of the Dirac equation. These solutions can be used to define quantum field operators, and the hamiltonian density for the system can be constructed in the canonical fashion. (For the details of these procedures, see ref. 1.) The results are

$$\hat{H} = \hat{H}_{\text{MFT}} + \delta H\,, \tag{11}$$

$$\hat{H}_{\text{MFT}} = g_v V_0 \hat{B} + \sum_{\underset{\sim}{k}\lambda} E^\star(k)(A^\dagger_{\underset{\sim}{k}\lambda} A_{\underset{\sim}{k}\lambda} + B^\dagger_{\underset{\sim}{k}\lambda} B_{\underset{\sim}{k}\lambda}) + V(\tfrac{1}{2}m_s^2\phi^2 - \tfrac{1}{2}m_v^2 V_0^2)\,, \tag{12}$$

$$\hat{B} = \sum_{\underset{\sim}{k}\lambda} A^{\dagger}_{\underset{\sim}{k}\lambda} A_{\underset{\sim}{k}\lambda} - B^{\dagger}_{\underset{\sim}{k}\lambda} B_{\underset{\sim}{k}\lambda} , \tag{13}$$

$$\delta H = -\sum_{\underset{\sim}{k}\lambda} \left[(\underset{\sim}{k}^2 + M^{*2})^{1/2} - (\underset{\sim}{k}^2 + M^2)^{1/2} \right] . \tag{14}$$

Here $A^{\dagger}_{\underset{\sim}{k}\lambda}$, $B^{\dagger}_{\underset{\sim}{k}\lambda}$, $A_{\underset{\sim}{k}\lambda}$, and $B_{\underset{\sim}{k}\lambda}$ are creation and destruction operators for baryons and antibaryons with shifted mass and energy, and $\hat{B}$ is the baryon number operator, which clearly counts the number of baryons minus the number of antibaryons. (The index λ denotes both spin and isospin projections.) The correction term δH arises from placing the operators in $\hat{H}_{\mathrm{MFT}}$ in "normal order."[1] This correction is easily interpreted in the context of Dirac hole theory. The spectrum of the infinite Dirac sea of occupied states shifts in the presence of the surrounding nucleons at finite density. Since all energies are measured relative to the vacuum, the energy shift must be computed by subtracting the total energy of the Dirac sea in the vacuum, where the nucleons have their free mass M. This leads to the result in eq. (14). This "zero-point energy" correction can be calculated straightforwardly in the renormalizable Walecka model,[1] but here we will concentrate on the MFT hamiltonian defined by eq. (12).

Since $\hat{H}_{\mathrm{MFT}}$ is diagonal, this model mean-field problem has been solved *exactly*. The solution retains the essential features of QHD: relativistic covariance, explicit meson degrees of freedom, and the incorporation of antiparticles. Furthermore, it yields a simple solution to the field equations that should become increasingly valid as the baryon density increases. This solution and model problem thus provide a useful starting point for describing the nuclear many-body system as well as a consistent basis for computing corrections using relativistic quantum field theory and standard many-body techniques.

3. THE NUCLEAR MATTER EQUATION OF STATE

For uniform nuclear matter in its rest frame, the ground state of the hamiltonian (12) is obtained by filling states with wavenumber $\underset{\sim}{k}$ and spin-isospin degeneracy γ up to Fermi level k_{F}. For nuclear matter, $\gamma = 4$ (neutrons and protons with spin up and down). The baryon density ρ_{B} and energy density $\mathcal{E}$ may be readily evaluated by taking ground-state expectation values of the operators in eqs. (13) and (12):

$$\rho_{\rm B} = \frac{\gamma}{(2\pi)^3} \int_0^{k_{\rm F}} d^3k = \frac{\gamma}{6\pi^2} k_{\rm F}^3 , \tag{15}$$

$$\mathcal{E} = \frac{g_{\rm v}^2}{2m_{\rm v}^2} \rho_{\rm B}^2 + \frac{m_{\rm s}^2}{2g_{\rm s}^2} (M - M^\star)^2 + \frac{\gamma}{(2\pi)^3} \int_0^{k_{\rm F}} d^3k \, E^\star(k) , \tag{16}$$

and the pressure p follows from similar manipulations on the trace of the stress tensor ($\frac{1}{3}T_{ii}$):

$$p = \frac{g_{\rm v}^2}{2m_{\rm v}^2} \rho_{\rm B}^2 - \frac{m_{\rm s}^2}{2g_{\rm s}^2} (M - M^\star)^2 + \frac{1}{3} \frac{\gamma}{(2\pi)^3} \int_0^{k_{\rm F}} d^3k \, \frac{\underset{\sim}{k}^2}{E^\star(k)} . \tag{17}$$

Here V_0 has been eliminated in terms of the conserved baryon density using the mean-field equation [see eqs. (3) and (7)]

$$V_0 = \frac{g_{\rm v}}{m_{\rm v}^2} \langle \psi^\dagger \psi \rangle = \frac{g_{\rm v}}{m_{\rm v}^2} \rho_{\rm B} , \tag{18}$$

and eq. (9) serves to eliminate ϕ. The first two terms in eqs. (16) and (17) arise from the classical meson fields. The final terms in these equations are those of a relativistic gas of baryons of mass $M^\star$. These expressions give the nuclear matter equation of state in parametric form: $\mathcal{E}(\rho_{\rm B})$ and $p(\rho_{\rm B})$.

The effective mass $M^\star$ can be determined by minimizing $\mathcal{E}(M^\star)$ with respect to $M^\star$, leading to the *self-consistency condition*

$$M^\star = M - \frac{g_{\rm s}^2}{m_{\rm s}^2} \frac{\gamma}{(2\pi)^3} \int_0^{k_{\rm F}} d^3k \, \frac{M^\star}{E^\star(k)} = M - \frac{g_{\rm s}^2}{m_{\rm s}^2} \langle \bar{\psi}\psi \rangle \equiv M - \frac{g_{\rm s}^2}{m_{\rm s}^2} \rho_{\rm s} , \tag{19}$$

which also defines the scalar density $\rho_{\rm s}$. This equation is equivalent to the MFT scalar field equation for ϕ. Note that the scalar density is smaller than the baryon density [eq. (15)] due to the factor $M^\star/E^\star(k)$, which is an effect of Lorentz contraction. Thus the contribution of rapidly moving baryons to the scalar source is significantly reduced. Most importantly, eq. (19) is a *transcendental self-consistency equation* for $M^\star$ that must be solved at each value of $k_{\rm F}$. This illustrates the *nonperturbative* nature of the mean-field solution.

An examination of the analytic expression for the energy density (16) shows that the system is unbound ($\mathcal{E}/\rho_{\rm B} > M$) at either very low or very high densities.[6]

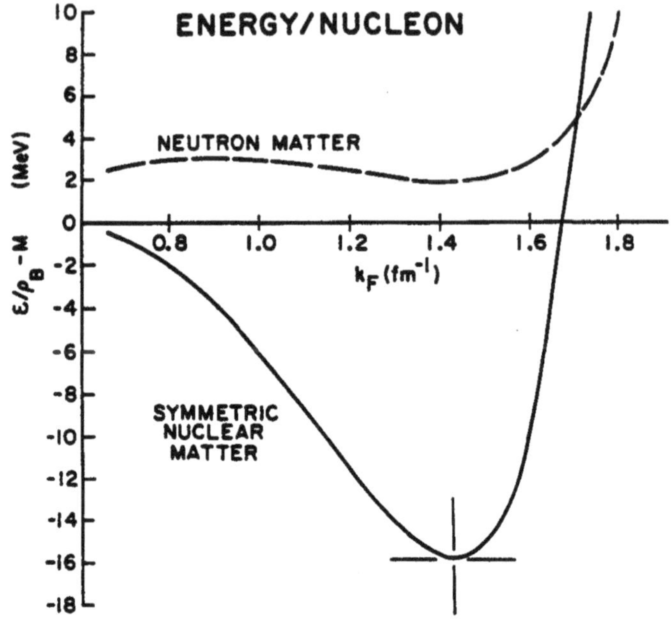

Fig. 1. Saturation curve for nuclear matter. These results are calculated in the relativistic mean-field theory with baryons and neutral scalar and vector mesons (QHD–I). The coupling constants are chosen to fit the value and position of the minimum. The prediction for neutron matter ($\gamma = 2$) is also shown.

At intermediate densities, the attractive scalar interaction will dominate if the coupling constants are chosen properly. The system then *saturates*. The empirical equilibrium properties of nuclear matter are reproduced if the couplings are chosen as

$$C_s^2 \equiv g_s^2\left(\frac{M^2}{m_s^2}\right) = 267.1\,, \qquad C_v^2 \equiv g_v^2\left(\frac{M^2}{m_v^2}\right) = 195.9\,, \tag{20}$$

which leads to an equilibrium Fermi wavenumber $k_F^0 = 1.42$ fm^{-1} and an energy/nucleon $(\mathcal{E}/\rho_B - M) = -15.75$ MeV. (I use this somewhat large saturation density merely for illustration, since it yields results consistent with those in refs. 1 and 6.) Note that only the *ratios* of coupling constants to masses enter in eqs. (16), (17), and (20). The resulting saturation curve is shown in fig. 1. In this approximation, the relativistic properties of the scalar and vector fields are responsible for saturation; a Hartree–Fock variational estimate built on the nonrelativistic (Yukawa) potential limit of the interaction shows that such a system is unstable against collapse.

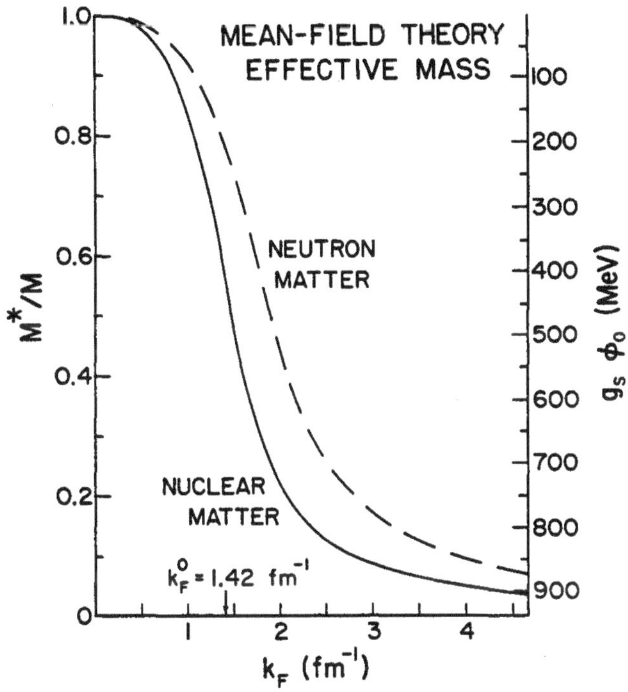

Fig. 2. Effective mass as a function of density for nuclear ($\gamma = 4$) and neutron ($\gamma = 2$) matter based on fig. 1.

The solution of the self-consistency condition (19) for M^* yields an effective mass that is a decreasing function of the density, as illustrated in fig. 2. Note that M^*/M becomes small at high density and is significantly less than unity at ordinary nuclear density. This is a consequence of the large scalar field $g_s\phi$, which is approximately 400 MeV and produces a large attractive contribution to the energy/baryon. There is also a large repulsive energy/baryon from the vector field $g_v V_0 \approx 350$ MeV. Thus *the Lorentz structure of the interaction leads to a new energy scale in the problem*, and the small nuclear binding energy ($\approx$ 16 MeV) arises from the cancellation between the large scalar attraction and vector repulsion. As the nuclear density increases, the scalar source ρ_s becomes small relative to the vector source ρ_B, and the attractive forces saturate, producing the minimum in the binding curve. Clearly, because of the sensitive cancellation involved near the equilibrium density, corrections to the MFT must be calculated before the importance of this new saturation mechanism can be assessed. Nevertheless, the Lorentz structure of the interaction provides an *additional saturation mechanism that is not present in the nonrelativistic potential limit*, as this limit ignores the distinction between ρ_s and ρ_B.

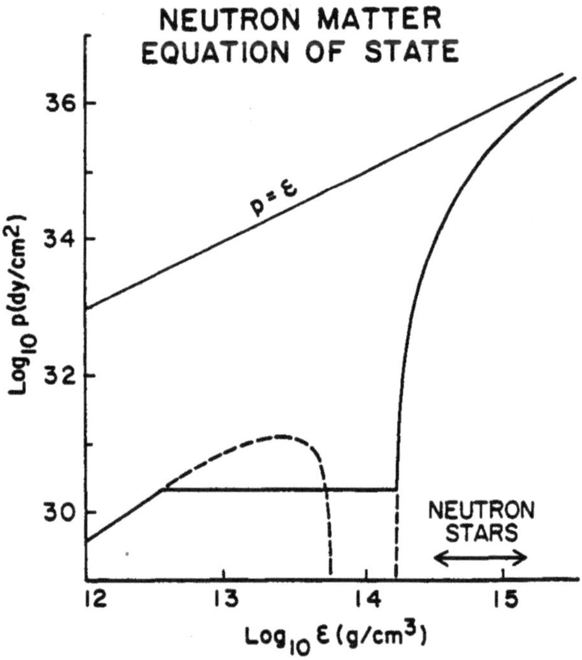

Fig. 3. Predicted equation of state for neutron matter. The solid and dashed curve shows the result for QHD–I based on the dashed line in fig. 1. A Maxwell construction determines the equilibrium (horizontal) curve in the region of the phase transition. The density regime relevant for neutron stars is also shown.

The corresponding curves for neutron matter obtained by setting $\gamma = 2$ are also shown in figs. 1 and 2, and the equation of state (pressure *vs.* energy density) for neutron matter is given in fig. 3. In this mean-field model, there is a van der Waals (liquid–gas) phase transition, and the properties of the two phases are deduced through a Maxwell construction. At high densities, the system approaches the "causal limit" $p = \mathcal{E}$ representing the stiffest possible equation of state.

The neutron matter equation of state shown in fig. 3 can be used in the Tolman–Oppenheimer–Volkoff equation for the general-relativistic metric to compute the masses of neutron stars as a function of their central density.[17] This MFT gives a maximum neutron star mass of roughly 2.6 solar masses at a central density approximately six times larger than the central density in ^{208}Pb. Note that the asymptotic approach of the equation of state to the causal limit is already relevant in this regime. (Normal nuclear density is $\approx 2.5 \times 10^{14}$ g/cm^3.) I emphasize that although the low-density behavior of nuclear matter is sensitive to the nearly exact cancellation between attractive scalar and repulsive vector components, the stiff high-density equation of state is determined simply by the Lorentz structure of the interaction. (The scalar attraction saturates completely at high densities, producing an essentially massless gas of baryons interacting through a strong vector repulsion, which leads to a stiff equation of state.) Moreover, because the individual Lorentz components are comparable to the nucleon mass, the onset of the asymptotic regime occurs at densities similar to those in the interiors of neutron stars.

4. HOT NUCLEAR MATTER

To describe the system at a finite temperature T, calculate the partition function Z and the thermodynamic potential Ω using the standard expressions from statistical mechanics:[18]

$$Z = \mathrm{Tr}\exp\left\{-(\hat{H} - \mu\hat{B})/T\right\} \equiv \exp\left\{-\Omega(T,V,\mu)/T\right\}. \qquad (21)$$

Here μ is the baryon chemical potential. Since the operators appearing in eq. (21) are all diagonal, Ω can be evaluated exactly in this MFT. The results are analogous to those for noninteracting fermions, and one finds

$$\Omega(T,V,\mu) = \tfrac{1}{2}(m_s^2\phi^2 - m_v^2V_0^2)V - T\sum_{\underset{\sim}{k}\lambda}\left\{\ln\left(1 + e^{-[E^*(k)-\nu]/T}\right)\right.$$
$$\left. + \ln\left(1 + e^{-[E^*(k)+\nu]/T}\right)\right\}. \qquad (22)$$

The sum runs over all single-particle states labeled by momentum $\underset{\sim}{k}$ and intrinsic quantum numbers λ. The *effective chemical potential* ν is defined by

$$\nu \equiv \mu - g_{\rm v} V_0 \,. \tag{23}$$

The ensemble average of an operator $\hat{A}$ is given by

$$A = \langle\!\langle \hat{A} \rangle\!\rangle = Z^{-1}\, \mathrm{Tr}\big[\hat{A}\; \mathrm{e}^{-(\hat{H}-\mu\hat{B})/T}\big] \,. \tag{24}$$

For example, the baryon density is

$$\rho_{\rm B} \equiv \frac{\langle\!\langle \hat{B} \rangle\!\rangle}{V} = -\frac{1}{V}\Big(\frac{\partial\Omega}{\partial\mu}\Big) = \frac{1}{V}\sum_{\underset{\sim}{k}\lambda}(n_k - \overline{n}_k) \,. \tag{25}$$

Here the partial derivative is taken with all other thermodynamic parameters and field variables held fixed, and I have identified the baryon and antibaryon occupation numbers

$$n_k \equiv n_k(T,\nu) \equiv \langle\!\langle A^{\dagger}_{\underset{\sim}{k}\lambda} A_{\underset{\sim}{k}\lambda} \rangle\!\rangle = \Big\{1 + \mathrm{e}^{[E^*(k)-\nu]/T}\Big\}^{-1} \,, \tag{26}$$

$$\overline{n}_k \equiv \overline{n}_k(T,\nu) \equiv \langle\!\langle B^{\dagger}_{\underset{\sim}{k}\lambda} B_{\underset{\sim}{k}\lambda} \rangle\!\rangle = \Big\{1 + \mathrm{e}^{[E^*(k)+\nu]/T}\Big\}^{-1} \,. \tag{27}$$

With these results, I can derive the equations that determine the meson fields. For a system in equilibrium, these should be chosen to make the thermodynamic potential Ω stationary. For example, $\partial\Omega/\partial\phi = 0$ leads to

$$\phi = \frac{g_{\rm s}}{m_{\rm s}^2}\frac{1}{V}\sum_{\underset{\sim}{k}\lambda}\frac{M^*}{E^*(k)}(n_k + \overline{n}_k) = \frac{g_{\rm s}}{m_{\rm s}^2}\langle\!\langle \hat{\rho}_{\rm s} \rangle\!\rangle \,. \tag{28}$$

Similarly, the vector field equation becomes

$$V_0 = \frac{g_{\rm v}}{m_{\rm v}^2}\frac{1}{V}\sum_{\underset{\sim}{k}\lambda}(n_k - \overline{n}_k) = \frac{g_{\rm v}}{m_{\rm v}^2}\langle\!\langle \hat{\rho}_{\rm B} \rangle\!\rangle \,. \tag{29}$$

Thus, by making the thermodynamic potential stationary with respect to the fields, they automatically satisfy the ensemble averages of the normal-ordered field equations resulting from the lagrangian (1).

The relations (28) and (29) are extremely useful, for they imply that ϕ and V_0 can be held *fixed* in computing thermodynamic functions as derivatives of the thermodynamic potential through the relations

$$\begin{aligned} B &= -\Big(\frac{\partial}{\partial \mu}\Omega(T,V,\mu)\Big)_{T,V}, \\ S &= -\Big(\frac{\partial}{\partial T}\Omega(T,V,\mu)\Big)_{V,\mu}, \\ p &= -\Big(\frac{\partial}{\partial V}\Omega(T,V,\mu)\Big)_{T,\mu}. \end{aligned} \tag{30}$$

Thus the calculation of the thermodynamic functions is straightforward in this MFT, and one obtains

$$\rho_s = \frac{\gamma}{(2\pi)^3}\int d^3k\, \frac{M^*}{E^*(k)}(n_k + \overline{n}_k), \tag{31}$$

$$\rho_B = \frac{\gamma}{(2\pi)^3}\int d^3k\,(n_k - \overline{n}_k), \tag{32}$$

$$\mathcal{E} = \frac{g_v^2}{2m_v^2}\rho_B^2 + \frac{m_s^2}{2g_s^2}(M - M^*)^2 + \frac{\gamma}{(2\pi)^3}\int d^3k\, E^*(k)(n_k + \overline{n}_k), \tag{33}$$

$$p = \frac{g_v^2}{2m_v^2}\rho_B^2 - \frac{m_s^2}{2g_s^2}(M - M^*)^2 + \frac{1}{3}\frac{\gamma}{(2\pi)^3}\int d^3k\, \frac{\underset{\sim}{k}^2}{E^*(k)}(n_k + \overline{n}_k), \tag{34}$$

$$\begin{aligned} \sigma = -\frac{\gamma}{(2\pi)^3}\int d^3k\, \{ & n_k \ln n_k + (1 - n_k)\ln(1 - n_k) \\ & + \overline{n}_k \ln \overline{n}_k + (1 - \overline{n}_k)\ln(1 - \overline{n}_k)\}. \end{aligned} \tag{35}$$

Here γ is the spin-isospin degeneracy, σ is the entropy density, and the baryon and antibaryon occupation number distributions are given by eqs. (26) and (27). Note the crucial difference in signs in these expressions: $\mathcal{E}$ and p depend on the *sum* $n_k + \overline{n}_k$, while ρ_B depends on the *difference* $n_k - \overline{n}_k$. Moreover, eq. (28) can be recast as

$$M^* \equiv M - g_s\phi = M - \frac{g_s^2}{m_s^2}\rho_s = M - \frac{g_s^2}{m_s^2}\frac{\gamma}{(2\pi)^3}\int d^3k\, \frac{M^*}{E^*(k)}(n_k + \overline{n}_k). \tag{36}$$

This is a transcendental *self-consistency condition* that determines ϕ (or M^*) at finite temperature and is a generalization of eq. (19).

To compute the thermodynamic functions and the equation of state, one first specifies T and ν. The self-consistency condition (36) is then solved to determine M^*. (There may be several solutions for fixed T and ν.) These solutions specify the distribution functions n_k and $\overline{n}_k$, and the remaining integrals in eqs. (32) to

(35) can be evaluated directly. At the end of the calculation, one can (in principle) invert these relations to find μ in terms of ρ_B, but in practice, this is usually done by searching on values of ν.

The resulting thermodynamic surface $p(\mathcal{E}, T)$ defines the equation of state. In fig. 4, the isotherms $p(\mathcal{E}, T$ = constant) are shown for neutron matter with $\gamma = 2$. As we have already seen, there is a phase transition in this model, and the region of the phase coexistence is indicated. (The isothermal spinodal curves can also be calculated easily.[15]) The predicted value of the critical temperature is $T_c = 9.1 \pm 0.2$ MeV, and the effective mass at the critical point is $M_c^* = 0.88M$.

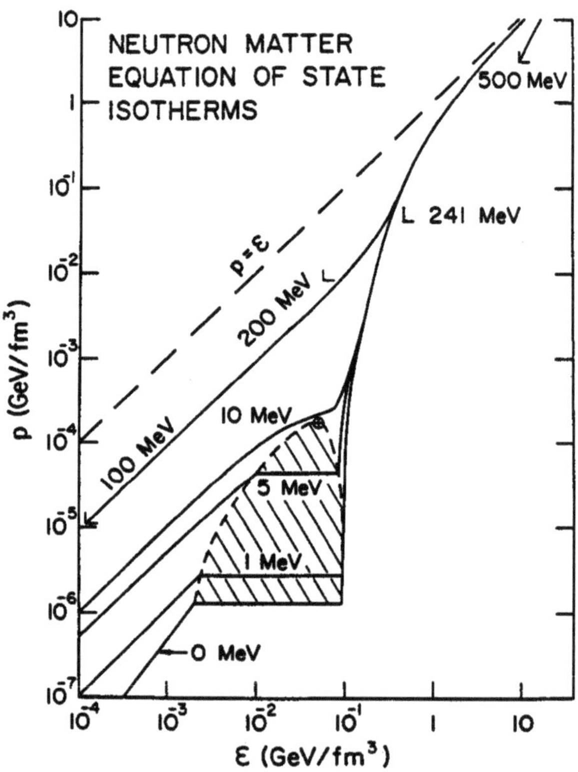

Fig. 4. Isotherms of the neutron matter equation of state, as calculated in the mean-field theory of QHD–I. The curves are labeled by the temperature, and the left hand "notches" correspond to zero baryon density. The shaded area shows the region of the phase separation, and the critical point is indicated by $\oplus$.

Several limiting cases of the equation of state are of interest:

- As $T \to 0$ at any $\rho_B > 0$, the baryon distribution becomes a step function: $n_t \to \theta(k_F - t)$. The results of eqs. (15) to (17) are then reproduced. This conclusion also obtains as $\rho_B \to \infty$ at any T.

- For $T \ll M$ and $\rho_B \to 0$, the equation of state is that of a classical nonrelativistic gas: $\mathcal{E} = \rho_B(M + 3T/2)$, $p = \rho_B T = 2(\mathcal{E} - M\rho_B)/3$.

- As $T \to \infty$ at any ρ_B, baryon-antibaryon pairs will be produced, and the equation of state resembles that of a black body: $\mathcal{E} = 7\pi^2\gamma T^4/120$, $p = \mathcal{E}/3$.

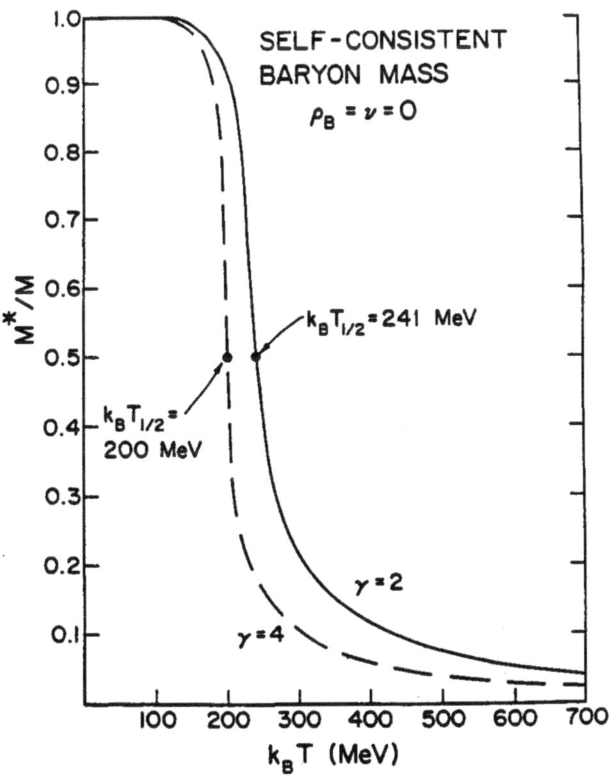

Fig. 5. Self-consistent nucleon mass as a function of temperature at vanishing baryon density. Results are indicated for both neutron matter ($\gamma = 2$) and nuclear matter ($\gamma = 4$) in QHD–I.

The solution to the self-consistency equation (36) shows that the effective mass of the nucleon decreases as the temperature is raised due to $N\overline{N}$ pair formation. The values of M^* at vanishing baryon density are shown in fig. 5. The most striking feature is the sudden decrease in the nucleon mass well below $T = M$. Thus, at high temperatures (as at high density), the baryons are essentially massless. As the temperature is lowered, the baryons acquire mass suddenly from the self-consistent "freezing out" of the vacuum pairs and the corresponding decrease in ϕ.

As a consequence of pair formation, the high-temperature isotherms shown in fig. 4 will terminate as the density is decreased. Thus, at a given temperature, there is a finite, limiting value of the energy density as the baryon density goes to zero. One simply has a vanishingly dilute solution of baryons in a sea of pairs. These limiting points on the isotherms are indicated for a representative set of temperatures in fig. 4. The isotherms start at these limiting values (at $\rho_B = 0$) and approach the causal limit $p = \mathcal{E}$ as the density is increased.

We have now studied the properties of nuclear matter in the simplest approximation to the Walecka model. As I emphasized in the Introduction, however, it is crucial to extend these calculations to more sophisticated approximations, in order to produce reliable predictions and to verify that the simple physical picture obtained in the MFT is a true consequence of the model QHD lagrangian.[20] Unfortunately, these extensions take me beyond the scope of this lecture, so I will end the discussion here.

5. SUMMARY

Quantum hadrodynamics is a consistent framework for studying the relativistic nuclear many-body problem. By specifying the interactions with a local, renormalizable lagrangian density, we may include the effects of meson exchange, relativistic propagation, retardation, causality, and the dynamical quantum vacuum. In addition, the correct Lorentz structure of the NN interaction can be maintained. Renormalizable theories keep the number of parameters to a minimum, and since they require no additional *ad hoc* cutoffs, they are minimally sensitive to input from short-distance physics.

In this lecture, I examined some simple results in the Walecka model, which incorporates the dominant features of the observed NN force relevant for bulk nuclear systems through the exchange of neutral scalar and vector mesons.

This Lorentz structure leads naturally to nuclear saturation in the mean-field approximation, and the model parameters were chosen to reproduce the empirical saturation properties. The small binding energy of nuclear matter arises from a sensitive cancellation between large attractive and repulsive components in the potential energy. These large components introduce a new energy scale into the nuclear matter problem and lead to new physical effects from the shifted mass of the nucleon in nuclear matter. They also imply a stiff nuclear equation of state at energy densities greater than approximately 0.5 GeV/fm^3.

The mean-field theory is a simple approximation to the full quantum field equations of QHD. Because QHD is a consistent many-body theory, corrections to the MFT may be examined systematically. In particular, our experience in nuclear physics suggests that there are important nucleon-nucleon correlations that must also be included.[19] There are many techniques available for studying relativistic, interacting many-body systems with quantum field theory. For example, Feynman diagrams allow corrections to the MFT to be included systematically, as described for zero temperature in chapter 5 of ref. 1. Clearly, since QHD is a strong-coupling theory, perturbative approaches are of no use, and various classes of diagrams must be summed to infinite order in the couplings. Unfortunately, diagrammatic techniques are of limited utility in carrying out this program: while the diagrams tell us *how* to calculate, they give no indication of *what* diagrams are important. We must therefore invent more powerful approaches that allow us to include various infinite classes of diagrams in a well-defined fashion, grouped in order of decreasing importance. The search for practical, reliable schemes for calculating in the strongly coupled, relativistic quantum field theory of QHD is an important problem that is currently under active investigation.

This work was supported in part by DOE contract DE–FG02–87ER40365.

REFERENCES

1. B. D. Serot and J. D. Walecka, Adv. Nucl. Phys. **16**, 1 (1986).
2. G. Baym and L. P. Kadanoff, Phys. Rev. **124**, 287 (1961).
3. G. Baym, Phys. Rev. **127**, 1391 (1962).
4. W. E. Caswell and A. D. Kennedy, Phys. Rev. **D25**, 392 (1982).
5. J. C. Collins, *Renormalization* (Cambridge University Press, New York, 1984).
6. J. D. Walecka, Ann. Phys. (N.Y.) **83**, 491 (1974).
7. S. A. Chin and J. D. Walecka, Phys. Lett. **52B**, 24 (1974).
8. S. A. Chin, Ann. Phys. (N.Y.) **108**, 301 (1977).
9. C. J. Horowitz and B. D. Serot, Nucl. Phys. **A368**, 503 (1981).
10. C. J. Horowitz and B. D. Serot, Phys. Lett. **140B**, 181 (1984).
11. J. A. McNeil, J. R. Shepard, and S. J. Wallace, Phys. Rev. Lett. **50**, 1439 (1983).
12. J. R. Shepard, J. A. McNeil, and S. J. Wallace, Phys. Rev. Lett. **50**, 1443 (1983).
13. B. C. Clark, S. Hama, R. L. Mercer, L. Ray, and B. D. Serot, Phys. Rev. Lett. **50**, 1644 (1983).
14. D. G. Boulware, Ann. Phys. (N.Y.) **56**, 140 (1970).
15. R. J. Furnstahl and B. D. Serot, Indiana University Nuclear Theory Center Preprint IU/NTC 89–2, submitted to Phys. Rev. C.
16. J. D. Bjorken and S. D. Drell, *Relativistic Quantum Mechanics* (McGraw-Hill, New York, 1964).
17. C. W. Misner, K. S. Thorne, and J. A. Wheeler, *Gravitation* (Freeman, San Francisco, 1973).
18. A. L. Fetter and J. D. Walecka, *Quantum Theory of Many-Particle Systems* (McGraw–Hill, New York, 1971).
19. C. J. Horowitz and B. D. Serot, Nucl. Phys. **A464**, 613 (1987); **A473**, 760 (1987) (E).
20. R. J. Furnstahl, R. J. Perry, and B. D. Serot, Phys. Rev. **C40**, 321 (1989).

A USER'S GUIDE TO COMPUTATIONAL CHEMISTRY

Marcy E. Rosenkrantz

University of Dayton Research Institute
Air Force Astronautics Laboratory
Edwards Air Force Base, California 93523-5000

INTRODUCTION

This talk is divided into several parts. In the first part I will discuss how I came to prepare this talk and what I hope you will learn from it. In part two, I will present a brief introduction to the kinds of investigations in computational chemistry we are engaged in at the Astronautics Lab. Specific examples throughout the rest of the presentation are taken from the work we do at the lab. Part three is meant to be a brief introduction to molecular orbital (MO) theory. Finally we will deal with three specific methods commonly employed in computational chemistry.

PART 1. MOTIVATION

This talk started as a result of an eight hour drive between Boston and Aberdeen, Maryland which I took with a colleague from the Harvard Smithsonian Center for Astrophysics. He is an experimental atomic physicist. As a result of several "shop talk" questions he asked, I became aware of the communication gap that sometimes exists between experimentalists and theorists even in closely related fields. I decided shortly thereafter that there was a need for a talk that dealt with this gap in understanding. I have since learned that a similar gap exists even between theoretical atomic physicists and theoretical chemists. Hence, this particular talk.

There are two major points I would like you to take away from this dicusssion.

1. An appreciation for the methods used in computational chemistry.

2. A critical mind when reading the computational chemistry literature.

If I fulfill these goals you will also be able to appreciate some of the new developments in computational chemistry.

PART 2. COMPUTATIONAL CHEMISTRY AT THE ASTRONAUTICS LABORATORY

At the Astro Lab we are primarily interested in atoms and molecules with high potential energy content and high density. Such systems are most useful due to their energy storage capabilities. In addition to these high energy density materials (HEDM) we study the precursers to and products of the combustion of our HEDM candidates. So our studies also include diagnostics of small molecular fragments. It is interesting to note that many of these fragments are also of interest to astronomers and astrophysicists since the fragments are often found in space.

One study we engaged in was that of the high spin states of carbon monoxide, CO(1). Our calculations led us to conclude that the lowest $^5\Sigma^+$ state of CO is metastable. And since even at its potential minimum it is about 11 eV above the minimum in the ground electronic $X^1\Sigma^+$ state of CO, the excited state is capable of releasing (without combustion) essentially all 11 eV of energy upon realignment of the electron spins. We still must determine the decay mechanisms for the quintet state which might preclude storing energy in the form of $^5\Sigma^+$ CO. I shall use examples from this CO work throughout this talk.

In order to characterize new molecular states we need to obtain electronic potential energy curves and wavefunctions. The potential energy curves can of course be obtained from molecular spectroscopy and/or scattering experiments. However, both of these methods have drawbacks. Molecular spectroscopy often limits us to a few vibrational energy levels. The potential curves are then obtained from an inversion of the data using, for example, the RKR technique. As a result we obtain incomplete potential curves. Scattering experiments are sensitve only to certain regions of the potential curves and require a fit to the cross section data to obtain electronic potential curves which are also incomplete.

However, from a well designed *ab initio* (from first principles) calculation we can obtain a wealth of information. We can use our electronic potential energy curves to obtain all the vibrational energy levels as well as the left and right classical turning points. In addition, rotational analyses can be performed. The dipole and quadrupole moments can be obtained as well as transition moments between states. Since we are not limited to bound electronic states, information about both bound and dissociating states and states whose selection rules preclude their facile investigation are readily available. Thus it is "easy" to explore regions difficult to obtain experimentally.

In principle, the methods are quite straighforward. We wish to apply the variational principle and solve the Schrodinger equation

$$H\Psi_{el}=E_{el}\Psi_{el} \tag{1}$$

for E_{el} and Ψ_{el} by diagonalizing the Hamiltonian matrix. E_{el} and Ψ_{el} are the electronic energy and wavefunction respectively. H is the Hamiltonian of the system.

In practice it is difficult to obtain the wavefunction for more that one state at the same time to the same degree of accuracy. To do so requires large amounts of computer time, large wavefunctions and thus large financial resources. Computer time is rarely free.

PART 3. BRIEF INTRODUCTION TO MOLECULAR ORBITAL THEORY

Linear Combination of Atomic Orbitals

In atomic structure we allocate electrons to atomic orbitals of lowest energy subject to the constraints of the Pauli Principle. Excited states are obtained by promoting electrons to higher lying atomic orbitals.

A similar approach is used in molecular structure. Each molecular orbital (MO) is a linear combination of atomic orbitals (LCAO). We allocate electrons to MO's of lowest energy subject to the Pauli Principle constraints. Molecular excited states are obtained by promoting electrons to higher lying MO's.

Each AO can be expressed as a linear combination of appropriately normalized functions. In practice these functions are usually either Slater functions (STF)

$$f(\text{STF}) = N\, r^{n-\ell}\, e^{-\xi r}\, Y_{\ell}^{m}(\theta,\phi) \tag{2}$$

or Gaussian functions (GTF)

$$f(\text{GTF}) = N\, r^{\ell}\, e^{-\zeta r^{2}}\, Y_{\ell}^{m}(\theta,\phi) \tag{3}$$

where n, ℓ, m are the well known quantum numbers and a set of ξ or ζ are chosen for each atom. N is the normalization constant for the function.

The wavefunction for the ith orbital then has the form

$$\phi_i = \Sigma_j a_{ij} f_j \tag{4}$$

where each f_j has a unique ξ_j or ζ_j. Examples of basis sets can be found in most thorough papers in the literature and in ref. 2. The latter is a compendium of ζ values for use in Gaussian basis sets. For example, suppose I needed a basis set for the hydrogen atom. The set of ζ values in Table 1 might be chosen.

Table 1. A simple Gaussian basis set for hydrogen.

ℓ	ζ
0	1.3325
0	0.2015

The AO wavefunction which corresponds to the basis set in Table 1 is

$$\psi_{1s} = a_1 N_1 e^{-1.3325r^2} Y_0^0(\theta,\phi) + a_2 N_2 e^{-0.2015r^2} Y_0^0(\theta,\phi) \quad (5)$$

One way to understand the importance of a basis set is to examine the term in a literal manner. The basis set is the foundation of the entire calculation. **The errors of a bad basis set can never be redeemed.** A basis set that is too small (i.e. the set is not even remotely complete) can lead to an error known as basis set superposition error (BSSE). In effect, this means that functions missing from one atom are compensated for by the basis set of another atom in the molecule. This can lead to potential curves that are artificially deep, with minima at distances that are artificially small.

In our study of the high spin states of CO we expected the molecules to be bound only at large interatomic distances. So we chose STF's to help us reproduce the atomic dipole (α_d), quadrupole (α_q) and octupole (α_o) polarizabilities as closely as possible. These polarizabilities give rise to such long range forces as induced dipole-induced dipole interactions, etc. Table 2 gives the ξ values we used for our carbon and oxygen Slater basis sets. Table 3 gives the polarizabilities of the fragment atoms. As you can see we do quite well in obtaining the polarizabilities of these atoms as judged by comparison with the best literature values[3]. We have sufficient confidence in our basis sets for the atoms as a result of these polarizabilities that appropriate linear combinations of them should describe very well the long range interactions which dominate the high spin states of CO.

Table 2a. Slater type basis sets for carbon and oxygen.

n	ℓ	ξ	
		carbon	oxygen
1	0	9.48260	13.75740
1	0	5.43600	7.61410
2	0	4.20100	5.86600
2	0	2.68440	4.31200
2	0	1.52430	2.48020
2	0	1.05750	1.69820
2	1	6.51000	7.56480
2	1	2.60050	3.44990
2	1	1.43360	1.81730
2	1	0.98070	1.14390
3	2	3.64070	4.82990
3	2	2.02110	2.54420
3	2	0.87500	1.24837
4	3	2.59850	3.27110
4	3	1.76530	2.05900

Table 2b. Rydberg and polarizing functions for carbon and oxygen.

n	ℓ	ξ carbon	ξ oxygen
3	0	2.01431	2.14726
3	0	0.59072	0.71638
3	1	2.03978	2.81850
3	1	0.48277	0.53369
4	3	0.92500	1.18750
5	4	0.94375	1.16250

Table 3. Dipole, quadrupole and octupole polarizabilities of carbon and oxygen (in a.u.), obtained with the basis sets of table 2, compared with the best literature values from ref. 3.

		Carbon		Oxygen	
		present	Best	present	Best
α_d	$M_\ell=0$	10.102	10.10	5.005	5.14
α_d	$M_\ell=1$	12.990	13.05	4.540	4.58
$\bar{\alpha}_d$		12.027	12.07	4.695	4.77
α_q	$M_\ell=0$	41.60	41.73	20.32	20.61
α_q	$M_\ell=1$	61.80	61.28	14.74	15.05
$\bar{\alpha}_q$		55.07	54.76	16.60	16.90
α_o	$M_\ell=0$	449.57	-	141.54	-
α_o	$M_\ell=1$	697.73	-	98.35	-

The Hartree-Fock Method

This method is sometimes also called the single configuration self-consistent field (SCSCF or just plain SCF) method, for reasons which will become obvious shortly.

Consider the hydrogen molecule, H_2--a nice, easy, two electron system. At very large internuclear separations it can be considered to be a pair of barely ineracting hydrogen atoms. An approximate wavefunction for each atom might be $1s_a(1)$ and $1s_b(2)$, where we label the electrons 1 and 2 and associate them with the atoms a and b, respectively. Let us now take a linear combination of these functions:

$$g=N(1s_a+1s_b) \tag{6}$$

where I have used the g to denote gerade symmetry, i.e. the function is even with respect to inversion through the center of symmetry of the molecule. The simplest LCAO MO description of the H_2 molecule is the product of two of these g functions, one for each electron.

$$g(1)g(2)=N^2\{1s_a(1) + 1s_b(1)\}\{1s_a(2) + 1s_b(2)\} \tag{7}$$

$$\begin{aligned} g(1)g(2)=N^2\{&1s_a(1)1s_a(2) + 1s_b(1)1s_b(2) + \\ &1s_a(1)1s_b(2) + 1s_a(2)1s_b(1)\} \end{aligned} \tag{8}$$

Here we have omitted the multiplicative spin function $2^{1/2}\{\alpha(1)\beta(2) - \beta(1)\alpha(2)\}$.

Let's not be too hasty in accepting this description. Let's look first at the potential function E(R) obtained by solving Eq. (1) using the product wavefunction of Eq. (7). The potential function E(R) as a function of R, the internuclear distance, is given in fig. 1, Curve A. Note that the minimum is not too far from the correct equilibrium internuclear separation of the hydrogen molecule (r_e = 1.4 a_o) but at large R the curve does not approach the energy of two infinitely separated hydrogen atoms as our physical intuition would tell us. What does this function mean physically? Notice that the first two products in equation 8 have both electrons "sitting" on the same atom. This indicates that in the first term in Eq. (8) the molecule looks like $H_a^- + H_b^+$, while the second term in Eq. (8) looks like $H_a^+ + H_b^-$. Clearly, these ionic terms do not belong in our description at large separations and are the cause of the incorrect long range behavior of the potential curve in fig. 1, Curve A. This behavior is typical of a single configuration description of a system that dissociates to open shell atoms and is sometimes called the Hartree-Fock Catastrophe. How can we get rid of the offending electron distributions?

Consider the function

$$u = N'(1s_a - 1s_b) \qquad (9)$$

where the u denotes ungerade symmetry. The function is odd with respect to inversion through the center of symmetry of the molecule. Now let us take the product of two of these u functions, one for each electron. (We will ignore spin again.)

$$u(1)u(2) = N'^2\{1s_a(1)1s_a(2) + 1s_b(1)1s_b(2) - 1s_a(1)1s_b(2) - 1s_a(2)1s_b(1)) \qquad (10)$$

Note that this function alone still contains the offending electron distributions. Next consider the wavefunction

$$\Psi(^1\Sigma_g^+) = g(1)g(2) + \lambda u(1)u(2). \tag{11}$$

Let us apply the variational principle and minimize the energy with respect to the variational parameter λ. We will obtain two roots of the secular equation, the lowest of which will have the following property:

$$\lim_{R\to\infty} \lambda = -1. \tag{12}$$

Thus we have eliminated the offending ionic terms at large R by introducing a configuration in addition to the single one we started with. (This is the simplest form of configuration interaction which we will discuss shortly.) But we have kept an appropriate mix of the two configurations at other separations. Figure 1, Curve B results from having eliminated the offending configurations. In the process we have inproved the location of the potential minimum and obtained the correct asymptotic behavior of the energy.

In actual practice, we use the LCAO method with either a Slater or Gaussian basis set. The Hartree-Fock equations[(4)] are solved variationally for the MO's and their eigenvalues. The variational procedure is iterated until the energy changes by an amount which is less than a predetermined tolerance. In other words, the a_{ij}'s in Eq. 4 are varied until a self-consistent energy is reached.

Multiconfiguration Self Consistent Field Method

The MCSCF method uses more than one configuration. A configuration is an antisymmetrized product of MO's. Consider the simplest configuration for the Li_2 molecule and recall that a property of determinants is that they change sign upon interchange of two rows or columns. The Li_2 molecule has six electrons. For those who have forgotten, the lithium atom has the atomic configuration: $1s^2 2s$. Thus the possible MO's arising from two such lithium atoms are as follows:

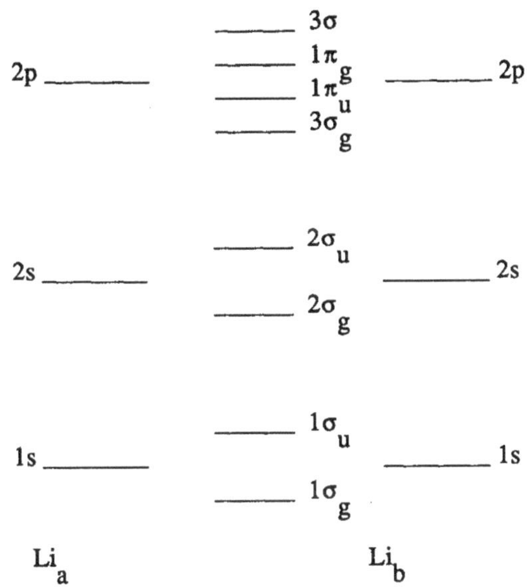

In this, so called, correlation diagram the notation, σ denotes a MO comprised of atoms with $m_\ell = 0$ and π denotes a MO comprised of atoms with $m_\ell = 1$.

The simplest configuration of these six electrons is

$$C_1 = | 1\sigma_g^2 \, 1\sigma_u^2 \, 2\sigma_g^2 | =$$

$$\begin{vmatrix} 1\sigma_g\alpha(1) & 1\sigma_g\beta(1) & 1\sigma_u\alpha(1) & 1\sigma_u\beta(1) & 2\sigma_g\alpha(1) & 2\sigma_g\beta(1) \\ \cdots & 1\sigma_g\beta(2) & \cdots & \cdots & \cdots & \cdots \\ \cdots & \cdots & 1\sigma_u\alpha(3) & \cdots & \cdots & \cdots \\ \cdots & \cdots & \cdots & 1\sigma_u\beta(4) & \cdots & \cdots \\ \cdots & \cdots & \cdots & \cdots & 2\sigma_g\alpha(5) & \cdots \\ \cdots & \cdots & \cdots & \cdots & \cdots & 2\sigma_g\beta(6) \end{vmatrix} \tag{13a}$$

Now consider the promotion of 2 valence electrons to the $2\sigma_u$ orbital.

$$C_2 = |\, 1\sigma_g^2\, 1\sigma_u^2\, 2\sigma_u^2\, | \tag{13b}$$

or to the $2\sigma_u$ and $3\sigma_u$

$$C_4 = |1\sigma_g^2\, 1\sigma_u^2\, 2\sigma_u^1\, 3\sigma_u^1|,\text{etc.} \tag{13c}$$

In general we have the following Slater determinant,

$$C_n = |\, \phi_1\, \phi_2\, \phi_3 \ldots |. \tag{13d}$$

Remember that each MO is a linear combination of functions in the basis set.

$$\phi_i = \Sigma_{t=1}^{q} a_{it} f_t \tag{14}$$

Here f_t is either a STO or a GTO. Now the wavefunction for the system is written as a linear combination of the Slater determinants, as shown in Eq. (15).

$$\Psi = \Sigma_{n=1}^{s} b_n C_n. \tag{15}$$

We must, therefore, have sufficient functions in the basis set to describe excited orbitals (the virtual orbitals) that are not occupied in the Hartree-Fock description but will be used in the MCSCF excitations. Look again at table 2 and notice the functions we added to account for the first excited states of carbon and oxygen--the Rydberg functions. The MCSCF method optimizes the energy with respect to the variational parameters, a_{it} and b_n. This can get very expensive. So in practice we limit the number of configurations allowed in the calculation; i.e. only single and double excitations within the collection of valence orbitals, or only single and double excitations into a limited number of virtual orbitals.

A complete active space self consistent field calculation (CASSCF) is a specific class of MCSCF calculation. A CAS allows all excitations within a limited number of orbitals, i.e. 4 electrons in 4 orbitals. Here the 4 orbitals can be a combination of valence and virtual orbitals.

Configuration Interaction (CI)

CI calculations are similar to MCSCF in so far as CI employs a linear combination of configurations. In a CI only the configuration coefficients are varied. The MO's (actually, the coefficients in the linear combinations of basis functions, in Eq. (14)) are kept fixed. We sacrifice the flexibility of the variation of the MO's, accomplished in the MCSCF procedure, for an increase in the number of configurations. ($10000 \le n \le 1{,}000{,}000$). The MO's for use in a CI calculation are usually obtained from a Hartree-Fock or MCSCF calculation.

There are several classes of CI calculations. The simplest is a First Order CI (FOCI). Here we allow single and double excitations among the valence orbitals, but only single excitations into the virtual orbitals. In a Second Order CI (SOCI) we allow all possible single and double excitations. In a full CI all possible excitations including single to quadruple excitations among all the orbitals are allowed. Multi-Reference CI (MRCI) calculations allow only single and double excitations into the virtuals from a number of different reference configurations. Often these references are from a CAS calculation. But we might also choose a set of MCSCF references, say those that have their b_i greater than some value, ε.

Our calculations of the potential energy curves of the quintet states of CO employed both MCSCF and CI techniques extensively. The characteristic constants of some of these curves are presented in Table 4. The only other calculations of these states were performed by O'Neil and Schaefer in 1970[5]. Their pioneering work included a full CI approach with a minimal basis set. This talk should give you enough insight to distinguish the virtues and shortcomings of that approach. If I have succeeded in acheiving my goals you should be confident that our results, obtained with a large SOCI calculation and a much larger basis set, specifically tuned to the calculations we planned, are more nearly correct. We leave it to experimentalists to prove us correct.

Table 4. Characteristic constants of five of the quintet states of CO. The numbers in parentheses are the constants reported in ref. 5. (Only the $1^5\Sigma^+$ and $1^5\Pi$ states were determined to be bound in that study.)

State	$R_e(a_o)$	$D_e(cm^{-1})$	$\omega_e(cm^{-1})$	$\omega_e x_e(cm^{-1})$
$1^5\Sigma^+$	4.71	749.7	123.3	5.00
	(3.59)	(3226)	(500)	
$2^5\Sigma^+$	7.00	94.6	38.7	3.96
$^5\Delta$	6.98	93.5	41.4	4.56
$^5\Sigma^-$	7.18	68.7	34.0	4.21
$1^5\Pi$	7.91	17.2	16.9	4.17
	(3.08)	(3549)	(759)	

Acknowledgments

The author wishes to thank Dr. Peter L. Smith for the questions and discussions that prompted this talk.

References

1. M.E. Rosenkrantz, J.E. Bohr, and D.D. Konowalow, Forefronts 4, 2 (June 1988) Forefronts is a publication of the Center for Theory and Simulation in Science and Engineering, Cornell Theory Center, Cornell University, Ithaca, New York.

2. R. Poirier, R. Kari, and I.G. Csizmadia, Handbook of Gaussian Basis Sets: A compendium for Ab Initio Molecular Orbital Calculations, Physical Sciences Data V 24, Elsevier (Amsterdam 1985).

3. H.-J. Werner and W. Meyer, Phys. Rev. A, 13 13 (1976); α_q from E.-A. Reinsch and W. Meyer, Phys. Rev. A, 18 1793 (1978).

4. For a more complete discussion of the Hartree-Fock method see, for example, I.N. Levine, Quantum Chemistry, pp 373-380, Allyn and Bacon, Inc. (Boston, 1983).

5. S.V. O'Neil and H.F. Schaefer III, J. Chem. Phys. 53 3994 (1970).

Bibliography

1. A.C. Hurley, Introduction to the Electron Theory of Small Molecules, Academic Press (London, 1976).

2. A. C. Wahl and G. Das, "The Multiconfiguration Self-Consistent Field Method", in Modern Theoretical Chemistry, vol. 3, pp 51-78, H.F. Schaefer III, ed. Plenum Press (New York, 1977).

3. I. Shavitt, "The Method of Configuration Interaction", in Modern Theoretical Chemistry, vol. 4, pp 189-275, H.F. Schaefer III, ed. Plenum Press (New York, 1978).

4. P. Pulay, "Direct Use of the Gradient for Investigationg Molecular Energy Surfaces", in Modern Theoretical Chemistry, vol. 4. pp 153-185, H.F. Schaefer III, ed. Plenum Press (New York, 1978).

GAUGE THEORIES AND DYNAMICAL SYMMETRY BREAKING

by

Thomas Appelquist
Center for Theoretical Physics, Department of Physics
Yale University, New Haven, Connecticut 06511

ABSTRACT

Asymptotically free technicolor theories are reviewed and the impact of additional interactions at very high energies on the spontaneous breaking of electroweak symmetry is described. These new interactions can enhance the chiral condensate and even the Goldstone boson decay constant relative to the confinement scale. This can lead to important consequences in technicolor theories, including realistically large quark and lepton masses and the possibility of new physics at energies below the weak scale. In addition, small differences in strength of the four-fermion interactions, due say to isospin or flavor breaking, can be greatly magnified in the resulting quark and lepton mass spectrum.

One of the key features of asymptotically free gauge field theories is that they allow a natural understanding of the scale of dynamical chiral symmetry breaking. The running coupling $\alpha(p)$ is strong at momenta on the order of the confinement scale Λ_C. It then decreases like $1/\ln(p)$ until it reaches some smaller value $\alpha(\Lambda)$, where Λ is the cutoff beyond which the theory can no longer be used in isolation. The value of $\alpha(\Lambda)$ is determined by the unknown high energy physics beyond Λ. Dynamical chiral symmetry breaking can only take place in a gauge field theory when the coupling exceeds some critical value α_C [1], expected to be reasonably strong [2]. Thus if $\alpha(\Lambda) << \alpha_C$, the breaking must take place at scales well below the cutoff and perhaps on the order of the confinement scale Λ_C. In QCD this is precisely what happens.

Technicolor theories are expected to behave in a similar way. The Goldstone boson decay constant F is equal to $250\mathrm{GeV}/(n)^{1/2}$, where n is the number of $SU(2)_L$ doublets. The confinement scale and the dynamical technifermion mass are typically of the same order and the ultraviolet cutoff Λ, the so-called extended technicolor scale, must be much larger.

In either of these contexts, the new physics that exists beyond Λ can be described at low energies in terms of effective higher dimension operators. The dominant contribution at low energies should arise from dimension six, four-fermion operators. In technicolor theories, the effect of these operators (ETC interactions) has been considered extensively [3]. They are treated as perturbations to the dominant technicolor gauge interactions, giving rise to quark and lepton masses, pseudo-Goldstone boson masses, and typically to flavor-changing neutral currents among the quarks and leptons. Without some GIM-type mechanism, they must be suppressed by taking the cutoff Λ to be at least several hundred TeV, and this then leads to a potential problem for technicolor theories: Quark and lepton masses are given by $m(q,l) \sim (g^2/\Lambda^2)<\bar{\psi}\psi>$ where g^2 is a dimensionless coupling strength and $<\bar{\psi}\psi>$ is the technicolor condensate, naively expected to be of order $\Lambda_C{}^3$. These masses then turn out to be no more than a fraction of an MeV.

Another embarrassing fact is that some of the pseudo Goldstone bosons generically present in these theories seem to have masses that are too small. They are the ones that do not get mass from QCD interactions, but instead must rely on the four-fermion (ETC) interactions. These masses are then also proportional to $<\bar{\psi}\psi>$ and turn out naively to be no larger than a few GeV. The electrically charged pseudos in this mass range should probably have been seen by now.

The recent revival of interest in technicolor theories is partly due to the observation [4] that some of them contain within themselves a partial solution to these problems. The large number of technifermions that typically populate these theories can slow the running of the coupling while maintaining the asymptotic freedom. The technicolor condensate, and therefore the quark, lepton and pseudo masses, are quite sensitive to this modest enhancement of the high momentum components. In many candidate technicolor theories ("walking theories"), the slowing of the running is sufficient to raise the masses by up to nearly two orders of magnitude [5]. Even this effect, however, does not seem adequate to produce a realistic quark and lepton mass spectrum.

In this talk I will summarize a study of the additional effect of the four-fermion operators on the dynamical breaking of chiral symmetry in asymptotically free theories, both walking and normally running. The essential point is that if these interactions are strong enough, they cannot necessarily be treated as perturbations to the gauge interactions. Instead, they should be included in the gap equation that governs the breaking and they can dramatically affect the mass scales that emerge. That four-fermion interactions can drive dynamical breaking has of course been known since the classic paper of Nambu and Jona-Lasinio [6]. If they begin to dominate the dynamics in the present framework, however, the entire low energy expansion could be breaking down and even higher dimension operators could then play a role. As we present our results, we will discuss this point. For economy, the discussion from here on will be in the context of technicolor.

To begin, flavor symmetry violation will be neglected. The technifermions together with their technicolor and four-fermion interactions will then respect a global $SU(N_f)xSU(N_f)$ symmetry. Spontaneous chiral symmetry breaking will lead to a residual $SU(N_f)_{L+R}$ symmetry with a technicolor singlet fermion mass determined by an appropriate gap equation for the dynamical technifermion mass $\Sigma(p)$. With the gauge sector treated in dressed ladder approximation [7], the gap equation will take the form

$$\Sigma(p) = \frac{1}{4}\int_0^{\Lambda^2} \frac{k^2dk^2}{M^2(k,p)} \frac{\alpha(M)}{\alpha_C} \frac{\Sigma(k)}{k^2 + \Sigma^2(k)} + 2\frac{g^2}{\Lambda^2}\left[\frac{1}{4\pi^2}\int_0^{\Lambda^2} k^2dk^2 \frac{\Sigma(k)}{k^2 + \Sigma^2(k)}\right] \qquad (1)$$

The first term is the technicolor contribution in Landau gauge. M(k,p) is the maximum of k and p, and $\alpha_C = \pi/3C_2$ is the critical gauge coupling required for spontaneous breaking in the absence of other interactions. C_2 is the quadratic Casimir of the technifermion representation. The second term comes from the chirally invariant four-technifermion interaction. Its coupling strength is defined to be g^2/Λ^2, and it is assumed that the physics giving rise to this interaction damps rapidly beyond Λ. It is also assumed that the four-fermion interaction is attractive in the technicolor-singlet channel. Factors arising from the closed loop sum over N_f and technicolor have been absorbed into g^2. The quantity in square brackets, when multiplied by the dimensionality of the tecnifermion representation, is the technifermion condensate.

We want to know how $\Sigma(p)$, and also F and the condensate $<\bar{\psi}\psi>$, depend on the four fermion strength $\lambda \equiv g^2/2\pi^2$. If there are interesting effects, a key question is whether they occur for a large enough range of λ so that they can be expected to be generic to the technicolor framework.

For very small λ, the physics will be dominated by the technicolor interaction. $\Sigma(p)$ will be of order Λ_C for $p \sim \Lambda_C$. For $p >> \Lambda_C$, it will fall with increasing p at a rate roughly between 1/p and $1/p^2$ depending on the rate of running of the coupling. The integral expression giving F in terms of $\Sigma(p)$ [8] converges rapidly for any rate of fall of $\Sigma(p)$ in the above range, and leads to a value of F on the order of Λ_C. The condensate depends sensitively on the running and can vary from being roughly of order $\Lambda_C{}^3$ (ordinary running) to up to two orders of magnitude larger (very slow running).

Suppose next that λ is very large, corresponding to the existence of new strong interactions among the technifermions at energies of order Λ. If the gauge interaction is neglected, chiral symmetry breaking will take place only for $\lambda \equiv g^2/2\pi^2 > 1$. If λ is well above 1, the gauge interaction can be treated as a perturbation and $\Sigma(p)$ will turn out to be of order Λ ($>> \Lambda_C$) and essentially independent of p. In this limit, $F = O(\Lambda)$ and $<\bar{\Psi}\psi> = O(\Lambda^3)$. This of course is a completely unphysical limit from the point of view of technicolor. For the reasons described above, a heirarchy at least of order 10^3 must be maintained between F and Λ. It is also a limit in which the low energy expansion has broken down since the physics is taking place dominantly at the cutoff, and the above estimates are not reliable in detail.

It is the range of intermediate values of λ that is of interest for technicolor theories. We [9] have studied the evolution of F and $<\bar{\Psi}\psi>$ throughout this range, both analytically and numerically. For most of the range, $\Sigma(0) << \Lambda$, and an analytic treatment is made possible because the integral equation can be linearized for $p,k >> \Sigma(0)$. The analytic treatment can then be supplemented numerically for all λ. The numerical results for a typical case, an SU(4) gauge theory with the number of technifermions varied to adjust the rate of running of the gauge coupling [10], are displayed in Table I. It is of course only the orders of magnitude of these results and their sensitivity to λ that are important. The main features of both the analytic and numerical results for the SU(4) case as well as many others are the following:

1. For essentially any rate of running, the four-fermion term makes a relatively small contribution to $\Sigma(0)$ for most of the range of λ between 0 and 1. For normal running ($(b-\delta b)\alpha_C \sim 1$, where b and δb are the gauge and fermion contributions to the beta function $\beta(\alpha) = -(b-\delta b)\alpha^2$), it becomes comparable to the technicolor contribution only for $\lambda \cong 0.93$. For somewhat slower running ($(b-\delta b)\,\alpha_C \sim 0.4$), as might be expected in many technicolor theories, it becomes comparable for $\lambda \cong 0.85$. For very slow running ($(b-\delta b)\alpha_C \sim 0.1$), it becomes comparable for $\lambda \cong 0.7$.

2. For λ between 0 and the above values, $\Sigma(0)$ and F remain essentially unchanged and of order Λ_C. The condensate, however, increases substantially throughout each range. The increase from the $\lambda = 0$ value is at

first modest, but it is at least three orders of magnitude near the end of each range. In the intermediate running case, the condensate is $O(\Lambda_C{}^3)$ in the absence of the four-fermion interaction, corresponding to a quark or lepton mass on the order of one Mev. The condensate then increases by nearly an order of magnitude by the time λ reaches 0.7 and by more than another two orders of magnitude in the range between 0.7 and 0.85. At the end of the range, a quark or lepton mass exceeding 1 GeV is generated. In the case of very slow running, the additional effect of the four-fermion interaction can produce a dramatically enhanced condensate within the above range. Near the end, quark and lepton masses in the range of 5 - 10 Gev can be generated.

3. In each of the above cases, the contribution of the four-fermion term to $\Sigma(p)$ is small compared to Λ, that is, $<\bar{\psi}\psi>/(\Lambda)^3 << 1$. This suggests that higher dimension operators will lead to even smaller contributions to $\Sigma(p)$ and that the low energy expansion will remain valid [11].

4. For any rate of running, as λ is increased beyond the ranges described above, some dramatic effects set in. The changes are very rapid and by the time λ exceeds unity, the theory becomes unphysical for technicolor purposes. Consider the intermediate running case $(b-\delta b)\alpha_C \sim 0.4$. As λ increases from 0.85 to 0.92, F grows from being somewhat less than Λ_C to being several times larger. Since F is the fixed weak scale of the theory, this means that the technicolor confinement scale drops below the weak scale. The resulting light physics would consist of relatively narrow techni-glueballs which could be produced at existing or soon-to-exist accelerators. With $\lambda = 0.92$, for example, the technicolor confinement scale drops to roughly 20 GeV, compared to $F = 90$ GeV, $\Sigma(0) = 120$ GeV, and the condensate is large enough to produce a quark or lepton mass greater than 50 GeV. In general, a fermion mass large enough to accomodate the t quark emerges in the range of λ values that can lead to physics below the weak scale. While, the low energy expansion still appears to remain valid for λ in this range, the results are becoming more sensitive to physics at Λ. It will be important to explore this sensitivity in more detail.

A natural question is whether there is an expected range of values of λ in technicolor theories. It is clear from Table I that substantial condensate enhancement, due partly to the four-fermion interaction, occurs for a reasonably large range of λ especially if there are enough fermions to slow the running of the coupling somewhat. An enhancement large enough to produce a t quark mass and light techniglueballs, on the other hand, emerges only for a narrower range of λ values nearer to unity.

I will here make only a few general remarks on expectations for the size of λ. If the four-fermion interactions are due to ETC boson exchange, g will be the ETC boson coupling. It would then not be unreasonable to speculate that $(\pi/2)\lambda = g^2/4\pi$ should be on the order of $\alpha(\Lambda)$, the technicolor coupling at the ETC scale. This would be expected if these interactions become unified at Λ. In the SU(4) examples of Table I, $\alpha(\Lambda)$ ranges between 0.2 and 0.35, somewhat less than the "interesting" range of values of $(\pi/2)\lambda$. On the other hand, λ contains group theory factors depending on the unified theory and therefore not present in the technicolor coupling $\alpha(\Lambda)$. Because the unified group is larger than the technicolor group, these factors might naturally be expected to be larger than unity. A value for $(\pi/2)\lambda$ of order unity and larger than $\alpha(\Lambda)$ might also be expected in a tumbling scenario [12]. There, the unified group, its fermion particle content, and its coupling strength at Λ are such that the unified theory breaks in a nonsinglet channel. This could presumably leave $\alpha(\Lambda) < \alpha_c$ (the critical technicolor coupling), and yet allow $(\pi/2)\lambda = O(1)$.

Whatever the source of the four-fermion interactions, it should be clear from the above analysis that a viable electroweak theory does not seem possible if $\lambda > 1$. This could prove to be a stumbling block for theories in which the four-fermion interactions are imagined to arise from a composite structure for the fermions and technifermions.

While the appearance of fermion masses up into the GeV range is welcome, the problem of producing a realistic quark and lepton mass spectrum remains. In technicolor theories, the mysteries of the large hierarchies in these spectra get shifted into the effective four-fermion interactions among the technifermions and the quarks and leptons. It should be clear from the above analysis, however, that this mapping can be far from linear. In particular, a small breaking of weak isospin or flavor symmetry in the four-technifermion couplings can be considerably magnified in the quark and lepton mass spectrum. This breaking could be due, for example, to group factors arising from the spontaneous breaking of some ETC theory at scale Λ.

We [13] have studied this magnification effect in more detail in a simple two-coupling example constructed from the SU(4) theory by including $N_f = 16$ technifermions divided into two families of eight. With four-technifermion couplings λ_1 and λ_2 for the two families, a variety of hierarchies will appear depending on the relative size of λ_1 and λ_2. In particular, for $\lambda_1 << \lambda_2 < 1$, F_2 and $\Sigma_2(0)$ can be on the order of the weak scale while F_1, $\Sigma_1(0)$ and the confinement scale can be smaller. Thus, there could be some relatively light technimesons and baryons along with light techni-glueballs. A smaller difference between λ_1 and λ_2 will keep all the above scales on the order of the weak scale and yet still lead to a large ratio between the condensates of family one and family two. If each technifamily couples to an ordinary fermion (quark and lepton) family with strength of order λ_1 (λ_2), a large hierarchy for these masses will emerge. An important point is that large quark and lepton mass hierarchies can develop in this framework without a hierarchy of ETC scales, but instead from modest coupling differences at a single ETC scale.

The large amount of weak isospin breaking in the quark and lepton mass spectrum poses a special problem for technicolor theories. Not only must a sufficiently large t-b quark mass splitting be generated, but this must be done in a way that doesn't overly infect the experimentally measured ρ parameter [14]. The fact that $\rho - 1 \equiv M_W^2/M_Z^2 \cos^2\theta_W - 1 < 10^{-2}$ [15], means that very little weak isospin violation can be allowed in ρ. We have begun an analysis of this problem by including an isospin violating four-fermion coupling λ_3 along with the symmetric coupling λ. To lowest order in electroweak interactions, $\rho = F_\pm^2/F_0^2$ where $F_\pm$ and F_0 are the decay constants for the charged and neutral Goldstone bosons. A weak isospin violating deviation of this ratio from unity will be induced by the λ_3 interaction, but unless it becomes quite strong, it will not violate the above experimental constraint. Nevertheless, a somewhat smaller λ_3 can induce a healthy isospin violation in the condensate, leading to realistic up-down fermion mass splittings. The size of the effect is essentially the same as that of Table I. In a recent preprint [13], we have argued that it is possible to accommodate even the t-b mass splittings without upsetting the ρ-parameter.

Finally, it is worth remarking that small amounts of weak isospin breaking will be induced by the electroweak interactions. The magnification effect I have described could then mean that these small increments to the four-fermion couplings could produce large weak isospin splitting in the quark and lepton mass spectrum. QCD effects could be magnified in a similar way, leading to important contributions to quark-lepton mass differences [16]. There will be analogous contributions to the masses of possible pseudo-Goldstone bosons. In general, the new interactions at scale Λ, if they are strong enough, can add to and magnify the contributions of standard model gauge interactions to the mass parameters of the standard model and its extensions.

TABLE I CAPTION

The results of a numerical investigation of chiral symmetry breaking, in an SU(4) theory with $N_f = 8, 16, 20$ fermion flavors, are tabulated for various values of the four-fermion coupling $\lambda \equiv g^2/2\pi^2$. The gauge and fermion contributions to the β-function are given by $b = 44/6\pi$ and $\delta b = N_f/3\pi$ respectively. Since there are $N_f/2$ weak doublets, $F = 250\ \text{GeV}/(N_f/2)^{1/2}$. The confinement scale is Λ_C and Λ is the (ETC) cutoff. With $\Lambda \cong 2500$ TeV, $\Lambda/g > \Lambda/(2\pi^2)^{1/2} \cong 600$ TeV. The dynamical technifermion mass is $\Sigma(0)$ and the size of $\Sigma(\Lambda)$ is a measure of the rate of fall of $\Sigma(p)$. To normalize F relative to $\Sigma(0)$ and Λ_C, we have used the relation appearing in Ref. 10, which is accurate for the high momentum components. We note that this expression underestimates f_π in QCD by roughly a factor of two. In the numerical solution of Eq. 1, confinement dynamics is approximated crudely by allowing $\alpha(p)$ to plateau at $\alpha(\Lambda_C) = 2.5\ \alpha_C$ for $p \leq \Lambda_C$. For $\Lambda_C < p < 2\Sigma(0)$, $\alpha(p)$ evolves according to $\beta(\alpha) = -b\alpha^2$, and for $p > 2\Sigma(0)$ it evolves according to $\beta(\alpha) = -(b-\delta b)\ \alpha^2$. The output numbers, in particular the dependence on the rate of running and on λ, are high momentum phenomena and not terribly sensitive to the low momentum approximation for $\alpha(p)$. For values of λ small enough so that Λ_C and $\Sigma(0)$ remain essentially unchanged relative to Λ, $\alpha(\Lambda)$ remains constant. For larger values of λ, our procedure leads to a small decrease in $\alpha(\Lambda)$. This, however, doesn't play an important role in the results for fermion masses or for the F/Λ_C hierarchy.

TABLE I

$N_f = 8$ Case. ($F = 125$ GeV, $b\alpha_C = 1.304$, $\delta b\alpha_C = 0.474$)

$\lambda \equiv g^2/2\pi^2$	$\Sigma(0)/\Lambda_C$	$\Sigma(\Lambda)/\Lambda_C$	F/Λ_C	$<\bar{\psi}\psi>/\Lambda^2\Lambda_C$	Λ_C	m_f/λ_{Ff}
0.00	0.48	8.2×10^{-9}	0.13	2.3×10^{-8}	970 GeV	0.4 MeV
0.50	0.48	2.7×10^{-7}	0.13	5.0×10^{-8}	970 GeV	1.0 MeV
0.70	0.48	6.9×10^{-7}	0.13	9.6×10^{-8}	970 GeV	1.8 MeV
0.80	0.48	1.4×10^{-6}	0.13	1.8×10^{-7}	970 GeV	3.4 MeV
0.90	0.48	9.5×10^{-6}	0.13	1.0×10^{-6}	970 GeV	20 MeV
0.91	0.48	1.9×10^{-5}	0.13	2.0×10^{-6}	970 GeV	39 MeV
0.92	0.48	3.6×10^{-4}	0.13	3.9×10^{-5}	960 GeV	730 MeV
0.93	0.84	6.3×10^{-2}	0.26	6.6×10^{-3}	480 GeV	62 GeV
0.94	1.6	0.30	0.77	3.2×10^{-2}	160 GeV	100 GeV
0.95	4.1	1.5	3.2	0.16	39 GeV	120 GeV

$N_f = 16$ Case. ($F = 88$ GeV, $b\alpha_C = 1.304$, $\delta b\alpha_C = 0.948$)

$\lambda \equiv g^2/2\pi^2$	$\Sigma(0)/\Lambda_C$	$\Sigma(\Lambda)/\Lambda_C$	F/Λ_C	$<\bar{\psi}\psi>/\Lambda^2\Lambda_C$	Λ_C	m_f/λ_{Ff}
0.00	0.55	3.3×10^{-8}	0.15	4.6×10^{-8}	590 GeV	0.5 MeV
0.50	0.55	6.4×10^{-7}	0.15	1.1×10^{-7}	590 GeV	1 MeV
0.70	0.55	2.1×10^{-6}	0.15	2.8×10^{-7}	590 GeV	3 MeV
0.80	0.55	8.8×10^{-6}	0.15	1.0×10^{-6}	590 GeV	12 MeV
0.82	0.55	1.9 x 10-5	0.15	2.2×10^{-6}	590 GeV	25 MeV
0.84	0.57	1.0×10^{-3}	0.15	1.2×10^{-4}	570 GeV	1.3 GeV
0.86	0.78	1.9×10^{-2}	0.23	2.1×10^{-3}	390 GeV	17 GeV
0.88	1.1	8.2×10^{-2}	0.42	8.9×10^{-3}	210 GeV	37 GeV
0.90	1.9	0.31	1.0	3.3×10^{-2}	85 GeV	55 GeV
0.91	2.8	0.62	1.8	6.6×10^{-2}	49 GeV	64 GeV
0.92	4.6	1.3	3.5	0.14	26 GeV	72 GeV

$N_f = 20$ Case. ($F = 79$ GeV, $b\alpha_C = 1.304$, $\delta b\alpha_C = 1.185$)

$\lambda \equiv g^2/2\pi^2$	$\Sigma(0)/\Lambda_C$	$\Sigma(\Lambda)/\Lambda_C$	F/Λ_C	$<\bar{\psi}\psi>/\Lambda^2\Lambda_C$	Λ_C	m_f/λ_{Ff}
0.00	0.63	4.9×10^{-7}	0.17	3.4×10^{-7}	470 GeV	3 MeV
0.50	0.63	9.6×10^{-6}	0.17	1.5×10^{-6}	460 GeV	14 MeV
0.55	0.63	1.6×10^{-5}	0.17	2.3×10^{-6}	460 GeV	21 MeV
0.60	0.63	3.6×10^{-5}	0.17	4.9×10^{-6}	460 GeV	45 MeV
0.65	0.65	3.1×10^{-4}	0.18	3.9×10^{-5}	450 GeV	350 MeV
0.70	0.75	3.2×10^{-3}	0.21	4.0×10^{-4}	370 GeV	3 GeV
0.75	0.95	1.6×10^{-2}	0.30	1.9×10^{-3}	270 GeV	10 GeV
0.80	1.3	6.8×10^{-2}	0.54	7.7×10^{-3}	150 GeV	22 GeV
0.90	8.4	2.6	7.0	0.27	11 GeV	61 GeV

REFERENCES

[1] T. Masakawa and H. Nakajima, Prog. Theor. Phys. 52, 1326 (1974); 54, 860 (1976); R. Fukuda and T. Kugo, Nucl. Phys. B117, 250 (1976); K. Higashijima, Phys. Rev. D29, 1228 (1984); P. Castorina and S.Y. Pi, Phys. Rev. D31, 411 (1985); R. Casalbuoni, S. De Curdis, D. Dominici and R. Gatto, Phys. Lett. 150B, 295 (1985).

[2] There is some evidence, however, that the ladder expansion may be reasonably convergent for $\alpha = O(\alpha_c)$. See T. Appelquist, K. Lane and U. Mahanta, Phys. Rev. Lett. 61, 1553 (1988).

[3] S. Dimopoulos and L. Susskind, Nucl. Phys. B155, 237 (1979); E. Eichten and K. Lane, Phys. Lett. 90B, 125 (1980). For a recent view of technicolor see R. Kaul, Rev. Mod. Phys. 55, 449 (1983).

[4] T. Appelquist, D. Karabali and L.C.R. Wijewardhana, Phys. Rev. Lett. 57, 957 (1986); T. Appelquist and L.C.R. Wijewardhana, Phys. Rev. D35, 774 (1987); Phys. Rev. D36, 568 (1987). The idea of raising the fermion condensate was first discussed by Bob Holdom [Phys. Rev. D24, 1441 (1981)]. To achieve this he imagined that the theory might have an ultraviolet stable fixed point. In an effort to understand monojet events once suggested by the UA1 group at CERN, he also examined slowly running theories [Phys. Lett. 150B, 301 (1985)] to try to produce a hierarchy between the confinement and chiral breaking scales. He noted that such theories can also enhance fermion masses.

[5] T. Appelquist, D. Carrier, L.C.R. Wijewardhana and W. Zheng, Phys. Rev. Lett. 60, 1114 (1987). Higher condensate enhancements can be obtained if one assumes the coupling to remain constant all the way to the ETC cutoff; T. Akiba and Y. Yanagida, Phys. Lett. B169, 432 (1986); K. Yamawaki, M. Bando and K. Matumoto, Phys. Rev. Lett. 56, 1335 (1986) and Phys. Lett. B178, 308 (1986).

[6] Y. Nambu and G. Jona-Lasinio, Phys. Rev. 122, 345 (1961).

[7] In this improved version of the ladder approximation vertices and propagators are dressed by taking all the relevant one loop corrections into account.

[8] Pagels and Stokar, Phys. Rev. D20, 2947 (1979). The integral is convergent if $\Sigma(p)$ falls with p as any power. If $\Sigma(p)$ falls logarithmically or does not fall at all as is the Nambu-Jona-Lasinio case, $F \sim \log \Lambda$. Thus as $\lambda \to 1$, F will increase relative to $\Sigma(0)$.

[9] T. Appelquist, M. Einhorn, T. Takeuchi and L.C.R. Wijewardhana, Physics Letters B220, 223 (1989).

[10] The lowest order expression for the β function is used here simply to illustrate the effect of different rates of running. Higher order effects are considered in Ref. 4. For the analysis of the role of a four-fermion interaction in a gauge theory with a non-running coupling, see T. Appelquist, M. Soldate, T. Takeuchi, and L.C.R. Wijewardhana, YCTP-P19-88, (Proceedings of the 12th Johns Hopkins Workshop on Current Problems in Particle Theory, Batimore, Maryland, June 8-10, 1988).

[11] This will presumably require only that the dimensionless coupling strength of the higher dimension operators, normalized to Λ, is no larger than λ.

[12] H. Georgi, Nucl. Phys. B156, 126 (1979); S. Raby, S. Dimopoulos and L. Susskind, Nucl. Phys. B169, 373 (1980).

[13] T. Appelquist, M. Einhorn, T. Takeuchi and L.C.R. Wijewardhana, Yale University preprint YCTP-P11-89, June, 1989.

[14] This point has been stressed in a recent paper by R.S. Chivukula, [Phys. Rev. Lett. 61, 2657 (1988)].

[15] For a recent review, see P. Langacker, Invited talk at the XXIV International Conference on High Energy Physics, Munich, August, 1988; and DPF-88, Storrs, Connecticut, August, 1988.

[16] It has also been pointed out that in the presence of a slowly running technicolor coupling, QCD contributions to quark-lepton mass differences can be enhanced; B. Holdom, Phys. Rev. Lett. 60, 1233 (1988). He also noted that symmetry breaking effects at the ETC scale could produce condensate hierarchies for technifermions when the gauge coupling runs slowly. Our analysis shows that this can happen for normally running theories as well, when the four fermion coupling is sufficiently strong.

THE PARTICLE-NUCLEAR INTERFACE

F. E. Close
Oak Ridge National Laboratory*
Oak Ridge, TN 37831-6373
and
University of Tennessee
Knoxville, TN 37996-1200

Once upon a time nuclear physics was the study of nucleons and pions vibrating and oscillating at the center of the atom; particle physics was the study of nucleons and pions interacting and producing resonances. From the latter, people gradually realized that hadrons are built from quarks; the fundamental rules governing their interactions were deduced ("QCD" — quantum chromodynamics) and the similarities with QED suggested the possibility that all of the natural forces can be described in a grand unified theory. Today, particle physics deals with questions ranging from the origins of matter in the first microseconds of the universe, whose experimental investigation requires the energies of the SSC, down to the quark structure of protons and nuclei that can be studied at relatively low energies.

Nuclear physics theory has taken QCD on board. There are still detailed studies going on in nuclear excitations and there are important overlaps with nuclear astrophysics. The field is very rich. The attempt to understand nuclear phenomena at the quark level causes many nuclear physicists to be concerned with the same sort of problems as their colleagues in particle physics.

During the last three years there has been a blossoming in this overlap area which I would like to call "hadron physics". I would like to make the following definitions.

*Operated by Martin Marietta Energy Systems, Inc. under Contract DE-AC05-84OR21400 with the U.S.D.O.E.

Nuclear Physics deals with the collective properties of the many-body nucleus consisting of nucleons. Hadron physics deals with the structure and interaction dynamics of those hadrons. Particle physics uses these particles as tools with which to elucidate deeper truths, seeking the origins of matter, the nature of mass, and the several other parameters which presently have to be invoked ad hoc (the weak mixing angles, fermion masses, etc.).

With tongue in cheek, one might contrast the two extremes. In extreme high-energy particle physics there is infinite theorizing but almost no data; in the nuclear structure field, we have copious data but no truly fundamentally useful theory. In hadron physics we have much data and the hope of confronting them with the fundamental QCD theory of interacting quarks and gluons. This is stimulating but also difficult. A major problem is confinement of colored quarks and gluons within the nucleons — this is simulated on computers where space-time is described as a discrete lattice, but its origins analytically are still rather poorly understood.

This confinement phenomenon is also the catch 22 of "deriving" nuclear physics from QCD. As Bob Jaffe once remarked, "Looking for evidence of quarks in nuclei is like looking for the mafia in Sicily: everyone knows they are there, but it's hard to find the evidence." The quarks are confined in nucleons, and so any successful description of the nuclear structure must reduce to that of quarks clustered inside individual nucleons. Thus we have first to understand the proton and neutron — hence the interest in hadron physics. There is the interesting possibility that the interactions and overlaps among closely packed nucleons in large or dense nuclei, or in high-energy collisions of heavy ions, may disturb the distributions of quarks — their spatial or momentum distributions — relative to their behavior when confined in isolated free nucleons. There are indeed hints of such behavior, e.g., the "EMC effect" where quarks in iron have a slightly different momentum distribution relative to that in the deuteron. Their mean momenta are reduced in iron and other heavy nuclei suggesting greater spatial freedom. Is this "liberation" a "cold" precursor to "hot" deconfinement? Are nucleons in nuclei physically "enlarged", or is this a manifestation of quarks exchanged between nucleons, tying the nucleus together and having more spatial mobility then when in a single free nucleon? Future experiments may help to answer these questions — questions raised, in large degree, by the underlying quark theory.

I would like now to draw some analogies between QED (electrical charges, atoms and molecules) on the one hand, and on the other, QCD (color, hadrons and nuclei).

QED	QCD
electric charge	3 colors
attraction of opposites	attraction of unlike colors
electrically zero atoms	colorless hadrons
radiation photon	radiation gluons
magnetic effects	chromomagnetic effects
hyperfine splitting $^3S_1 - {}^1S_0$	color hyperfine $m_\rho - m_\pi$, $m_\Delta - m_N$
Fermi-Breit in hydrogen	Fermi-Breit splittings in hadron spectroscopy

The similarity is such that one could rewrite Bjorken and Drell's QED text by inserting a traceless 3×3 matrix (λ of SU(3)) at the fermion gauge boson vertices and let QED become QCD (with α replaced by $\alpha_s \simeq 1/10$). However, the gluons themselves have color and so mutually interact via the color forces (contrast the photon of QED which transmits but does not directly "feel" the electromagnetic force). These new inter-gluon interactions give rise to vertices involving three or four gluons at a point and so a text on QCD requires more than just a coloring of Bjorken and Drell.

Now let's make a matrix to summarize how systems variously react to the forces.

		QED	QCD
Feel the force	Carry the charge	e^- Z^+ Na^+ Cl^-	Quarks Gluons
	Contain the charge	Atoms Molecules	Hadrons Nuclei
Do not feel the force Do not contain the charge		Neutrinos Photon	Leptons (ν,e)

Notice that the gluon and photon are in different slots. This small difference gives rise to the different long-range phenomena in QCD compared to QED (e.g. confinement versus ionization).

Within atoms and hadrons one finds analogues. The Coulomb potential of hydrogen has an analogue in quark systems: as the quarks' relative separations $r \to o$, $V(r) \sim 1/r$, but at large r, $V(r) \sim r$, presumably due to the detailed self-interactions among the gluons that are transmitting the force. In the ground state of hydrogen the magnetic interaction ("one photon exchange") splits the $^3S_1$ and $^1S_0$ levels. In the ground state quark conglomerates, the high-J combinations have enhanced masses relative to their low-J counterparts due to "one gluon exchange"; thus the $\Delta(1230)$ resonance, with $J = 3/2$, has greater mass than the $J = 1/2$ nucleon.

Now move up a layer in complexity to the world of molecules (QED) and nuclei (QCD).

At the risk of being accused of oversimplification by the atomic experts, I will divide the interatomic forces into three broad classes, then make analogy in the QCD world with interhadronic forces at the level of the quarks and gluons.

	Covalent	van der Waals	Ionic
Atoms Molecules	e^- exchange	"two photon"	Na^+Cl^-
Hadrons Nuclei	quark exchange	"two gluon"	no analogue if color confined in neutral clusters

If this was the whole story, then nuclear physics from QCD would be a rerun of molecules from QED. However, confinement of color breaks the simple analogy. The quark exchange at large distances ($>$ 1 fm) is contained within the confined packages, dominantly pions. The confinement of gluons in glueballs also breaks the analogy with van der Waals' forces. The hope that QCD would predict observable color van der Waals' forces in nuclei is most probably flawed, as the gluons will be confined within colorless glueballs. Computer simulations of QCD suggest that the lightest glueballs have masses in excess of 1 GeV and so transmit forces over much less than a nucleon radius. Thus their presence is hidden in nuclear physics.

It is an open question whether analogues of ionic forces occur in dense or hot nuclear systems; whether multiquark clusters occur within nuclei; whether quark-gluon plasma may form in hot-dense systems.

If color attractions among quarks are the source of internucleon forces, then there could exist analogous clusters of mesons — "meson molecules". The instability of most mesons prevents formation of these systems, but π, K, η are stable on the time scales of the strong interactions and may have the chance to bind. Indeed Weinstein and Isgur find that such attractions occur in S-wave. The $\pi\pi$ system has a strong enhancement above $2m_\pi$ which may be manifested in the $\psi \to \omega\pi\pi$ dipion spectrum. The KK system binds forming nearly degenerate I = 0, 1 systems 10 MeV below $2m_K$. The S^*(975 MeV) and δ(980 MeV), scalar "mesons", thus appear to be meson molecules; meson analogues of the I=0 dueteron (whose I=1 partner is above $2m_N$).

The color attractions among quarks and gluons lead to the prediction of glueballs and hybrid hadrons — the latter where gluons play a dynamical role, attracted to quarks to from hybrid mesons and baryons.

The problem in predicting the masses of these states is that we have to simulate the effects of confinement. Perhaps the simplest way of doing this is to suppose that the constituent quarks or gluons are free until they hit an infinitely high wall. This is the essence of cavity or bag models. Confine a massless J=1/2 quark in a radius, R, and it gains an energy that scales as 1/R. This energy becomes of the order of 350 MeV if R is of the order of the proton radius, hence the proton mass may be modeled. For gluons, one solves the eigenvalue equations for J = 1 rather than J = 1/2 confined fields. There are electric or magnetic modes (actually TE and TM in the language of classical electrodynamics) with different eigenvalues. If R is the same as for quark systems, the typical confined-mass-scale is some 500 MeV per TE mode and 750 MeV per TM mode. Thus follows the prediction that the lightest systems consisting of at least two confined gluons weigh in at O(1 GeV) and that the lightest hybrid baryons weigh in at O(1.5 GeV). A problem is that as soon as the hyperfine shifts in energy are taken into account (this involves one first calculating the propagators of confined quarks and gluons), the lowest spin-J systems are pulled down significantly in mass. The lightest hybrid baryon might thus appear to have a mass near that of the proton which suggests either a profound rethink of baryon spectroscopy or that we have unearthed a naievity.

I suspect it is the latter. No one yet has convincingly set up a study of loop effects with renormalization within a cavity. These loop diagrams enter at the same order in perturbation theory to which the hyperfine shifts have been calculated and may alter the naive "effective" energies per confined gluon. In the case of quarks, their effects were subsumed in the MIT bag by an input mass parameter for the quark; this mass fitted to the overall mass scale of the spectroscopy. In the gluonic sector we have no mass scale to set the scale, and until we make sense of the (infinite!) self-energy diagrams, we cannot predict the absolute scale. So the mass separations among the various states may be reliable, but the absolute mass scale is beyond present analysis. To predict the masses of glueballs and hybrid hadrons, we have to resort to computer simulations — lattice QCD. This has proved to be a harder task than was originally thought.

The eventual discovery of the gluonic spectroscopy may give important insights into the nature of confinement of gluons. If lattice calculations, including quarks and gluons (to date, people work in the "quenched" approximation, which roughly translated means "ignore the quarks") merely point out masses of states that correspond to the particle data tables, we will confirm QCD but may still require much study to elucidate the analytic dynamics of confinement. The main outcome of such a success may be the advances that will have come in the art of computation and design of machines. Thus the significant questions posed by hadron physics are having a spinoff in the intellectual stimulation they provide to computational science and, in turn, the subsequent ability to encode problems in field theory, condensed matter, and other areas of science. I am reminded of the title of Tony Hey's talk at a recent meeting of the British Association for the Advancement of Science, and it provides an apt one-line summary of the multidisciplinary efforts flowing from computation at the nuclear-particle interface. It was: "Quarks, Supercomputers, and Oil Prospecting".

THE BASIS SPLINE METHOD AND ASSOCIATED TECHNIQUES

C. Bottcher and M. R. Strayer

Physics Division, Oak Ridge National Laboratory

Oak Ridge, TN 37831

ABSTRACT

We outline the Basis Spline and Collocation methods for the solution of Partial Differential Equations. Particular attention is paid to the theory of errors, and the handling of non-self-adjoint problems which are generated by the collocation method. We discuss applications to Poisson's equation, the Dirac equation, and the calculation of bound and continuum states of atomic and nuclear systems.

We can forgive a man for making a useful thing
as long as he does not admire it.
The only excuse for making a useless thing
is that one admires it intensely.
All art is quite useless.
OSCAR WILDE *(The Picture of Dorian Gray)*

1. THE BASIS SPLINE AND COLLOCATION METHODS

1.1 Basis Splines and their Generation

Splines of order $\mathcal{N}$ are functions $S^{\mathcal{N}}(x)$ of a single real variable belonging to the class $C^{\mathcal{N}-2}$ with continuous $(\mathcal{N}-2)^{\text{th}}$ derivatives. Each spline is associated with a set of points $\{x_k\}$, called knots; we take the knots to be distinct and ordered, $x_k < x_{k+1}$. Between each pair of knots, the spline is a polynomial of degree $\mathcal{N}-1$ (the order refers to the number of coefficients); at each knot, the function and derivatives up to the $(\mathcal{N}-2)^{\text{th}}$ are continuous. The $(\mathcal{N}-1)^{\text{th}}$ derivative is bounded but discontinuous. We shall consider only orders $\mathcal{N} \geq 3$, though piecewise linear or tent functions ($\mathcal{N}=2$) were useful at an earlier stage in development of numerical analysis, particularly as the simplest implementation of the finite element method.

Splines have long been used for curve fitting, but the introduction of basis splines converted the concept into a power technique. Given a set of knots $\{x_k\}$, the basis splines of order $\mathcal{N}$ are a set of functions $B_k^{\mathcal{N}}(x)$ such that any spline $S^{\mathcal{N}}(x)$ is identically a linear superposition

$$S^(x) \equiv \sum_k a_k B_k^{\mathcal{N}}(x) \, . \tag{1}$$

It is easy to see that $B_k^{\mathcal{N}}(x)$ is uniquely defined by the condition that it is zero outside the range of $\mathcal{N}+1$ consecutive knots $x_k, x_{k+1}, \ldots, x_{k+\mathcal{N}}$. Such a function of order $\mathcal{N}=3$ is illustrated in Fig. 1. At either end of the range of knots, the basis splines can be modified to incorporate boundary conditions. As an example, Fig. 2 shows how a set of splines can be fitted together to represent functions

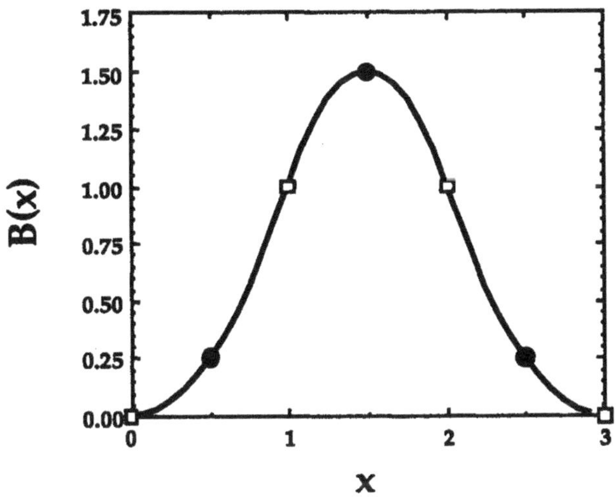

Fig. 1. Basis spline of order $\mathcal{N} = 3$. The knots are denoted by open squares, and the collocation points by filled circles.

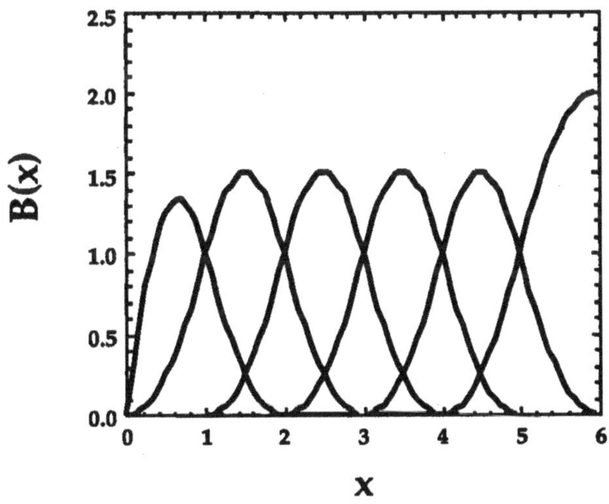

Fig. 2. Basis splines of order 3 satisfying the boundary conditions (2).

satisfying the boundary conditions

$$\psi(x_1) = 0, \quad \psi'(x_N) = 0 \; , \tag{2}$$

where the prime denotes differentiation. Several algorithms are available to construct basis splines from the continuity conditions at the knots, and the boundary conditions at the endpoints [1,2,4,5], and we refer the reader to the literature rather than pursue this rather technical aspect of the subject here.

Given the basis splines, interpolation is straightforward. A function $\psi(x)$ is approximated by the interpolant $\hat{\psi}$, where

$$\hat{\psi}(x) = \sum_{k=1}^{N} \psi^k B_k(x) \; , \tag{3}$$

and the ψ^k are determined if $\hat{\psi}(x) \equiv \psi(x)$ at a set of N data, or *collocation*, points $\{\xi_\alpha\}$. We have only to solve

$$\sum_{k=1}^{N} B_{\alpha k} \psi^k = \psi_\alpha \; , \tag{4}$$

where $B_{\alpha k} = B_k(\xi_\alpha)$ and $\psi_\alpha = \psi(\xi_\alpha)$; the order $\mathcal{N}$ is omitted for simplicity. It is necessary and sufficient that the matrix $\mathbf{B}$ be non-singular: this will always be the case if we limit our considerations to odd orders, and choose the collocation points halfway between the knots. Further, as is clearly seen in Fig. 1, the elements of $\mathbf{B}$ are positive and largest on the diagonal, implying that the inversion of (4) is very stable. It is convenient to have a notation for the inverse of $\mathbf{B}$: we write the elements as $B^{k\alpha}$ so that

$$\sum_{k=1}^{N} B^{k\alpha} \hat{\psi}^\alpha = \psi^k \; . \tag{5}$$

Formulae for derivatives and integrals are readily derived from (3)-(5) [4].

1.2 Collocation Methods

The collocation method for solution of an operator equation, formally written as

$$L[\psi] = 0 , \tag{6}$$

is intimately related to the interpolation procedure of (3) and (4). We obtain N equations for the unknowns $\{\psi^k\}$ by requiring that

$$L[\sum \psi^k B_k] = 0 \text{ at } x = \xi_\alpha . \tag{7}$$

Though the method is perfectly applicable to non-linear problems, we are mostly concerned with linear operators L, in which case (7) becomes

$$\sum_{k=1}^{N} \psi^k L[B_k]_{x=\xi_\alpha} = 0 . \tag{8}$$

If we use (4) and (5) to eliminate the coefficients $\{\psi^k\}$ in favour of the values of the solution $\{\psi^k\}$ at the collocation points,(8) is replaced by

$$\sum_{\alpha=1}^{N} L_{\beta\alpha}\psi_\alpha = 0 , \tag{9}$$

where

$$L_{\beta\alpha} = \sum_{k=1}^{N} B^{k\alpha} L[B_k]_{x=\xi_\beta} = 0 . \tag{10}$$

As a simple example, consider the non-relativistic Schrödinger equation in one dimension,

$$(T+V)\psi = \left[-\frac{1}{2}\frac{d^2}{dx^2} + V(x)\right] = 0 , \tag{11}$$

where V is a local potential. An agreeable feature of the method is that expressions such as $V\psi$ are replaced by $V_\alpha\psi_\alpha$, where $V_\alpha = V(\xi_\alpha)$. In other words, local operators come to be represented simply as diagonal matrices of their values at the collocation points. Non-local operators, notably the kinetic energy, appear complicated at first sight. In summary, (11) is replaced by

$$\sum_{\beta=1}^{N}(T_{\alpha\beta} + V_\alpha\delta_{\alpha\beta})\psi_\beta = E\psi_\alpha \ , \tag{12}$$

where

$$T_{\alpha\beta} = -\frac{1}{2}\sum_{k=1}^{N} B''_{\alpha k}B^{k\beta} \ ,$$
$$B''_{\alpha k} = \left[\frac{d^2 B_k}{dx^2}\right]_{x=\xi_\alpha} \ . \tag{13}$$

We shall see in Section 2 that the analysis of (6)-(13) is readily extended to problems in two and three dimensions.

1.3 Completeness and Convergence

Given a formal interpolation procedure, such as that implied by (3)- (5), the issue of completeness refers to how well the interpolant $\hat{\psi}(x)$ represents $\psi(x)$ at points other than the exactly fit points $\{\xi_\alpha\}$. In other words, how does the error

$$\mathcal{E}(x) = |\psi(x) - \hat{\psi}(x)| \tag{14}$$

behave globally? A simple estimate is obtained from the Dirichlet functions $B_k(x)$, which arise when $\hat{\psi}$ is expressed directly in terms of $\{\psi_\alpha\}$. From (3)-(5), we find that

$$\hat{\psi}(x) = \sum_{k=1}^{N}\psi_\alpha\hat{B}_\alpha(x) \ ,$$

$$\hat{B}_\alpha(x) = \sum_{k=1}^{N} B^{k\alpha} B_k(x) \,. \tag{15}$$

A typical Dirichlet function is shown in Fig. 3. It is evident that $\hat{B}_\alpha(\xi_\beta) = \delta_{\alpha\beta}$ by construction, but at other points, $\hat{B}_\alpha$ is non-zero and oscillating in sign. The values of $|\hat{B}_\alpha|$ at successive turning points ξ'_β fall off rapidly, in fact approximately as

$$0.3 \exp\left[-\frac{(\xi_\alpha - \xi'_\beta)^2}{2\mathcal{N}\Delta^2}\right],$$

for knots with a uniform spacing Δ [4]. At the first minimum, however, $\hat{B}_\alpha \simeq -0.3$. Thus the maximum of the error function (14) is estimated to be

$$\mathcal{E}_{\text{max}} \simeq 0.3 \max_\alpha |\psi_{\alpha+1} - \psi_\alpha| \,. \tag{16}$$

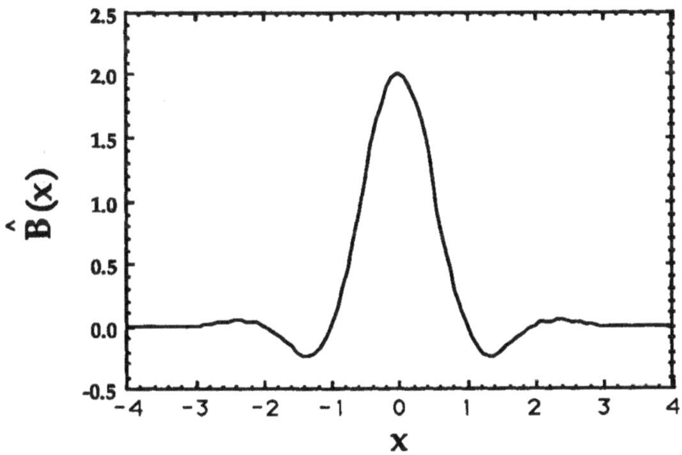

Fig. 3. Dirichlet function of order 3: $\hat{B}(x)$ is defined in (15).

A large error is encountered if ψ has a discontinuity, as seen in Fig. 4, which depicts the result of trying to fit a step function. However, in virtue of the rapid rate of falling off of $\hat{B}_\alpha(x)$ with $|x - \xi_\alpha|$, the overshooting oscillations in the fit also die off rapidly with distance from the discontinuity. This contrasts with the notorious behaviour of Fourier or orthogonal polynomial interpolation.

When splines are used to fit smooth functions (of class C^M, $M > 1$, say), the solutions to operator problems usually converge uniformly as the number of points is increased. Uniformity implies, in quantum mechanical problems such as (11), that the wavefunction converges at the same rate as the eigenvalue. A further point, of great practical significance, is that the rate of convergence improves as the order of the splines increases: usually $\mathcal{E} \sim \Delta^{-\mathcal{N}+1}$.

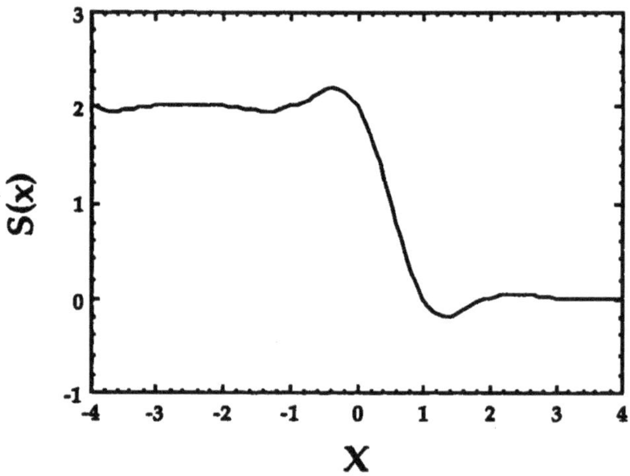

Fig. 4. Result of fitting a step function $S(x)$ with splines of order 3.

This is illustrated by Fig. 5 which shows the errors in the lowest eigenvalue of (11), where V is chosen to be a Morse potential [4]. Usually $\mathcal{N} = 7$ is adequate for practical purposes. Though spline interpolation remains stable for any order, a trend of diminishing returns is usually observed for $\mathcal{N} > 7$; for increasing $\mathcal{N}$ more splines may be needed to accommodate the boundary conditions instead of representing the solution.

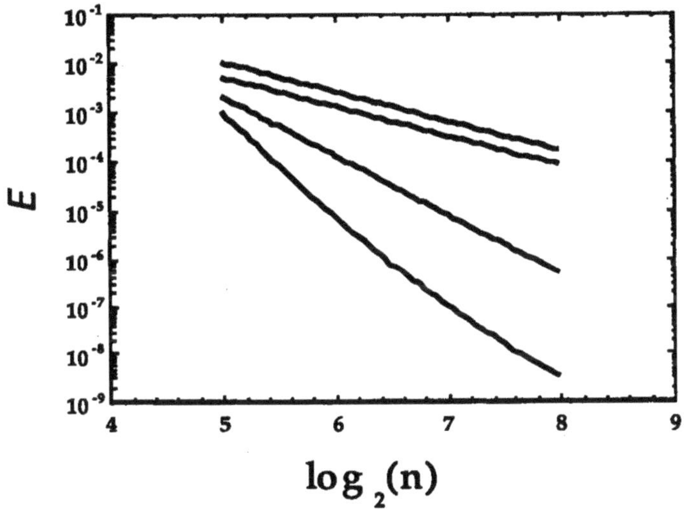

Fig. 5. Errors in the lowest eigenvalue of a Morse potential *vs.* the number of grid points n for different discretization algorithms. Reading from the top down, the curves refer to the 3-point finite difference method, and the basis spline-collocation method with splines of orders 3, 5 and 7 respectively. The model is described in more detail in Ref. [4].

1.4 Techniques for Non-Self Adjoint Problems

The most general collocation-variational method for the linear operator equation $L\psi = 0$ is obtained by expanding ψ in a righthand basis u_k, and projecting the equation onto a lefthand basis v_α,

$$\sum_k (v_\alpha, Lu_k)\psi^k = 0\,. \tag{17}$$

The collocation method results from the choice $v_\alpha = \delta(x - \xi_\alpha)$, and the self-adjoint variational method from the choice $v_\alpha = u_\alpha$. For a general choice of v_α, including that of the collocation method, (17) is not self adjoint. Though problems which are not self adjoint are not usually emphasised in textbooks, they pose no great difficulties. In some branches of physics, such as transport phenomena and quantum optics, non-self-adjoint formulations may be more natural.

We are interested in the stationary eigenvalue problem of quantum mechanics, $L = H - E$, for which the collocation procedure results in a non-self-adjoint matrix problem,

$$\mathbf{H}\psi^{(\lambda)} = \epsilon^{(\lambda)}\psi^{(\lambda)} ,$$

$$\mathbf{H}^\dagger\phi^{(\lambda)} = \epsilon^{(\lambda)}\phi^{(\lambda)} , \tag{18}$$

where the right and lefthand eigenvectors are biorthogonal $\phi_\lambda^\dagger\psi_\mu = \delta_{\lambda\mu}$. If we define a matrix $\boldsymbol{\Psi}$ whose elements are $\Psi_{\beta\lambda} = \psi_\beta^{(\lambda)}$, a similar matrix $\boldsymbol{\Phi}$, and a diagonal matrix of the eigenvalues $\boldsymbol{\epsilon}$, (18) becomes

$$\mathbf{H}\boldsymbol{\Psi} = \boldsymbol{\Psi}\boldsymbol{\epsilon}, \;\; \mathbf{H}^\dagger\boldsymbol{\Phi} = \boldsymbol{\Phi}\boldsymbol{\epsilon} . \tag{19}$$

The two equations (19) are only consistent if

$$\boldsymbol{\Psi}^{-1} = \boldsymbol{\Phi}^\dagger, \;\; \boldsymbol{\Phi}^{-1} = \boldsymbol{\Psi}^\dagger . \tag{20}$$

It is necessary and sufficient that the eigenvectors be linearly independent. From (19) and (20) the reconstruction of the matrix from its eigenvectors is given by

$$\mathbf{H} = \boldsymbol{\Psi}\boldsymbol{\epsilon}\boldsymbol{\Psi}^{-1} = \boldsymbol{\Psi}\boldsymbol{\epsilon}\boldsymbol{\Phi}^\dagger . \tag{21}$$

The significance of (21) is that H may be factored into a self-adjoint operator $\hat{\mathbf{H}}$, and a positive, definitive self adjoint operator $\mathbf{S}^{-1}$,

$$\mathbf{H} = \mathbf{S}^{-1}\hat{\mathbf{H}} . \tag{22}$$

where

$$\hat{\mathbf{H}} = \mathbf{\Phi}\epsilon\mathbf{\Phi}^{\dagger}, \quad \mathbf{S} = \mathbf{\Psi}\mathbf{\Psi}^{\dagger} . \tag{23}$$

If the eigenvectors are linearly independent, $\mathbf{S}$ is non-singular and can be interpreted as an overlap matrix between a set of linearly independent basis functions. The factorization (23) converts (19) into a *generalized eigenvalue problem*,

$$\mathbf{H\Psi} = \mathbf{S\Psi}\epsilon . \tag{24}$$

The factorization (22) is necessary and sufficient for $\mathbf{H}$ to possess a real spectrum.

It is instructive to pursue the physical interpretation of the lefthand eigenvectors. Given (18), the only consistent prescription for the expectation value of a local operator $a(x)$ in the eigenstate λ is

$$< a >= \boldsymbol{\phi}^{(\lambda)\dagger}\mathbf{a}\boldsymbol{\psi}^{(\lambda)} = \sum_{\alpha} \phi_{\alpha}^{(\lambda)*} a(\xi_{\alpha}) \psi_{\alpha}^{(\lambda)} . \tag{25}$$

This agrees intuitively with the usual formula of wave mechanics,

$$< a >= \int dx \psi(x)^* a(x) \psi(x),$$

if we identify

$$\psi_{\alpha}^{(\lambda)} \longrightarrow \psi(\xi_{\alpha}), \quad \phi_{\alpha}^{(\lambda)} \longrightarrow W_{\alpha}\phi(\xi_{\alpha}) , \tag{26}$$

and $\{W_\alpha\}$ are the quadrature weights associated with the points $\{\xi_\alpha\}$. The identification (26) suffers from the drawback that the weights have a dependence on λ and do not coincide identically with the conventional definition of numerical analysis,

$$W_\alpha = \int dx \hat{B}(x) \, .$$

Nonetheless (26) can sometimes be used to good effect, as we shall see in Sections 1.5 and 2.2.

1.5 Representation Theory

The concept of a *faithful* representation of a differential operator is taken over from the mathematical literature [1]. For our purposes, it means that the representation faithfully simulates the familiar intuitive properties of the differential calculus. No approximation can reproduce all properties exactly, but we can base our development on the following choice. Given the operator $D = d/dx$, we require that its matrix representation $\mathbf{D}$ satisfies a subset of identities

$$\mathbf{D}\boldsymbol{\chi}_M = \boldsymbol{\chi}_{M-1} \, , \tag{27}$$

analogous to those for differentiating the monomials

$$\chi_M(x) = \frac{x^M}{M!} \, . \tag{28}$$

Insofar as the basis splines belong to the class $C^{\mathcal{N}-2}$, (27) holds identically for $M \leq \mathcal{N} - 2$. In general, the identity is modified near the boundaries, since the polynomial representation must be constrained to accommodate the boundary conditions.

We remarked above that differential operators in the basis-spline-collocation representation had a rather recondite appearance. However, when the numerical presentation is examined, it becomes obvious that the operators have the qualitative and intuitive structure expected from elementary finite difference considerations. An example will make the point more clearly. If the second derivative operator is constructed as in (13), with periodic boundary conditions on a mesh with unit spacing and spline order $\mathcal{N} = 3$, the elements of the resulting matrix are given by

$$D^{(2)}_{\alpha,\alpha+\mu} = (\cdots, -0.0049, 0.0286, -0.1665, 0.9705, -1.657,$$
$$0.9705, -0.1665, 0.0286, \cdots) .$$

It is easy to see that the sum across a row is zero to good accuracy, and that the pattern of elements resembles the finite difference representation

$$D^{(2)}_{\alpha,\alpha+\mu} = (\cdots, 0, 1, -2, 1, 0, \cdots) .$$

These properties of $D^{(2)}$ make it a faithful representation of d^2/dx^2.

To develop these ideas more quantitatively, consider the simplest representation of a derivative, by a single backward difference,

$$\mathbf{D} = \begin{bmatrix} 1 & 0 & 0 & 0 & \cdots & & \\ -1 & 1 & 0 & 0 & \cdots & & \\ 0 & -1 & 1 & 0 & \cdots & & \\ & & & & \cdots & 1 & 0 \\ & & & & \cdots & -1 & 1 \end{bmatrix} . \tag{29}$$

This representation satisfies an identity of the form (27),

$$\mathbf{D}\boldsymbol{\chi}_0 = \mathbf{e} , \tag{30}$$

where $\chi_{0\alpha} = 1 \; \forall \alpha$ and $e_\alpha = \delta_{\alpha,1}$. In general $\boldsymbol{\chi}$ is the eigenvector of the modified operator $\tilde{\mathbf{D}}$ corresponding to a zero eigenvalue, where

$$\tilde{D}_{\alpha\beta} = D_{\alpha\beta} - \delta_{\alpha,1}\delta_{\beta,1} . \tag{31}$$

The accompanying righthand eigenvector $\boldsymbol{\omega}_0$ defines a set of quadrature weights in accordance with the prescription (26),

$$\boldsymbol{\omega}_0^\dagger \mathbf{D} = \mathbf{e}^\dagger , \tag{32}$$

where $\boldsymbol{\omega}_0$ is so normalized that $\omega_{01} = 1$. The significance of (32) is seen by applying both sides to an arbitrary vector $\boldsymbol{\psi}$,

$$\boldsymbol{\omega}_0^\dagger \mathbf{D} \boldsymbol{\psi} = \psi_1, \tag{33}$$

which is the familiar formula for the integral of a derivative. Higher moments are obtained by repeatedly operating on $\mathbf{e}$ with $\mathbf{D}^{-1}$,

$$\boldsymbol{\chi}_M = \mathbf{D}^{-M-1}\mathbf{e} . \tag{34}$$

In summary, the formal requirement of fidelity (30) leads to a quadrature formula which is identically inverse to differentiation on the lattice. This, in turn, leads to an analogue of Green's lemma as an identity on the lattice, and finally to exact conservation laws on the lattice [4].

2. APPLICATIONS

2.1 Atomic and Nuclear Collision Problems

One of the most powerful techniques for studying atomic and nuclear collisions is the numerical Time-Dependent Hartree-Fock (TDHF) method. This amounts to solving a set of partial differential equations in three spatial dimensions, in addition to time. Most current work is based on cartesian, rather than curvilinear coordinates. For a system with A independent particle orbits, the orbital wavefunctions satisfy

$$H\psi_K = i\frac{\partial\psi_K}{\partial t} \quad (K = 1,\ldots,A)\,,$$
$$H = T_x + T_y + T_z + V(x,y,z|t)\,, \tag{35}$$

where

$$T_x = -\frac{1}{2}\frac{\partial^2}{\partial x^2}\,,\quad \ldots$$

The potential energy is the sum of the self-consistent field $\mathcal{V}$, and an external time-dependent part U, which in atomic physics would be provided by the nuclear motion. The self-consistent field $\mathcal{V}$ depends on the orbitals through Poisson's equation,

$$\nabla^2 \mathcal{V} = -4\pi \sum_{K=1}^{A} n_K |\psi_K|^2 , \tag{36}$$

where n_K is the occupancy of the Kth orbital. Thus the equations are coupled through nonlinear interactions, though the nonlinearity is very weak in atomic physics. The foregoing presentation is rather schematic — the reader is referred to the literature [6,7,8,9] for an explanation of the complications arising from spin and the Pauli exclusion principle.

The coupled set (35) and (36) are ideally suited to the basis spline-collocation approach. The procedure of (7) is generalized by expanding the wavefunction in products

$$\psi(x, y, z) = \sum_{ijk} \psi^{ijk} B^i(x) B^j(y) B^k(z) . \tag{37}$$

For simplicity of exposition, we assume that the same set of splines is associated with each coordinate. In collocation space, we find that the representation of the Hamiltonian analogous to (12) is given by

$$\begin{aligned} H_{\alpha\beta\gamma,\alpha'\beta'\gamma'} = {} & T^x_{\alpha,\alpha'} \delta_{\beta,\beta'} \delta_{\gamma,\gamma'} + T^y_{\beta,\beta'} \delta_{\gamma,\gamma'} \delta_{\alpha,\alpha'} + T^z_{\gamma,\gamma'} \delta_{\alpha,\alpha'} \delta_{\beta,\beta'} \\ & + V(\xi_\alpha, \xi_\beta, \xi_\gamma) \delta_{\alpha,\alpha'} \delta_{\beta,\beta'} \delta_{\gamma,\gamma'} , \end{aligned} \tag{38}$$

where $T^x_{\alpha,\alpha'}$ is given by (13).

The equations (35) now have the form

$$i \frac{\partial}{\partial t} \psi = \mathbf{H} \psi , \tag{39}$$

where $\mathbf{H}$ is the sum of block matrices of a rather simple form. Thus algorithms which decompose into canonical operations of the form

$$\mathbf{H}\psi \longrightarrow \psi' , \tag{40}$$

can be implemented efficiently on supercomputers with vector and parallel capabilities. Such an algorithm is used for time propagation. If time is divided into small steps τ,

$$\psi(t+\tau) = \exp[-i\mathbf{H}(t+\tau/2)]\psi(t) , \tag{41}$$

with an error of order $\mathcal{O}(\tau^3)$. The exponential can be expanded in a power series using only operations of the form (40).

Propagation in time is only one aspect, and probably the simplest, of the complete solution of a collision problem. In the following sections, we treat the other aspects in turn:

- Solution of Poisson's equation (36).
- Construction of initial conditions by solution of the stationary eigenvalue problem.
- Extraction of continuum amplitudes by projection on final states.

2.2 Poisson's Equation

It is convenient to rewrite Poisson's equation (36) as

$$(\epsilon - T_x - T_y - T_z)\phi = 2\pi\rho , \tag{42}$$

where T_x is the kinetic energy operator associated with the x-coordinate, as before, and ϵ is a small positive quantity. This equation is to be solved in a finite domain

Ω, bounded by a closed surface Σ, with the boundary condition that the normal derivative is known on Σ,

$$\frac{\partial \phi}{\partial n} = g(S), \quad S \in \Sigma \, . \tag{43}$$

The function $g(S)$ is usually constructed from a multipole expansion. It is important that (42) and (43) be consistent in the sense of satisfying Gauss's theorem for the total charge within Ω,

$$Q = \int_\Omega \rho d\tau = \int_\Sigma g dS \, . \tag{44}$$

To incorporate the boundary condition (43), we transform the source in (42),

$$(\epsilon - T_x - T_y - T_z)\hat{\phi} = 2\pi\hat{\rho} = 2\pi[\rho - g(S)\delta(S)] \, , \tag{45}$$

so that the righthand side integrates to zero over Ω. Introducing a basis spline expansion, (45) can be discretized by the methods of earlier sections,

$$(\epsilon - \mathbf{T}_x - \mathbf{T}_y - \mathbf{T}_z)\hat{\boldsymbol{\phi}} = 2\pi\hat{\boldsymbol{\rho}} \, , \tag{46}$$

The boundary delta-function $\delta(S)$ is written as a sum of Kronecker deltas with appropriate weights. The old and new solutions, ϕ and $\hat{\phi}$, differ only by a constant in their exact (or continuum) representation. However in their matrix representation on a lattice, the relationship is more recondite. Suppose there exists a pair of left and right eigenvectors corresponding to a zero eigenvalue of $T = T_x + T_y + T_z$,

$$(\epsilon - \mathbf{T})\boldsymbol{\theta} = 0, \quad , \tilde{\boldsymbol{\theta}}^\dagger(\epsilon - \mathbf{T}) = 0, \, . \tag{47}$$

Then $\boldsymbol{\phi}$ and $\hat{\boldsymbol{\phi}}$ differ by a multiple of $\boldsymbol{\theta}$, and (46) has a solution only if

$$\tilde{\theta}^{\dagger}\hat{\rho} = 0\,. \tag{48}$$

According to the principles of Section 1.5, in particular (26), the elements of θ are quadrature weights, and (48) is the lattice form of Gauss's theorem. The existence of a zero eigenvalue thus guarantees a well-behaved solution. A zero eigenvalue mandates a zero gradient on the boundary: periodic boundary conditions or $\partial\psi/\partial n = 0$ are acceptable, but $\psi = 0$ is not.

The eigenvalue spectrum of $\mathbf{T}$ is readily constructed from that of $\mathbf{T}_x$,

$$\mathbf{T}_x \boldsymbol{\eta}_k = \tau_k \boldsymbol{\eta}_k, \;\; \mathbf{T}_x^{\dagger} \tilde{\boldsymbol{\eta}}_k = \tau_k \tilde{\boldsymbol{\eta}}_k\;, \tag{49}$$

with the identifications,

$$\tau_0 = 0, \;\;, \eta_{0\alpha} = 1, \;\;, \tilde{\eta}_{0\alpha} = W_\alpha, \;. \tag{50}$$

We can transform the source and solution of (45) to the new basis provided by (50),

$$\begin{aligned} \hat{\rho}_{\alpha\beta\gamma} &= \sum_{ijk} \eta_{i\alpha}\eta_{j\beta}\eta_{k\gamma}\hat{\rho}_{[ijk]}, \\ \hat{\phi}_{\alpha\beta\gamma} &= \sum_{ijk} \eta_{i\alpha}\eta_{j\beta}\eta_{k\gamma}\hat{\phi}_{[ijk]}\;. \end{aligned} \tag{51}$$

In this representation, the solution of (45) is simply,

$$\hat{\phi}_{[ijk]} = (\epsilon - \tau_i - \tau_j - \tau_k)^{-1}\hat{\rho}_{[ijk]}\;. \tag{52}$$

It follows from (48) and (51) that $\hat{\rho}_{[000]} = 0$. Thus (51) and (52) define the solution within a constant, which can be fixed by invoking the multipole expansion again.

2.3 Bound State Eigenfunctions and Continuum State Amplitudes

As stated above, we require a means of setting up the initial conditions of the scattering problem (35), as expressed in matrix form (39). These usually state that ψ is the stationary ground state satisfying

$$\mathbf{H}\psi_0 = E_0\psi_0 \,, \tag{53}$$

with a similar equation for the adjoint vector ϕ_0. The matrix $\mathbf{H}$ is usually too large to store explicitly in core, so we seek iterative methods which require only "operator knowledge", meaning the implicit ability to carry out the canonical operation (40). Such a technique is the *damped relaxation* method, described in detail in a recent paper [10]. The ground state eigenvector of the Schrödinger equation is the limit of the iterates $\psi^{(I)}$, generated by

$$\psi^{(K+1)} - \psi^{(K)} = v\mathbf{R}[\mathbf{H} - E^{(I)}] \,, \tag{54}$$

where $E^{(I)} = \psi^{(K)\dagger}\mathbf{H}\psi^{(K)}$, and $\mathbf{R}$ is a relaxation operator, designed to filter out the high-frequency components of $\psi^{(K)}$. A convenient choice for $\mathbf{R}$ is

$$\mathbf{R} = \left(1 + \frac{\mathbf{T}_x}{\mu}\right)^{-1} \left(1 + \frac{\mathbf{T}_y}{\mu}\right)^{-1} \left(1 + \frac{\mathbf{T}_z}{\mu}\right)^{-1} . \tag{55}$$

An extensive theory has been developed on the optimum choice of μ and v to guarantee rapid convergence.

Excited states ψ_m can be computed by enforcing the constraint of orthogonality to lower lying states. Then bound state amplitudes can be extracted from the final state wavefunction $\psi(T)$ by projection,

$$P_m = |\phi_m^\dagger \psi(T)|^2 \,. \tag{56}$$

Amplitudes may be calculated in a frame moving with one of the nuclei by using the eigenstates of a translated Hamiltonian

$$H' = H - \vec{v} \cdot \vec{p} \,, \tag{57}$$

where $\vec{v}$ is the nuclear velocity.

Direct calculation of every eigenvector is not practical for continuum amplitudes, so we use another technique, that of the *Gaussian filter* [11,12]. Suppose we expand

$$\psi(T) = \int dE \sum_\lambda A_\lambda(E) \Psi_\lambda(E) \,, \tag{58}$$

where λ distinguishes substates of the same energy. The amplitudes can be picked out of (58) by operating on ψ with the filter

$$F(E, \Delta) = (\sqrt{\pi}\Delta)^{-1/2} \exp\left[\frac{(E-H)^2}{2\Delta^2}\right] . \tag{59}$$

Defining $\theta = F(E, \Delta)\psi(T)$, it is easily shown that for moderately small Δ

$$\sum_\lambda |A_\lambda(E)|^2 \simeq \theta^\dagger \theta \,. \tag{60}$$

With some refinements, it is possible to extract other information, such as angular distributions.

In more recent work [12], we do not use the Gaussian (59) as such, but replace the exponential by a rational function

$$\exp(-Z) \longrightarrow \left(1 + \frac{Z}{M}\right)^{-M} ,$$

where the inversion is performed by damped relaxation.

2.4 The Dirac Equation

Discrete representations of the Dirac equation are plagued by the pathology of "fermion doubling", namely the appearance of high momentum components at low energies. We briefly indicate how this problem is avoided in the basis spline-collocation method [4].

It is sufficient to consider the continuum states of a free particle in one dimension. The upper and lower components g, f satisfy

$$\begin{aligned} g - f' &= Eg \\ f + g' &= Ef \, . \end{aligned} \tag{61}$$

The pathology arises when (61) is discretized on a uniform mesh with the prescription

$$g'_n \simeq \frac{g_{n+1} - g_{n-1}}{2\Delta x} \, , \tag{62}$$

manifesting itself in a double valued dispersion relation, as illustrated in Fig. 6. If by contrast we use forward and backward differences

$$f'_n \simeq \frac{f_{n+1} - f_n}{\Delta x}, \quad g'_n \simeq \frac{g_n - g_{n-1}}{\Delta x} \, , \tag{63}$$

a single valued relation is recovered.

In the basis spline-collocation method, we generalize this procedure by factorizing the second derivative representation

$$\mathbf{D}^{(2)} = \mathbf{D}_- \mathbf{D}_+ \, , \tag{64}$$

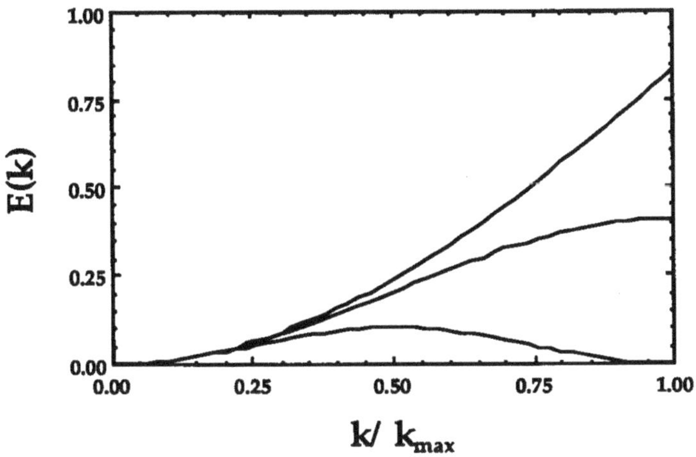

Fig. 6. Energy-momentum dispersion relation $E(k)$ for a free particle obeying the one dimensional Dirac equation. Reading from the top down, the curves refer to the exact continuum result, discretization using (62), and using (63).

and replacing

$$f' \longrightarrow \mathbf{D}_+\mathbf{f}, g' \longrightarrow \mathbf{D}_-\mathbf{g} .$$

This procedure can be justified from fidelity arguments [4]. If $\mathbf{D}^{(2)}$ satisfies an identity of the form (30), then (64) is the only factorization leading to operators which satisfy similar identities.

3. ACKNOWLEDGEMENTS

This research was sponsored by the Division of Nuclear Physics of the U.S. Department of Energy under contract No. DE-AC05-84OR21400 with Martin Marietta Energy Systems, Inc., and by the Division of Chemical Sciences, Office of Basic Energy Sciences.

REFERENCES

1. C. De Boor, *Practical Guide to Splines*, (Springer-Verlag, New York, 1978).

2. G. I. Marchuk, *Methods of Numerical Mathematics*, (Springer-Verlag, New York, 1975). (Springer-Verlag, New York, 1984).

4. C. Bottcher and M.R. Strayer, Ann. Phys. (N.Y.) **175**, 64 (1987).

5. A.S. Umar, J.S. Wu, M.R. Strayer and C. Bottcher, J. Comp. Phys., in press (1990).

6. A. S. Umar, M. R. Strayer, P. -G. Reinhard, K. T. R. Davies, and S. -J. Lee, Phys. Rev. C **40**, (1989); and references therein.

7. C. Bottcher, Invited Papers of XIII ICPEAC (North- Holland, 1984), p. 187.

8. C. Bottcher, N.I.M. Phys.Res. **B11**, 7 (1985).

9. M.S. Pindzola, G.J. Bottrell and C. Bottcher, J. Opt. Soc. *B*, in press (1990).

10. C. Bottcher, M.R. Strayer, A.S. Umar and P.-G. Reinhard, Phys. Rev. **A40**, 4182 (1989).

11. C. Bottcher, M.R. Strayer, A.S. Umar, and V. Oberacker, Phys. Rev. **C37**, 2487 (1988).

12. C. Bottcher and M.R. Strayer, *Electron correlation in the continuum*, invited talk at XVI Conference on the Physics of Electronic and Atomic Collisions, New York, August 1989, to appear in the APS series of Conference Proceedings.

BORN EXPANSIONS FOR CHARGED PARTICLE SCATTERING

J. H. Macek and R. O. Barrachina*

University of Tennessee, Knoxville, TN 37996-1200 and
Oak Ridge National Laboratory, Oak Ridge, TN 37831-6373

Abstract

High-order terms in Born expansions of scattering amplitudes in powers of charge are frequently divergent when long-range Coulomb interactions are present asymptotically. Expansions which are free from these logarithmic divergences have been constructed recently. We illustrate these expansions with the simplest example, namely the non-relativistic Rutherford scattering of two charged particles. This approach represents an adequate framework for the calculation of transition amplitudes and a comprehensive starting point for the development of consistent perturbation approximations in multi-channel descriptions of strongly interacting atomic systems.

Pacs. No. 34.10.+x

I. Introduction

The use of perturbation expansions is fundamental to all of atomic physics. The basic idea is that if a potential is regarded as a small perturbation disturbing the evolution of the system, a series expansion in powers of the strength of this potential may be employed. Such expansions have usually been called Born series[1], even though this name has been lately restricted to expansions involving free propagators. More general expansions then receive different names. For instance, a series in powers of the ratio of the strengths of the potentials in strongly asymmetric systems[2] has been termed the Strong Potential Born expansion. These expansions have been developed in many different forms and applied to a wide variety of problems.

It has been known for a long time that expansions in powers of charge are not well defined when Coulomb potentials are present asymptotically. In the case of relativistic quantum field theory, this dificulty is known as the infrared divergence[3,4,5]. Its counterparts on a non-relativistic level are divergences caused by the long-range of the Coulomb potential. For instance, in electron-capture reactions, where the proponderance of a double collision mechanism at high energy requires a second order description, high-order terms in some Strong Potential Born expansions are divergent[6]. Recently it was shown that the same kind of divergences is also present in excitation amplitudes[7]. In fact, the usual non-relativistic scattering theory is not strictly applicable to the collision of charged particles because of the long-range of the Coulomb interaction. The Coulomb potential falls off too slowly at large distances to satisfy the asymptotic conditions of the theory.

Since the pioneering papers by Dollard[8], many different ways of avoiding this difficulty have been proposed. In a time-independent version of Dollard's approach, the long-range effects due to the Coulomb potential have been included in asymptotically modified improper free states that lead to a convergent expression for the T-matrix[9]. Coulomb asymptotic states of the eikonal form have been introduced in other Distorted-Wave theories[10]. None of these methods expand the full amplitude in powers of charge. An alternative approach, previously introduced by Roberts[11] and Macek[7], employs an on-shell limit of a modified off-energy-shell transition matrix, which is well defined in terms of plane waves, even though there are long-range Coulomb interactions in the initial and final states. This approach is amenable to expansions in powers of charge.

In order to show how it is possible to develop expansions which avoid the appearance of logarithmic divergences we will analyse the simplest case, namely the non-relativistic Rutherford scattering of two charged particles. This approach represents an appropriate framework for the calculation of transition amplitudes and the development of consistent perturbation approximations in multi-channel scattering problems.

II. Born Expansion for the Transition Matrix

In order to provide a framework for our discussion, we first outline the basics of Born expansions in single-channel collisions. The on-shell ($E_{k'}=E_k=E$) transition matrix for the scattering of a particle of mass m and initial momentum **k** by a potential V(r) is given by

$$<\mathbf{k}'|T(E)|\mathbf{k}> = <\mathbf{k}'|V + V\,G(E+i\varepsilon)\,V|\mathbf{k}> \quad , \tag{1}$$

where $|\mathbf{k}'>$ and $|\mathbf{k}>$ are free-particle states and $G(z)=[z-H_0-V]^{-1}$ is the full Green's operator. The Born series results from the assumption that the transition matrix can be expanded in powers of the strength of the potential,

$$<\mathbf{k}'|T(E)|\mathbf{k}> = <\mathbf{k}'|V|\mathbf{k}> + <\mathbf{k}'|V\,G_0(E+i\varepsilon)\,V|\mathbf{k}> +$$
$$+ <\mathbf{k}'|V G_0(E+i\varepsilon)\,V G_0(E+i\varepsilon)\,V|\mathbf{k}> + \ldots \quad , \tag{2}$$

with $G_0(z)=[z-H_0]^{-1}$ the free Green's operator. The basic theory of this expansion -including convergence theorems and applications- has been profusely studied in the literature[12]. However, when standard textbooks apply this series to the

Coulomb potential V(r)=Z/r, they usually stop with the first term and fail to mention that higher-order terms do not exist[13]. The standard technique of evaluating these terms as the R $\rightarrow \infty$ limit of a Yukawa potential $V_R(r)= Z/r \exp(-r/R)$ leads to the appearance of logarithmic divergences . For R $\rightarrow \infty$ the first Born approximation yields the proper expression, while the second and third Born contributions are logarithmically divergent for large R [13,14,15]. Alternatively, the well known Rutherford scattering amplitude

$$t(E) = \frac{Z}{2\pi^2 h}\frac{\Gamma(1+in)}{\Gamma(1-in)}\frac{1}{|\mathbf{k}'-\mathbf{k}|^2}\left(\frac{4k^2}{|\mathbf{k}'-\mathbf{k}|^2}\right)^{in} \tag{3}$$

with n=mZ/hk the Sommerfeld's parameter, is analytic in the charge Z and thus has a well defined expansion in powers of Z with finite coefficients. This contrasts with Eq.(2) which has no well-defined expansion in powers of Z. This contradiction has been resolved only recently[7], yet this resolution provides the means to formulate expansions in powers of the charge generally.

III. Born expansions in the Rutherford scattering

The main shortcoming in the previous standard derivation of the Born expansion for Rutherford scattering is that the initial expression Eq.(1) for the transition matrix is not valid in the presence of Coulomb interactions. As mentioned in the Introduction, the usual non-relativistic scattering theory is not applicable to the collision of charged particles because of the long-range of the Coulomb interaction. Indeed, Rutherford's transition matrix Eq.(3) is not given by Eq.(1). Schwinger[16] shows that the transition matrix Eq. (1) does not approach a well-defined limit on the energy shell as $E_{k'}$ and E_k approach E, rather

$$\lim_{E_{k'},E_k\to E} <\mathbf{k}'|T(E)|\mathbf{k}> = g(E,E_{k'})\,t(E)\,g(E,E_k) \tag{4}$$

with

$$g(E,E') = \Gamma(1-in)\,e^{-\Pi n/2}\left(\frac{E'-E}{4E}\right)^{in} \quad . \tag{5}$$

and $E_k=k^2/2m$. Solving Eq.(4) gives an explicit expression for t(E) as the limit of the off-energy-shell transition matrix.

$$t(E) = \lim_{E_{k'},E_k\to E} g(E,E_{k'})^{-1} <\mathbf{k}'|T(E)|\mathbf{k}> g(E,E_k)^{-1} \tag{6}$$

We now have two expressions for t(E), namely Eq.(3) and Eq.(6). Eq.(3) is known to have a well-defined power series expansion in Z, while <**k**'|T(E)|**k**> in Eq.(6) does not. Rather it is known to have logarithmic divergences. The resolution of this apparent contradiction is fundamental to the formulation of expansions in powers of charges and charge ratios, and was achieved only recently[7,11] even though it requires only a simple observation. Note that g(E,E') involves the charge Z through the Sommerfeld parameter n=mZ/hk. Accordingly, expansions of the right hand side of Eq.(6) in powers of Z requires expansion of <**k**'|T(E)|**k**> and g(E,E') . When both factors are expanded in powers of Z it is found that the terms of a given order are finite and equal to those obtained by expanding Eq.(3).

To see this we expand the Coulomb transition matrix of Eq.(3) to second order in the charge Z, and obtain

$$t(E) \approx Z\left[\frac{1}{2\pi^2 h}\frac{1}{|\mathbf{k}'-\mathbf{k}|^2}\right] + Z^2\left[\frac{im}{hk}\frac{1}{2\pi^2 h}\frac{1}{|\mathbf{k}'-\mathbf{k}|^2}\left(\log(4k^2/|\mathbf{k}'-\mathbf{k}|^2) - 2\gamma\right)\right] + o(Z^3) \tag{7}$$

with γ=0.5772...

Alternatively, expanding the right hand side of Eq.(6) in powers of Z requires the expansion of the matrix element

$$<\mathbf{k}'|T(E)|\mathbf{k}> = <\mathbf{k}'|V|\mathbf{k}> + <\mathbf{k}'|V\,G_0(E+i\varepsilon)\,V|\mathbf{k}> + o(Z^3) \tag{8}$$

and g(E,E');

$$g(E,E')^{-1} = 1 + Z\left[\frac{m}{hk}\left(\frac{\pi}{2} - i\log\left(\frac{E'-E}{4E}\right) - i\gamma\right)\right] + o(Z^2) \tag{9}$$

so that the appropriate Born expansion of Eq.(6) reads

$$t(E) = \lim_{E_{k'},E_k\to E}\{<\mathbf{k}'|V|\mathbf{k}> + [<\mathbf{k}'|VG_o(E+i\varepsilon)V|\mathbf{k}> -$$

$$+Z<\mathbf{k}'|V|\mathbf{k}>\frac{m}{hk}(\pi - i\log\frac{(E_{k'}-E)(E_k-E)}{16E^2} - 2i\gamma)] + o(Z^3)\} \tag{10}$$

Close to the energy shell, the Born expansion of the off-shell T-matrix yields[17]

$$<\mathbf{k}'|V|\mathbf{k}> \approx Z\frac{1}{2\pi^2 h}\frac{1}{|\mathbf{k}'-\mathbf{k}|^2} \tag{11}$$

and

$$<\mathbf{k}'|VG_o(E+i\varepsilon)V|\mathbf{k}> \approx -Z^2\frac{1}{2\pi^2 h}\frac{1}{|\mathbf{k}'-\mathbf{k}|^2}\frac{m}{hk}\left[\pi - i\log\left(m\frac{(E_{k'}-E)(E_k-E)}{2E|\mathbf{k}'-\mathbf{k}|^2}\right)\right] \tag{12}$$

Upon substituting Eqs.(11) and (12) into Eq.(10), the proper Born expansion Eq.(7) is recovered.

We thus see that the factors g(E,E') play an essential role in formulating correct expansions of transition matrix elements in powers of charge. This role has largely gone unnoticed in the non-relativistic domain since it is not apparent in first order approximations.

IV. Conclusions

The standard Born expansion of the Rutherford scattering amplitude yields divergent high-order terms. Here, we have constructed perturbation expansions which are free from these logarithmic divergences. The main point is that the expansion has to be evaluated in a framework where the physical amplitude is represented as the on-shell limit of a modified off-shell T-matrix. This approach represents an appropriate starting point for the development of consistent perturbation approximations applicable to multi-channel scattering problems. If the collision involves, both in the initial and final channels, two agregates of particles which asymptotically interact through Coulomb potentials, the modified on-shell transition matrix[7] reads

$$t_{fi}(E) = \lim_{E_f, E_i \to E} g_f(E,E_f)^{-1} < \Psi_f | T(E) | \Psi_i > g_i(E,E_i)^{-1} \tag{13}$$

with

$$g_j(E,E_j) = \Gamma(1 - i n_j)\, e^{-\pi n_j/2} \left(\frac{E_j - E}{4\,(E - \xi_j)} \right)^{i n_j} , \tag{14}$$

where $n_j = (Z_j/h)\,[m_j/2\,(E - \xi_j)]^{1/2}$, Z_j and m_j are the Sommerfeld's parameter, the charge and the reduced mass between the two agregates in the channel j, with ξ_j the energy of the corresponding bound state. As in the single-channel case, a perturbation expansion of the transition matrix requires expansion of the factors $g_j(E,E_j)$ as well as the off-shell T-matrix element. However it is worthwhile to mention that the systematic development of such expansion methods -although possible- is not as simple as for Rutherford scattering. Firstly, in some situations an expansion of a transition matrix element in powers of a weak potential is not possible since the asymptotic aggregate states may not be bound in lowest order. Secondly, mainly in rearrangement processes, there can be more than one natural expansion parameter. However, as we have shown, the common assumption that wave functions satisfying correct asymptotic boundary conditions are needed to eliminate divergent terms is not true. It suffices to take the on-shell limit correctly. Thus we regard Eq.(13) as an appropriate starting point to formulate perturbation-type approximation schemes when only two aggregates are present in the initial and final states.

Acknowledgements: This research was sponsored by the U. S. Department of Energy, Office of Basic Energy Sciences, Division of Chemical Sciences, under contract No. DE-AC05-840R21400 with Martin Marientta Energy Systems, Inc.

References

* Permanent Address: Centro Atomico Bariloche (Comision Nacional de Energia Atomica), 8400 S. C. dc Bariloche, Rio Negro, Argentina.

1. N. F. Mott and H. S. W. Massey, *The Theory of Atomic Collisions* (Oxford University Press, London, 1965).
2. J. S. Briggs, J. Phys. B **10**, 3075 (1977).
3. J. M. Jauch and F. Rohrlich *The Theory of Photons and Electrons* (Springer-Verlag, New York, 1976)
4. P. P. Kulish and L. D. Fadde'ev, Teor. Matem. Fiz **4**, 153 (1970).
5. D. Zwanziger, Phys. Rev. D **11**, 3481 and 3504 (1975).
6. P. Dewangan and J. Eichler, J. Phys. B **18**, L65 (1985).
7. J. H. Macek, Phys. Rev. A **37**, 2365 (1988).
8. J. D. Dollard , J. Math. Phys. **5**, 729 (1964); Rocky Mount. J. Math. **1**, 5 (1971).
9. H. van Haeringen, J. Math. Phys. **17**, 17 (1976).
10. Dz. Belkic, R. Gayet and A. Salin, Phys. Rep. **56**, 279 (1979).
11. M. J. Roberts, J. Phys. B **18**, L707 (1985).
12. J. R. Taylor, *Scattering Theory* (John Wiley, New York, 1972).
13. R. H. Dalitz, Proc. Roy. Soc. A **206**, 509 (1951).
14. C. Kacser, Nuov. Cim. **13**, 303 (1959).

15. A. R. Holt and B. L. Moiseiwitsch, Adv. Atom. Mol. Phys. **4**, 143 (1968).

16. J. Schwinger, J. Math. Phys. **5**, 1606 (1964).

17. J. C. Y. Chen and A. C. Chen, Adv. Atom. Mol. Phys. **8**, 71 (1972).

COMPUTATIONAL ASPECTS OF BOUND-STATE QED

Peter J. Mohr
National Institute of Standards and Technology
Gaithersburg, MD 20899

ABSTRACT

Bound-state quantum electrodynamics (QED) is reviewed with an emphasis on the computational aspects of the subject. The path between the basic theory and the experimental predictions is outlined, with a discussion of some of the numerical methods that provide the basis for calculations.

INTRODUCTION

Quantum electrodynamics makes accurate predictions for phenomena that arise from electromagnetic interactions. Various techniques are needed to do calculations in applications that cover a wide spectrum that ranges from the perturbative domain of weak fields and high energy phenomena to the nonperturbative domain of strongly bound particles.

The perturbative region is characterized by the fact that the intermediate states in the perturbation expansion are free. This class of problems includes the very precise calculations of the anomalous magnetic moment of the electron where high orders (so far up to eighth order or four photons) in perturbation theory have been carried out. It also includes high-energy tests of QED where lower-order diagrams are sufficient to describe experiments.

The non-perturbative bound-state domain can be roughly divided into problems where the interaction is weak and recoil effects are important, such as in positronium, or problems where the interaction is strong and is well represented by a static external field, such as in highly-ionized atoms. The latter category includes studies of the Lamb shift at high Z and the relativistic structure of heavy atoms. The focus of this paper will be on the study of QED in a strong Coulomb field.

APPLICATION TO EXPERIMENT

An example of an application of the theory being considered here to experiment is the calculation of energy levels of two-electron uranium.[1] A recent

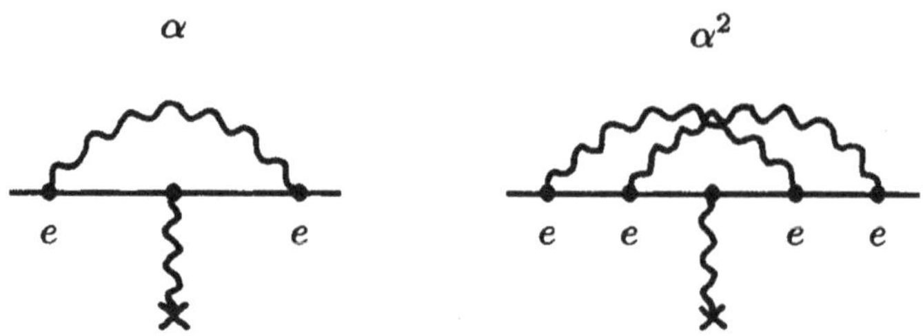

Figure 1: Examples of free-electron Feynman diagrams for the anomalous magnetic moment of the electron.

measurement at Lawrence Berkeley Laboratory has provided an indirect determination of the 2^3S_1 - 2^3P_0 energy separation in this atom. It was accomplished by observing photons emitted from the 2^3S_1 state which reflect the transition rate from the 2^3P_0 state. This rate is a sensitive function of the 2^3S_1 - 2^3P_0 energy separation. With reasonable theoretical assumptions, the measured value of the decay rate provides information on this energy separation. The experimentally derived value for the splitting is 260.0 ± 7.9 eV. This leads to a value for the one electron Lamb shift (2P - 2S splitting) of 70.4 ± 8.3 eV which is in good agreement with the theoretical value of 75.3 ± 0.4 eV. The quoted uncertainty in the deduced one-electron Lamb shift is partly due to the uncertainty in the experimental determination of the 2^3S_1 - 2^3P_0 energy separation and partly due to uncertainties in the theory linking the measured separation and the deduced one-electron Lamb shift. The theoretical interpretation of the experiment can be based on bound-state perturbation theory which is the subject of the following sections.

BOUND-STATE FEYNMAN DIAGRAMS

The theoretical basis for the discussion of strongly-bound electrons can be given in terms of bound-state Feynman diagrams, as contrasted with the more familiar free-particle diagrams. An example of free-particle perturbation theory is given by the QED corrections to the magnetic moment of the electron. The moment is defined in terms of the energy of an electron in a weak magnetic field. Only Feynman diagrams that are linear in the external field are included in the perturbation expansion for this quantity. Typical terms in the expansion are shown in Fig. 1. In that figure, the internal lines for the electron correspond to free electrons, so the only dependence on α arises from the vertices. This leads to the familiar series expansion in powers of α for the correction

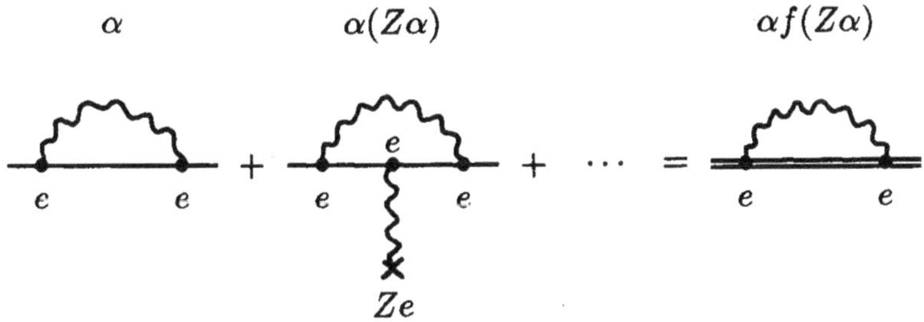

Figure 2: A bound-state Feynman diagram shown as a summation over all orders of the interaction with the external field.

$$a_e = \frac{\alpha}{\pi} a_e^{(2)} + \left(\frac{\alpha}{\pi}\right)^2 a_e^{(4)} + \left(\frac{\alpha}{\pi}\right)^3 a_e^{(6)} + \cdots \tag{1}$$

In the case of bound-state Feynman diagrams, the internal lines for the electrons correspond to electrons bound in the external field of the nucleus with coupling strength $Z\alpha$. This introduces an additional dependence in each term in the perturbation expansion on the coupling strength. In this case, the series expansion for the QED correction to the energy level takes the form

$$\Delta E = \frac{\alpha}{\pi} f^{(2)}(Z\alpha) + \left(\frac{\alpha}{\pi}\right)^2 f^{(4)}(Z\alpha) + \left(\frac{\alpha}{\pi}\right)^3 f^{(6)}(Z\alpha) + \cdots \tag{2}$$

Electrons in the bound-state Feynman diagrams are denoted by a double line which represents the sum over all orders of interactions with the external field, as indicated in Fig. 2.

BOUND-STATE QED

The bound-state Feynman diagram in Fig. 2 represents a term in perturbation theory in the bound-interaction picture (Furry picture) in QED.[2] In this picture, zero order corresponds to noninteracting electrons with wave functions that are solutions of the Dirac equation in an external Coulomb potential V (units are chosen such that $\hbar = c = m_e = 1$)

$$\left[-i\vec{\alpha}\cdot\vec{\nabla} + V(\vec{x}) + \beta - E_n\right]\phi_n(\vec{x}) = 0 \tag{3}$$

with the corresponding time-dependent solutions given by

$$\phi_n(x) = \phi_n(\vec{x})e^{-iE_n t} \tag{4}$$

In terms of these solutions, the electron-positron field operator is written as

$$\psi(x) = \sum_{E_n>0} a_n\phi_n(x) + \sum_{E_m<0} b_m^\dagger \phi_m(x) \tag{5}$$

where a_n is the annihilation operator for an electron in the positive energy state n, and $b_m^\dagger$ is the creation operator for a positron in the negative energy state m.

The interaction Hamiltonian is

$$H_I(x) = j^\mu(x)A_\mu(x) - \delta M(x) \tag{6}$$

where $j^\mu(x)$ is the current operator

$$j^\mu(x) = -\frac{e}{2}\ [\bar{\psi}(x)\gamma^\mu, \psi(x)] \tag{7}$$

$\delta M(x)$ is the mass renormalization operator

$$\delta M(x) = \frac{\delta m}{2}\ [\bar{\psi}(x), \psi(x)] \tag{8}$$

and $A_\mu(x)$ is the photon field operator. In (7) and (8), $\bar{\psi}(x) = \psi^\dagger(x)\gamma^0$.

The perturbation expansion of the S-matrix is

$$S_{\epsilon,\lambda} = 1 + \sum_{j=1}^{\infty} S_{\epsilon,\lambda}^{(j)} \tag{9}$$

with

$$S_{\epsilon,\lambda}^{(j)} = \frac{(-i\lambda)^j}{j!}\int d^4x_j \ldots \int d^4x_1\ e^{-\epsilon|t_j|}\ldots e^{-\epsilon|t_1|}T\,[H_I(x_j)\ldots H_I(x_1)] \tag{10}$$

where T is the time-ordering symbol. Level shifts are given by the Gell-Mann Low Sucher formula[3,4]

$$\Delta E = \lim_{\substack{\epsilon\to 0\\ \lambda\to 1}} \frac{i\epsilon}{2}\frac{\frac{\partial}{\partial\lambda}\langle S_{\epsilon,\lambda}\rangle_c}{\langle S_{\epsilon,\lambda}\rangle_c} \tag{11}$$

where c denotes the fact that only connected diagrams are included. This formulation of the theory leads to the self-energy level-shift expression, corresponding to the Feynman diagram in Fig. 2, for state 0

$$\begin{aligned} E_{SE}^{(2)} &= -4\pi i\alpha \int d(t_2-t_1)\int d\vec{x}_2 \int d\vec{x}_1 D_F(x_2-x_1) \\ &\times \bar{\phi}_0(x_2)\gamma_\mu S_F(x_2,x_1)\gamma^\mu\phi_0(x_1) - \delta m \int d\vec{x}\ \bar{\phi}_0(\vec{x})\phi_0(\vec{x}) \end{aligned} \tag{12}$$

Evaluation of this expression yields a numerical result that can be written in the form

$$E_{SE}^{(2)} = \frac{\alpha}{\pi}\frac{(Z\alpha)^4}{n^3}F(Z\alpha)m_e c^2 \tag{13}$$

where F is a slowly varying dimensionless function of Z that depends on the bound state, and n here denotes the principal quantum number of the state.

Equation (12) expresses the level shift in terms of the bound-electron propagation function and photon propagation function. The bound-electron propagation function is defined by

$$\begin{aligned} S_F(x_2, x_1) &= \langle 0|T[\psi(x_2)\bar{\psi}(x_1)]|0\rangle \\ &= \begin{cases} \sum\limits_{E_n>0} \phi_n(x_2)\bar{\phi}_n(x_1) & t_2 > t_1 \\ -\sum\limits_{E_n<0} \phi_n(x_2)\bar{\phi}_n(x_1) & t_2 < t_1 \end{cases} \\ &= \frac{1}{2\pi i}\int_{-\infty}^{\infty} dz \sum_n \frac{\phi_n(\vec{x}_2)\bar{\phi}_n(\vec{x}_1)}{E_n - z(1+i\delta)}\, e^{-iz(t_2-t_1)} \\ &= \frac{1}{2\pi i}\int_{-\infty}^{\infty} dz\ G(\vec{x}_2, \vec{x}_1, z(1+i\delta))\gamma^0 e^{-iz(t_2-t_1)} \end{aligned} \tag{14}$$

where the Green's function G is the solution of the equation

$$\left[-i\vec{\alpha}\cdot\vec{\nabla}_2 + V(\vec{x}_2) + \beta - z\right] G(\vec{x}_2, \vec{x}_1, z) = \delta(\vec{x}_2 - \vec{x}_1) \tag{15}$$

It is an analytic function of z except for the bound-state poles on the real axis in the interval (0,1), branch points at $z^2 = 1$, and cuts corresponding to the condition $\mathrm{Re}\big((1-z^2)^{1/2}\big) > 0$, as depicted in Fig. 3. The photon propagation function D_F defined by

$$\langle 0|T\left[A_{\mu_2}(x_2)A_{\mu_1}(x_1)\right]|0\rangle = g_{\mu_2\mu_1}D_F(x_2 - x_1) \tag{16}$$

is

$$\begin{aligned} D_F(x_2 - x_1) &= -\frac{i}{(2\pi)^4}\int d^4q\,\frac{e^{-iq\cdot(x_2-x_1)}}{q^2 + i\delta} \\ &= \frac{1}{2\pi i}\int_{-\infty}^{\infty} dq_0\ H(\vec{x}_2 - \vec{x}_1, q_0)e^{-iq_0(t_2-t_1)} \end{aligned} \tag{17}$$

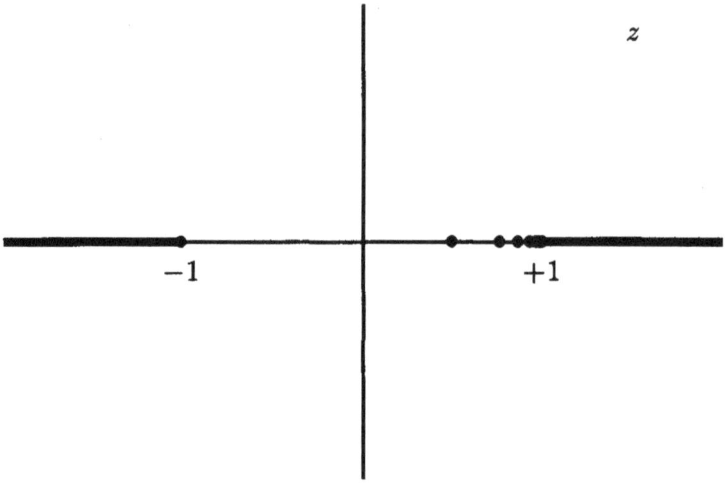

Figure 3: Singularities of the Dirac Green's function.

The function H, which is obtained by carrying out the integration over the space components of q in the above expression, is

$$H(\vec{x}_2 - \vec{x}_1, q_0) = -\frac{e^{-bx_{21}}}{4\pi x_{21}} \tag{18}$$
$$x_{21} = |\vec{x}_2 - \vec{x}_1| \ ; \ b = -i(q_0^2 + i\delta)^{1/2} \ , \ \mathrm{Re}(b) > 0$$

It is analytic in the complex q_0 plane except for branch points and cuts corresponding to the condition $\mathrm{Re}(b) > 0$, as shown in Fig. 4.

Carrying out the integration over $t_2 - t_1$ that appears in (12) and the integration over the variable q_0 that appears in (17) yields an expression for the level shift in terms of the Dirac Green's function given by[5]

$$E_{SE}^{(2)} = -\frac{i\alpha}{2\pi}\int d\vec{x}_2 \int d\vec{x}_1 \phi_0^\dagger(\vec{x}_2)\alpha_\mu \int_{C_F} dz\ G(\vec{x}_2, \vec{x}_1, z)\alpha^\mu \phi_0(\vec{x}_1)\frac{e^{-bx_{21}}}{x_{21}} - \delta m \int d\vec{x}\ \phi_0^\dagger(\vec{x})\beta\phi_0(\vec{x}) \tag{19}$$

where

$$b = -i\left[(E_0 - z)^2 + i\delta\right]^{1/2}; \qquad \mathrm{Re}(b) > 0 \tag{20}$$

Fig. 5 shows the contour C_F and the singularities of the integrand as a function of z in Eq. (19).

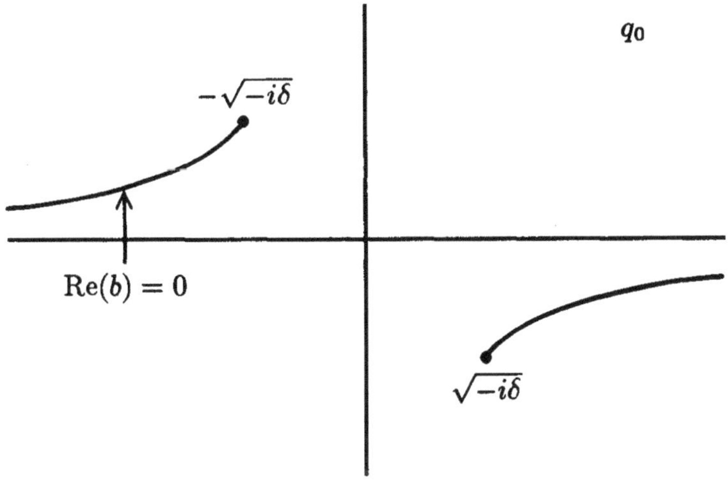

Figure 4: Singularities of the photon Green's function.

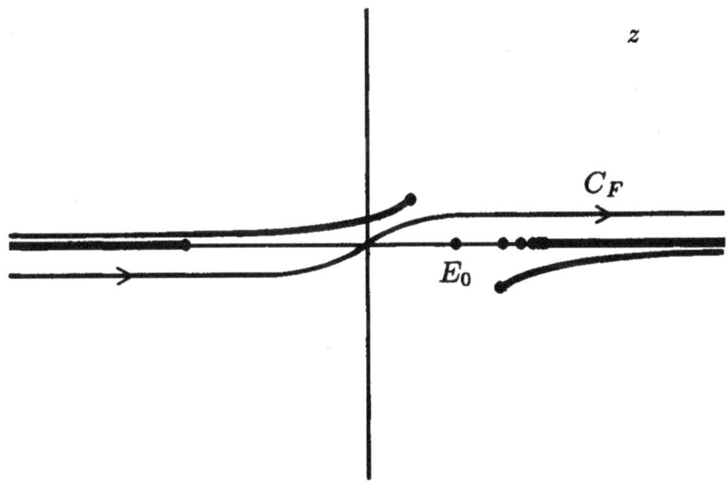

Figure 5: The contour of integration C_F and the singularities of the integrand in the complex z plane in Eq. (19).

In calculations, the contour of integration C_F in Fig. 5 is deformed to a contour C'_F that lies along the imaginary z axis with a segment along the positive real axis as shown in Fig. 6. In that figure, the limit $\delta \to 0$ has been taken for b

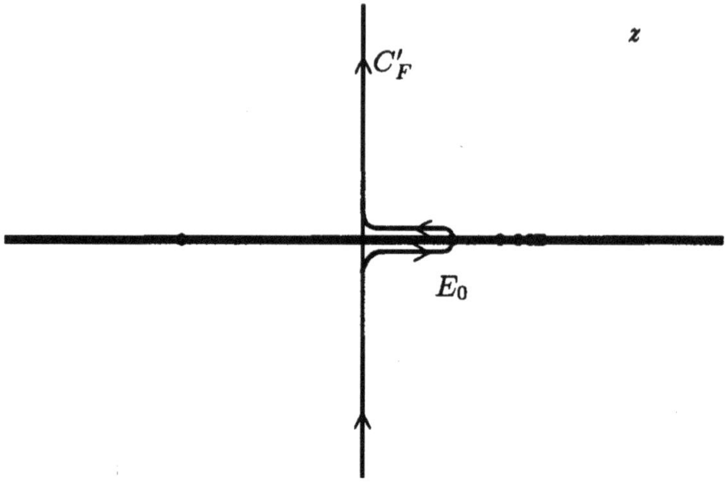

Figure 6: The contour of integration C'_F employed in numerical calculations.

in (20), so the photon branch points are located at the point $z = E_0$. This change of contour is convenient for calculations, because the Dirac Green's function is exponentially damped along the new contour C'_F while it oscillates along the old contour C_F.

DIRAC COULOMB GREEN'S FUNCTION

One of the principal challenges in the calculation of QED level shifts in a strong Coulomb field is the numerical evaluation of the Dirac Green's function $G(\vec{x}_2, \vec{x}_1, z)$. There is an added difficulty arising from the fact that the mass renormalization subtraction term is infinite, so that suitable counter terms need to be introduced to make numerical evaluation possible. This fact does not affect the essential features of the numerical calculations, so the subtraction methods will not be discussed here.

Numerical calculations with the Dirac Green's function are facilitated by introducing radial Green's functions in analogy with the familiar two-component radial wave functions that are defined by

$$\phi_n(\vec{x}) = \begin{bmatrix} f_1(x)\chi^{\mu}_{\kappa}(\hat{x}) \\ if_2(x)\chi^{\mu}_{-\kappa}(\hat{x}) \end{bmatrix} \tag{21}$$

for solutions ϕ_n that are eigenfunctions of the Dirac spin-angular-momentum operator with eigenvalue $-\kappa$

$$\beta(\vec{\sigma} \cdot \vec{L} + 1)\phi_n(\vec{x}) = -\kappa\phi_n(\vec{x}) \tag{22}$$

and are eigenfunctions of the z-component of the total angular momentum operator J_z with eigenvalue μ

$$(L_z + \tfrac{1}{2}\sigma_z)\phi_n(\vec{x}) = \mu\phi_n(\vec{x}) \tag{23}$$

Substitution of these functions in the eigenfunction expansion of the Dirac Green's function

$$G(\vec{x}_2, \vec{x}_1, z) = \sum_n \frac{\phi_n(\vec{x}_2)\phi_n^\dagger(\vec{x}_1)}{E_n - z} \tag{24}$$

yields

$$\begin{aligned} G(\vec{x}_2, \vec{x}_1, z) &= \sum_{n\kappa\mu} \frac{1}{E_{n,\kappa} - z} \\ &\times \begin{bmatrix} f_1(x_2)\chi_\kappa^\mu(\hat{x}_2)f_1(x_1)\chi_\kappa^{\mu\dagger}(\hat{x}_1) & -if_1(x_2)\chi_\kappa^\mu(\hat{x}_2)f_2(x_1)\chi_{-\kappa}^{\mu\dagger}(\hat{x}_1) \\ if_2(x_2)\chi_{-\kappa}^\mu(\hat{x}_2)f_1(x_1)\chi_\kappa^{\mu\dagger}(\hat{x}_1) & f_2(x_2)\chi_{-\kappa}^\mu(\hat{x}_2)f_2(x_1)\chi_{-\kappa}^{\mu\dagger}(\hat{x}_1) \end{bmatrix} \end{aligned} \tag{25}$$

The radial Green's functions are simply defined as the radial portion of this expression, *i.e.*,

$$G_\kappa^{ij}(x_2, x_1, z) = \sum_n \frac{1}{E_{n,\kappa} - z} f_i(x_2)f_j(x_1) \tag{26}$$

and, in analogy with the full Green's function equation in (15), the radial Green's functions satisfy the 2×2 matrix radial Green's function equation

$$\begin{bmatrix} V(x_2) + 1 - z & -\frac{1}{x_2}\frac{d}{dx_2}x_2 + \frac{\kappa}{x_2} \\ \frac{1}{x_2}\frac{d}{dx_2}x_2 + \frac{\kappa}{x_2} & V(x_2) - 1 - z \end{bmatrix} \begin{bmatrix} G_\kappa^{11} & G_\kappa^{12} \\ G_\kappa^{21} & G_\kappa^{22} \end{bmatrix} = I\frac{1}{x_2 x_1}\delta(x_2 - x_1) \tag{27}$$

Substitution of the right-hand-side of Eq. (25) for G in (19) yields an expression in which the integration over angles can be carried out completely to obtain a result expressed in terms of radial Green's functions and Bessel functions

$$\begin{aligned} E_{SE}^{(2)} &= -\frac{i\alpha}{2\pi}\int_{C_F} dz \int_0^\infty dx_2 x_2^2 \int_0^\infty dx_1 x_1^2 \\ &\times \sum_\kappa \sum_{ij} f_i(x_2)G_\kappa^{ij}(x_2, x_1, z)f_j(x_1)A_\kappa(x_2, x_1) \quad + \quad \text{similar terms} \end{aligned} \tag{28}$$

where the summation over κ runs over all nonzero integers, and the summation over i and j runs over 1, 2. For S-states, and for $\kappa > 0$, for example, the function A is

$$A_\kappa(x_2,x_1) = -\kappa b j_\kappa(ibx_<)h_\kappa^{(1)}(ibx_>) \tag{29}$$

To numerically evaluate Eq. (28), it is useful to take advantage of the fact that solutions of Eq. (27) may be constructed from appropriately normalized two-component solutions $F_<$ and $F_>$ of the homogeneous radial equation, regular at $x = 0$ and $x = \infty$, respectively, by writing[6]

$$G_\kappa(x_2,x_1,z) = \theta(x_2 - x_1)F_>(x_2)F_<^T(x_1) + \theta(x_1 - x_2)F_<(x_2)F_>^T(x_1) \tag{30}$$

In the case of an external Coulomb potential, the solutions are known functions given by

$$F_<(x) = \begin{bmatrix} \frac{\sqrt{1+z}}{2ax^{3/2}}\left[(\lambda-\nu)M_{\nu-(1/2),\lambda}(2ax) - \left(\kappa - \frac{\gamma}{a}\right)M_{\nu+(1/2),\lambda}(2ax)\right] \\ \frac{\sqrt{1-z}}{2ax^{3/2}}\left[(\lambda-\nu)M_{\nu-(1/2),\lambda}(2ax) + \left(\kappa - \frac{\gamma}{a}\right)M_{\nu+(1/2),\lambda}(2ax)\right] \end{bmatrix}$$

$$F_>(x) = \frac{\Gamma(\lambda-\nu)}{\Gamma(1+2\lambda)}\begin{bmatrix} \frac{\sqrt{1+z}}{2ax^{3/2}}\left[\left(\kappa + \frac{\gamma}{a}\right)W_{\nu-(1/2),\lambda}(2ax) + W_{\nu+(1/2),\lambda}(2ax)\right] \\ \frac{\sqrt{1-z}}{2ax^{3/2}}\left[\left(\kappa + \frac{\gamma}{a}\right)W_{\nu-(1/2),\lambda}(2ax) - W_{\nu+(1/2),\lambda}(2ax)\right] \end{bmatrix} \tag{31}$$

where M and W are the Whittaker functions,[7] and where $a = (1 - z^2)^{1/2}$, with $\mathrm{Re}(a) > 0$, $\gamma = Z\alpha$, $\nu = \gamma z/a$, and $\lambda = (\kappa^2 - \gamma^2)^{1/2}$.

The remainder of the calculation to evaluate Eq. (28) requires numerical integration and evaluation of the special functions. Some of the techniques that have been useful for this purpose are described in the following sections. In particular, methods for evaluation of M and W, evaluation of j and h, dealing with large numbers, summation over angular momentum, numerical integration, and principal value integration are discussed.

EVALUATION OF M AND W

Numerical evaluation of the functions F has been done by power series and asymptotic expansions of the Whittaker functions $M_{\alpha,\beta}(x)$ and $W_{\alpha,\beta}(x)$.[8] The relevant range of parameters is α complex with $|\alpha| = \mathcal{O}(1)$, β real and in the range $0 < \beta < 20{,}000$, and x real and in the range $0 < x < 40{,}000$. The methods described here can be applied to obtain approximately 12 significant figure accuracy.

The function M is given by the power series

$$M_{\alpha,\beta}(x) = x^{\beta+(1/2)}e^{-x/2}\sum_{n=0}^{\infty}\frac{\Gamma(n+\beta+\frac{1}{2}-\alpha)\Gamma(2\beta+1)}{\Gamma(\beta+\frac{1}{2}-\alpha)\Gamma(n+2\beta+1)}\frac{x^n}{n!} \tag{32}$$

or

$$M_{\alpha,\beta}(x) = x^{\beta+(1/2)} e^{-x/2} \sum_{n=0}^{\infty} T(n) \tag{33}$$

where the terms $T(n)$ can be calculated by the simple algorithm

$$\begin{aligned} T(0) &= 1 \\ T(n+1) &= \frac{(n+\beta+\frac{1}{2}-\alpha)\,x}{(n+2\beta+1)\,(n+1)}\, T(n) \end{aligned} \tag{34}$$

based on the gamma function identity $\Gamma(z+1) = z\Gamma(z)$. For sufficiently large n, the terms in (34) decrease rapidly as n increases

$$\frac{T(n+1)}{T(n)} \rightarrow \frac{x}{n} \tag{35}$$

so the sum over n converges rapidly. In order to have control over the calculation, one can have a rigorous upper bound on the remainder after the sum is truncated at a finite limit $n = N_2$. To be specific, let an error term E_2 be defined by writing

$$S = \sum_{n=0}^{\infty} T(n) = \sum_{n=0}^{N_2} T(n) + E_2 \tag{36}$$

If

$$\theta_n = \left|\frac{T(n+1)}{T(n)}\right| = \left|\frac{(n+\beta+\frac{1}{2}-\alpha)x}{(n+2\beta+1)(n+1)}\right| \tag{37}$$

then for sufficiently large n,

$$\begin{aligned} \theta_n &< 1 \\ \theta_n &> \theta_{n+1} \end{aligned} \tag{38}$$

Hence for large enough N_2

$$\begin{aligned} |E_2| \leq \sum_{n=N_2+1}^{\infty} |T(n)| &= |T(N_2+1)|\,(1 + \theta_{N_2+1} + \theta_{N_2+1}\theta_{N_2+2} + \cdots) \\ &< |T(N_2+1)| \sum_{i=0}^{\infty} (\theta_{N_2+1})^i \\ &= |T(N_2+1)|\, \frac{1}{1-\theta_{N_2+1}} \end{aligned} \tag{39}$$

when $\theta_{N_2+1} < 1$. This straightforward approach works for moderate values of x, but for large values, the magnitude of the terms $T(n)$ can be too large for the computer.

The case of large x may be dealt with as follows. To simplify the discussion, consider the special case of $\alpha = -\frac{1}{2}$. The terms $T(n)$ begin with $T(0) = 1$ and increase rapidly corresponding to $\theta \gg 1$, then level off corresponding to $\theta \approx 1$, and eventually decrease rapidly corresponding to $\theta \ll 1$. The main contribution to the sum comes from the plateau region where $T(n)$ is a maximum and $\theta \approx 1$. More precisely, the maximum term occurs for $n \approx u$ where u is a continuous variable that is the solution of

$$\left|\frac{(u+\beta)x}{(u+2\beta)u}\right| = 1 \tag{40}$$

given by

$$u = \left(\beta^2 + \frac{x^2}{4}\right)^{1/2} + \frac{x}{2} - \beta \tag{41}$$

The magnitude $T(n)$ near the maximum is then given by $|T(n_0)|$, where n_0 is the integer part of u, $n_0 = [u]$. In this region, the magnitude of the terms is of order e^x if $\beta = 0$, for example. For the maximum x under consideration, $x = 40,000$, the terms have a magnitude of roughly $10^{17,000}$ which is out of the range of constants for most computers.

The procedure that is employed in calculations is to normalize the terms in the sum so that the magnitude of the maximum term is close to 1

$$\bar{T}(n) = \frac{T(n)}{|T(n_0)|} \tag{42}$$

The sum S is just

$$S = |T(n_0)| \sum_{n=0}^{\infty} \bar{T}(n) \tag{43}$$

Summation begins at n_0, terms for increasing n are summed, the summation is terminated with the aid of the error bound discussed previously, and the summation resumes at $n_0 - 1$ in the direction of decreasing n. The sum is terminated in the decreasing n direction when the error is sufficiently small. For terms decreasing in the direction of decreasing n, the error E_1 in terminating the sum at $n = N_1$ satisfies

$$|E_1| = \left|\sum_{n=0}^{N_1-1} \bar{T}(n)\right| < N_1 |\bar{T}(N_1 - 1)| \tag{44}$$

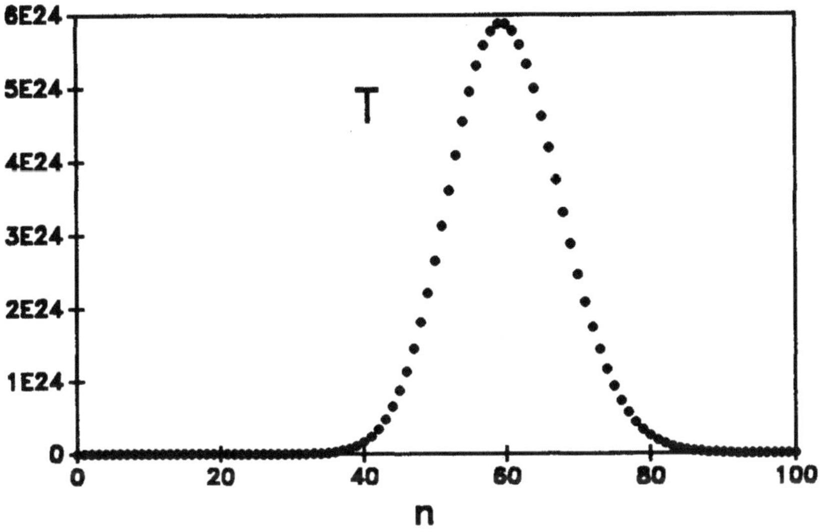

Figure 7: Terms in the power series for the function M for the case where $\alpha = -\frac{1}{2}$, $\beta = 0$, and $x = 60$. In this case, $n_0 = 60$.

In this way, the magnitude of the calculated terms is of order 1. The overall factor $|T(n_0)|$ is calculated separately. Fig. 7 shows terms in the sum for a particular set of values of the parameters.

For $x < 30$, evaluation of the Whittaker function W can be done with the aid of the relation

$$W_{\alpha,\beta}(x) = \frac{\Gamma(2\beta)}{\Gamma(\beta + \frac{1}{2} - \alpha)} M_{\alpha,-\beta}(x) + \frac{\Gamma(-2\beta)}{\Gamma(-\beta + \frac{1}{2} - \alpha)} M_{\alpha,\beta}(x) \tag{45}$$

and the methods described above for evaluation of the M functions are applied with slight modifications in the error estimates. However for $x > 30$, this approach leads to numerical difficulties. In particular, for large x

$$W_{\alpha,\beta}(x) = x^{\alpha} e^{-x/2} \left[1 + \mathcal{O}\left(\frac{1}{x}\right)\right] \tag{46}$$

while

$$M_{\alpha,\beta}(x) = C(\alpha,\beta) x^{-\alpha} e^{x/2} \left[1 + \mathcal{O}\left(\frac{1}{x}\right)\right] \tag{47}$$

so that

$$\frac{W_{\alpha,\beta}(x)}{M_{\alpha,\beta}(x)} = \frac{x^{2\alpha}e^{-x}}{C(\alpha,\beta)}\left[1+\mathcal{O}\left(\frac{1}{x}\right)\right] \tag{48}$$

Hence, there are severe cancellations between the two terms in (45), roughly of order e^{30} or about 13 significant figures. For $x \approx 30$, the individual terms are calculated in double precision (128 bits) to obtain the desired accuracy.

For $x > 30$, the asymptotic expansion of W in powers of x^{-1} gives roughly 12 significant figure accuracy for the specified range of parameters. The expansion is given by[7]

$$W_{\alpha,\beta}(x) = x^{\alpha}e^{-x/2}S' \tag{49}$$

where

$$S' = \sum_{n=0}^{N} T'(n) + E' \tag{50}$$

The terms $T'(n)$ are defined by

$$\begin{aligned} T'(0) &= 1 \\ T'(n+1) &= \frac{(n+\beta+\frac{1}{2}-\alpha)(\beta-\frac{1}{2}+\alpha-n)}{(n+1)x}T'(n) \end{aligned} \tag{51}$$

A rigorous bound on the error E' for α real is given by

$$|E'| \leq |T'(N+1)| \quad \text{for} \quad N+1 > \beta - \tfrac{1}{2} + \alpha \tag{52}$$

For large n, the terms, and consequently the error estimate, increase without bound due to the asymptotic nature of the series. Despite this fact, highly accurate values for the sum may be obtained provided the summation is suitably terminated. In this calculation it is often efficient to sum terms starting from the peak term as in the calculation of M. Fig. 8 shows the terms in the asymptotic series for W for a particular choice of the parameters.

CALCULATION OF BESSEL FUNCTIONS

In strong-field QED calculations a frequently occurring quantity is the product of spherical Bessel and Hankel functions $j_\ell(x)\, h_\ell^{(1)}(y)$ that arise from the expansion of the photon propagator in spherical harmonics. A relevant parameter range is given by x/y real with $0 < x/y < 1$, $0 < \mathrm{Re}(y) < 200$, $0 < \mathrm{Im}(y) < 20{,}000$, and $0 \leq \ell \leq 20{,}000$.

From the recurrence relation for the Bessel functions

$$(2\ell+1)j_\ell(x) = x[j_{\ell-1}(x) + j_{\ell+1}(x)] \tag{53}$$

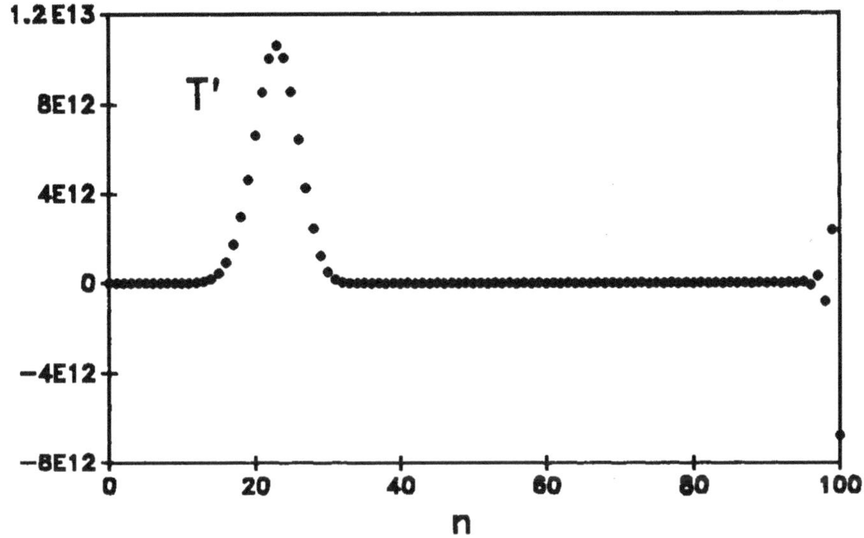

Figure 8: Terms in the asymptotic series for the function W for the case where $\alpha = 0.25$, $\beta = 35$, and $x = 30$. The scatter in the value of the terms near $n = 100$ is the onset of rapid divergence of the asymptotic expansion.

and the definition

$$r_\ell(x) = \frac{2\ell+3}{x}\frac{j_{\ell+1}(x)}{j_\ell(x)} \tag{54}$$

one can write

$$r_{\ell-1}(x) = \frac{1}{1 - \frac{x^2}{(2\ell+1)(2\ell+3)}r_\ell(x)} \tag{55}$$

to calculate the ratios $r_\ell(x)$ recursively. Such a calculation is stable numerically in the direction of decreasing ℓ. For large ℓ, the ratios satisfy

$$\lim_{\ell\to\infty} r_\ell(x) = 1 \tag{56}$$

The stability for decreasing ℓ can be seen from the fact that the error in $r_\ell(x)$ is diminished by a factor of order $\left(\frac{x}{2\ell}\right)^2$ in obtaining $r_{\ell-1}(x)$ from (55).

In order to calculate the functions $r_\ell(x)$ for the range $0 < \ell < L$, the value of $r_L(x)$ is needed. This problem can be solved with the method of J.C.P. Miller.

The calculation begins with the approximation $\tilde{r}_N(x) = 1$, with $N \gg L$. Then the succession of approximate values $\tilde{r}_N(x) = 1 \to \tilde{r}_{N-1}(x) \to \tilde{r}_{N-2}(x) \to \cdots \to \tilde{r}_L(x)$ are calculated by means of (55). Since the error decreases at each step of the calculation, the value of $\tilde{r}_L(x)$ will be an accurate approximation to $r_L(x)$ for a sufficiently large starting value of N. In particular, for the range of parameters under consideration, the choice

$$N = \max(L, [|x|]) + [15 + 0.1\mathrm{Re}(x)] \tag{57}$$

produces a value for $r_L(x)$ that is correct to about 12 figures.

Similarly, with the definition

$$t_\ell(y) = \frac{2\ell+1}{y}\frac{h_\ell^{(1)}(y)}{h_{\ell+1}^{(1)}(y)} \tag{58}$$

for which

$$\lim_{\ell\to\infty} t_\ell(y) = 1 \tag{59}$$

we have

$$t_{\ell+1}(y) = \frac{1}{1 - \frac{y^2}{(2\ell+1)(2\ell+3)}t_l(y)} \tag{60}$$

In this case, the recursion relation is stable in the direction of increasing ℓ, and the initial value is

$$t_0(y) = \frac{1}{1 - iy} \tag{61}$$

The products of interest can then be calculated from the relations

$$j_0(x)h_0^{(1)}(y) = \frac{\sin x}{x}\frac{e^{iy}}{iy} \tag{62}$$

and

$$j_{\ell+1}(x)h_{\ell+1}^{(1)}(y) = \frac{2\ell+1}{2\ell+3}\frac{x}{y}\frac{r_\ell(x)}{t_\ell(y)}j_\ell(x)h_\ell^{(1)}(y) \tag{63}$$

The calculation of the product as shown bypasses a potential problem, because the individual terms $j_\ell(x)$ and $h_\ell^{(1)}(x)$ are outside the allowed range of numbers in most computers for extreme values of the parameters under consideration here.

DEALING WITH LARGE NUMBERS

A practical difficulty in calculations involving Green's functions is the fact that it is sometimes necessary to deal with intermediate stages of calculation that involve numbers with magnitudes outside the range of the computer, even though the final results are not. For example, consider the asymptotic forms for $\beta \to \infty$

$$M_{\alpha,\beta}(ry) = (ry)^{\beta+(1/2)}\left[1 + \mathcal{O}\left(\frac{1}{\beta}\right)\right] \tag{64}$$

where $0 < r < 1$, and

$$W_{\alpha,\beta}(y) = \frac{\Gamma(2\beta)}{\Gamma(\beta + \frac{1}{2} - \alpha)} y^{-\beta+(1/2)}\left[1 + \mathcal{O}\left(\frac{1}{\beta}\right)\right] \tag{65}$$

The Green's function is roughly of order

$$G \sim (ry)^{\beta} y^{-\beta} \tag{66}$$

If $y \approx 1000$, $\beta \approx 10,000$, and $r \approx 0.99999$, then

$$G \sim 0.99999^{10,000} \cdot 10^{30,000} \cdot 10^{-30,000} = 0.90484 \cdot U \cdot S \tag{67}$$

where $U = 10^{30,000}$ and $S = 10^{-30,000}$. To avoid the problem of extreme magnitudes in this case, one can calculate $\ln(U)$ and $\ln(S)$ and form the product as

$$U \cdot S = e^{[\ln(U)+\ln(S)]} \tag{68}$$

For example logarithm of the large coefficient $|T(n_0)|$ in (43) is calculated rather than the number itself.

This procedure introduces a new problem, because there will be an amplification of the relative error when a large positive number, $\ln(U)$, is added to a large negative number, $\ln(S)$, in the exponent. To maintain sufficient accuracy, large magnitude numbers are calculated in double precision.

The loss of significance when multiplying a very large number by a very small number is a problem of a more general nature than this example indicates. In general, if a number y with a relative error δ is raised to a large power N, there will be an amplification of the error

$$y^N \to [y(1+\delta)]^N \approx y^N(1 + N\delta) \tag{69}$$

and similarly for a large negative power $-N$

$$y^{-N} \to [y(1+\delta')]^{-N} \approx y^{-N}(1 - N\delta') \tag{70}$$

The product is

$$y^N y^{-N} \to 1[1 + N(\delta - \delta')] \tag{71}$$

If the methods of calculating the two terms in (71) are different, so that the error terms δ and δ' are independent, then the order-of-magnitude of the relative error in the result will be amplified by a factor N.

SUMMATION OVER ANGULAR MOMENTUM

The expansion of the Dirac Green's function in terms of radial Green's functions introduces summations over angular momentum quantum numbers κ of the form

$$S = \sum_{\kappa=1}^{\infty} T_\kappa \tag{72}$$

where summation over signs of κ has already been carried out. In order to terminate the sum over κ with control over the error, some estimate of the remainder is needed. Two methods with differing levels of simplicity and precision are described here.

For sufficiently large κ, the terms T_κ approach an asymptotic form that can be found from asymptotic expansions of the various functions that make up T_κ. In Eq. (28), with suitable subtraction terms included, the asymptotic form of the terms is (for z on the imaginary axis)

$$T_\kappa \to \frac{r^{2\kappa}}{\kappa} P_\kappa \tag{73}$$

where P_κ is a slowly varying function of κ that approaches a constant as $\kappa \to \infty$ and $0 < r < 1$. Then if κ is sufficiently large for T_κ to be in the asymptotic region, an approximation for the remainder R_N after summing over N terms is

$$R_N = \sum_{\kappa=N+1}^{\infty} T_\kappa \approx \left(\sum_{\kappa=N+1}^{\infty} \frac{r^{2\kappa}}{\kappa} \right) P_{N+1} \tag{74}$$

A simple bound on the sum in (74) is provided by factoring out the maximum value of κ^{-1}

$$\sum_{\kappa=N+1}^{\infty} \frac{r^{2\kappa}}{\kappa} < \frac{1}{N+1} \sum_{\kappa=N+1}^{\infty} r^{2\kappa} = \frac{1}{N+1} \frac{r^{2(N+1)}}{1-r^2} \tag{75}$$

This estimate leads to the approximate bound on the remainder

$$|R_N| \le \left| \frac{1}{N+1} \frac{r^{2(N+1)}}{1-r^2} P_{N+1} \right| \approx \frac{1}{1-r^2} |T_{N+1}| \tag{76}$$

The estimate is approximate in the sense that the approximate equality in (74) may be an underestimate for R_N if P_κ grows in magnitude as κ increases. However, since κ is large enough that P_κ varies slowly, the bound can be expected to be a good approximation. This bound is convenient, since it is simply proportional to the next term in the series.

A tighter estimate that requires somewhat more programming to implement can be made for the remainder. For this estimate, the coefficient of P in (74) is summed exactly by writing

$$\sum_{\kappa=N+1}^{\infty} \frac{r^{2\kappa}}{\kappa} = \sum_{\kappa=1}^{\infty} \frac{r^{2\kappa}}{\kappa} - \sum_{\kappa=1}^{N} \frac{r^{2\kappa}}{\kappa} \tag{77}$$

In (77), the first term on the right-hand-side is just

$$\sum_{\kappa=1}^{N} \frac{r^{2\kappa}}{\kappa} = -\ln(1-r^2) \tag{78}$$

and the second term is obtained by explicit summation. This method requires some care in the case of r close to 1 and N large, because there can be substantial numerical cancellation between the two terms on the right-hand-side of (77).

NUMERICAL INTEGRATION

The final step in the calculation of the level shift as given in (28) is a three-dimensional numerical integration. This can be done to high precision by repeated one-dimensional Gaussian quadrature formulas.[9] The general features of the integrand, for the 1S state for example, relevant to choosing coordinate transformations are

$$\begin{aligned} f_i(x) &\sim e^{-\gamma x} \\ G_\kappa^{ij}(x_2, x_1, z) &\sim e^{-a|x_2-x_1|} \\ A_\kappa(x_2, x_1) &\sim e^{-b|x_2-x_1|} \end{aligned} \tag{79}$$

where a, b, and γ are as defined previously. A crude model function that represents the general features of the integration (for z on the real axis) is given by

$$\tilde{Q} \sim \int_0^1 dt \int_0^\infty dy \int_0^1 dr e^{-y} e^{-\frac{1}{2}(\frac{a}{\gamma}-1)(1-r)y} \frac{\sin\left[\frac{E_0 t y}{2\gamma}(1-r)\right]}{1-r} \tag{80}$$

in terms of the new variables

$$\begin{aligned} r &= \frac{\min(x_2, x_1)}{\max(x_2, x_1)} \\ y &= 2\gamma \max(x_2, x_1) \\ t &= 1 - \frac{z}{E_0} \end{aligned} \tag{81}$$

For the numerical evaluation, integrals over the interval (0,1) are evaluated with Gauss-Legendre quadrature formulas [with the appropriate variable change to the interval (-1,1)], and integrals over the interval (0,∞), with integrands that fall off exponentially, are evaluated with Gauss-Laguerre quadrature formulas. Additional changes of variables are made to deal with steep exponential falloff or nonanalytic features of the integrand. For example the integration over y is broken into two Legendre integrations in the range (0,1) and (1,5) and a Laguerre integration in the range (5,∞), and for large y, the integration over r is done with various variable transformations that take into account the steep falloff of the integrand. The nominal number of evaluation points for this example are 24, 32, and 8 to 19 for the t, y, and r integrations, for a total number of order 10^4. Sample results for the actual evaluation for $Z = 10$ are

$$Q = \begin{cases} 0.492604558261 & \text{for} \quad I = -2 \\ 0.492604558212 & \text{for} \quad I = 0 \\ 0.492604558219 & \text{for} \quad I = +2 \end{cases} \tag{82}$$

where I represents an increment in the number of integration points in every integration formula that is varied to check the precision of the integration.

PRINCIPAL VALUE INTEGRATION

Evaluation of the self energy of excited states with the approach described here involves numerical evaluation of principal value integrals, as in the calculation of the Bethe logarithm. A convenient method of doing this is as follows.[10] Consider a function $f(x)$ analytic on the interval (-1,1) except for a pole at $x = 0$

$$f(x) = \frac{R}{x} + g(x) \tag{83}$$

where R is the residue of the pole of $f(x)$, and $g(x)$ is analytic on the interval (-1,1). Then the principal value integral of $f(x)$ is given by

$$P\int_{-1}^{1} dx\ f(x) = P\int_{-1}^{1} dx\ \frac{R}{x} + \int_{-1}^{1} dx\ g(x) = \int_{-1}^{1} dx\ g(x) \tag{84}$$

since

$$P\int_{-1}^{1} dx\ \frac{R}{x} = P\int_{-1}^{1} dx\ \tfrac{1}{2}\left(\frac{R}{x+i\epsilon} + \frac{R}{x-i\epsilon}\right) = \int_{-1}^{1} dx\ \frac{xR}{x^2+\epsilon^2} = 0 \tag{85}$$

The principal value integral in (84) can be evaluated numerically with the following prescription:

1) Apply an N-point Gauss-Legendre quadrature formula, with N even, to the function $f(x)$.

2) Ignore the principal value symbol.

This prescription yields

$$P\int_{-1}^{1} dx\ f(x) \approx \sum_{i=1}^{N} w_i f(x_i) = \sum_{i=1}^{N} w_i g(x_i) \tag{86}$$

because

$$\sum_{i=1}^{N} w_i \frac{R}{x_i} \equiv 0 \tag{87}$$

which follows from the symmetry of the Gaussian weights and abscissas

$$\begin{aligned} x_i &= -x_{N+1-i} \\ w_i &= w_{N+1-i} \end{aligned} \tag{88}$$

The theoretical accuracy of this method is the same as that of the quadrature formula applied directly to the analytic function $g(x)$, except possibly for some additional roundoff error.

REFERENCES

1. C. T. Munger and H. Gould, Phys. Rev. Lett. **57**, 2927 (1986).
2. W. H. Furry, Phys. Rev. **81**, 115 (1951).
3. M. Gell-Mann and F. Low, Phys. Rev. **84**, 350 (1951).
4. J. Sucher, Phys. Rev. **107**, 1448 (1957).
5. P. J. Mohr, Ann. Phys. (N.Y.) **88**, 26 (1974).
6. E. H. Wichmann and N. M. Kroll, Phys. Rev. **101**, 843 (1956).
7. E. T. Whittaker and G. N. Watson, "A Course of Modern Analysis," Cambridge University Press, Cambridge, England, 1927.
8. P. J. Mohr, Ann. Phys. (N.Y.) **88**, 52 (1974).
9. A. H. Stroud and Don Secrest, "Gaussian Quadrature Formulas," Prentice-Hall, Englewood Cliffs, NJ, 1966
10. P. J. Mohr, Phys. Rev. A **26**, 2338 (1982).

THE THEORY OF ELECTRON-ION COLLISIONS: THE ROLE OF DOUBLY EXCITED STATES

D. C. Griffin
Department of Physics, Rollins College
Winter Park, Florida 32789

M. S. Pindzola
Department of Physics, Auburn University
Auburn, Alabama 36849

C. Bottcher
Physics Division, Oak Ridge National Laboratory
Oak Ridge, Tennessee 37831

ABSTRACT

The important contributions of doubly excited autoionizing states to electron-impact excitation, ionization, and recombination are reviewed. Various theoretical methods of treating these states are considered, and theoretical cross sections are presented and compared to experimental measurements where available.

1. INTRODUCTION

Doubly excited states play a major role in electron-ion collisions. If such states are above the first ionization limit of an N-electron ion, they interact with the adjacent continuum and can decay by the de-excitation of one electron and the simultaneous ejection of a second electron, resulting in the production of an (N-1)-electron ion. However, in addition to this process of autoionization, they can also decay to a bound state of the

N-electron ion, with the emission of a photon.

They may be populated from the N-electron ion by the excitation of an inner-shell electron or the excitation of two valence-shell electrons, or from the (N-1)-electron ion by the excitation of one bound electron and the simultaneous capture of a free electron. We refer to this last process as resonant recombination; it is just the reverse of autoionization. These processes, which are part of the indirect mechanisms in electron impact-excitation, ionization, and recombination, are depicted schematically in Fig. 1.

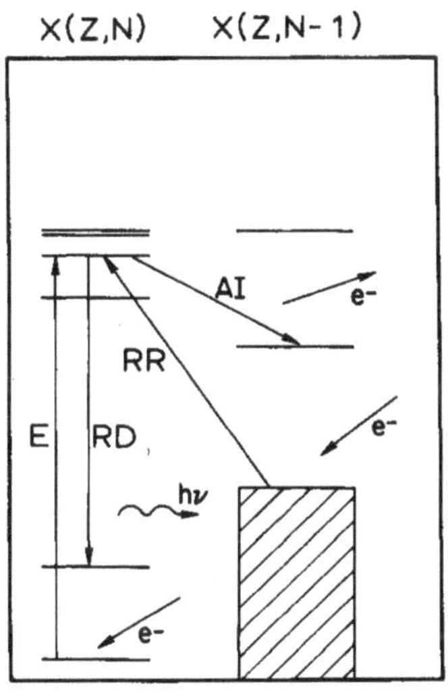

Fig. 1. Schematic drawing showing the population and decay of a doubly excited state of an ion X(Z,N) with atomic number Z and N electrons. It can be populated by excitation (E) of the N-electron ion or resonant recombination (RR) from the (N-1)-electron ion, and can decay by autoionization (AI) or radiative decay (RD).

Since these doubly excited states are coupled to the adjacent continuum, and through the continuum to each other, one should properly treat such processes by constructing a wavefunction which is some linear combination of states consisting of N bound orbitals with states consisting of N-1 bound orbitals and one continuum orbital. However, in many cases, one can employ an independent-processes model and treat the interaction with the continuum using perturbation theory.

2. ELECTRON-IMPACT EXCITATION

The contribution of doubly excited autoionizing states to electron-impact excitation can be included in a very natural way through the use of the close-coupling approximation. One begins by obtaining a set of bound-state eigenfunctions and eigenvalues which describe the target states of an (N-1)-electron ion

$$H_{N-1}\chi_i(x_1 \cdots x_{N-1}) = E_i\chi_i(x_1 \cdots x_{N-1}), \tag{1}$$

where x_i denotes the space and spin coordinates. The radial wavefunctions for the bound-state orbitals can be obtained from the Hartree-Fock approximation, and the energy eigenvalues, E_i, and the eigenvectors, χ_i, are determined by diagonalizing the Hamiltonian matrix for the (N-1)-electron system including the effects of configuration interaction.

The wavefunction for the N-electron collision problem is then constructed from vector-coupled antisymmetized products of the (N-1)-electron wavefunctions with a spin-orbital of the form

$$\theta(\ell_i m_{\ell_i} m_{s_i}| x) = Y_{\ell_i m_{\ell_i}}(\hat{r})\delta(m_{s_i},\sigma_i)\, \frac{1}{r}\, F_i(r), \tag{2}$$

where $Y_{\ell m}$ is a spherical harmonic, σ is a spin coordinate, and $F_i(r)$ describes the radial part of the wavefunction for the N'th electron when the ion is in the state i.

For atomic systems with low Z, we can ignore the spin-orbit interaction, as well as other relativistic effects, and the conserved quantum numbers are $(LS\pi)$, where L is the total orbital angular momentum, S is the total spin angular momentum, and π is the parity of the N-electron system. The state i of the (N-1)-electron ion in LS-coupling is specified by the quantum numbers $\alpha_i L_i S_i M_{L_i} M_{S_i}$, where α_i represents all other quantum numbers needed to completely specify the state. One then forms wavefunctions for each channel, i, from the vector-coupled antisymmetrized product functions

$$\Theta_i = \sum_{\substack{M_{L_i} m_{\ell_i} \\ M_{S_i} m_{s_i}}} C^{L_i \ell_i L}_{M_{L_i} m_{\ell_i} M_L} \; C^{S_i 1/2 S}_{M_{S_i} m_{s_i} M_S} \; A\chi(\alpha_i L_i S_i M_{L_i} M_{S_i} | x_1 \cdots x_{N-1}) \, \theta(\ell_i m_{\ell_i} m_{s_i} | x_N) \ , \tag{3}$$

where A is the operator that antisymmetrizes the total wavefunction.

In the close-coupling (CC) approximation, the total wavefunction is taken to be a finite expansion over the wavefunctions for each channel

$$\Psi(x_1 \cdots x_N) = \sum_i \Theta_i \ . \tag{4}$$

In the most widely used formulation of the CC approximation, one also explicity includes a set of N-electron bound-state functions in the above expansion. These so-called correlation functions are necessary to remove a constraint on the total wavefunction if one insists on orthogonality between the bound-state radial functions and the radial function $F_i(r)$. However, for simplicity, we will consider the form of the CC equations when one does not impose this condition.

Application of the variational principle leads to a set of coupled differential equations, which in atomic units are of the form

$$\left[-\frac{1}{2}\frac{d^2}{dr^2}+\frac{\ell_i(\ell_i+1)}{2r^2}-\frac{Z}{r}+V_{ii}-\frac{k_i^2}{2}\right]F_{ii'}(r)$$

$$+\sum_{j\neq i} V_{ij}F_{ji'}(r) = 0, \tag{5}$$

where we now employ a double index on the radial function $F_{ii'}(r)$ to denote the incident channel i', and the channel index i now signifies the collection of quantum numbers $\alpha_i L_i S_i k_i \ell_i LS\pi$. The potential function V_{ij} contains both direct and exchange electrostatic terms, as well as exchange overlap terms which arise because of the non-zero overlap between the bound orbitals and the radial functions $F_{ii'}(r)$ with the same orbital angular momentum.

For open channels for which $k_i^2 > 0$, the asymptotic form of the radial wavefunction may be written in the form

$$F_{ii'}(r)_{r\to\infty} \sim \frac{1}{\sqrt{k_i}}\left[\delta_{ii'}\sin k_i x + R_{ii'}\cos k_i x\right], \tag{6}$$

where $k_i x = k_i r - \ell_i\pi/2 + (z/k_i)\ln(2k_i r) + \arg\Gamma[\ell_i+1-i(z/k_i)]$, with $z=Z-N+1$, is the asymptotic phase of the regular Coulomb function, and $R_{ii'}$ is an element of the reactance matrix R. For closed channels with $k_i^2 < 0$, we require that $F_{ii'}(r) \longrightarrow 0$ as $r \longrightarrow \infty$.

The cross section for the excitation $\alpha_i L_i S_i \longrightarrow \alpha_f L_f S_f$ in atomic units is given by

$$\sigma_{if} = \frac{\pi}{k_i^2\ 2(2L_i+1)(2S_i+1)} \sum_{LS\ell_i\ell_f} (2L+1)(2S+1) \left| T_{fi} \right|^2 , \quad (7)$$

where T_{fi} is an element of the transition matrix T, which is related to the reactance matrix R by the equation

$$T = \frac{-2iR}{(1 - iR)} . \quad (8)$$

For initial electron energies where both open and closed channels are included in the wavefunction expansion, we obtain Feshbach or closed-channel resonances. For example, consider the idealized system shown in Fig. 2 with the (N-1)-electron ion initially in the ground state i. The incident electron energy ε_i is such that channels associated with level f are open, but those associated with level c are closed. The incident electron can be captured (resonant recombination) into the doubly excited autoionizing state j consisting of a Rydberg electron attached to the (N-1)-electron core in level c. This state can: (1) autoionize to the initial state i, and we have a resonance in the elastic cross section; (2) autoionize to the excited state f, with the emission of an electron of energy ε_f, and we have a resonance in the excitation cross section; or (3) it can radiatively decay to a bound state b and contribute to the recombination (dielectronic) cross section. In order to include this last process, which of course reduces the size of the resonance contributions to the scattering cross sections, the interaction with the radiation field would have to be included in the close-coupling formalism.

Resonances can make significant contributions to the excitation cross section. For example, the cross section resulting from a 6-state (4s, 3d, 4p, 5s, 4d, 5p) close-coupling calculation[1]) for the transition 4s⟶3d in Ca^+ is

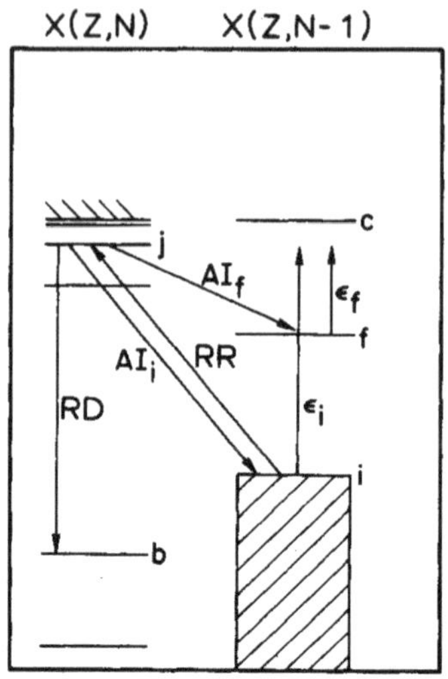

Fig. 2. Schematic drawing of an idealized resonant system. The initial electron with energy $\varepsilon_i=k_i^2/2$ can be captured (RR) into a doubly excited state j. This state can then autoionize to the initial state i [AI_i], or the excited state f [AI_f] (ejecting an electron of energy $\varepsilon_f=k_f^2/2$); or radiatively decay (RD) to a bound state b.

shown in Fig. 3. Resonances, especially from the doubly excited states of the type 4pnℓ, dominant the cross section for incident electron energies between the 3d and 4p thresholds.

As the net ionic charge increases, the spin-orbit interaction becomes large, and the LS representation is no longer valid. The close-coupling approximation must then be formulated in terms of channels associated with individual intermediate-coupled levels, rather than LS terms. For a complex ion there may be a large number of levels which should be included in a single calculation, and solving the coupled equations becomes a formidable task. However, for many cases in highly ionized species, the coupling between open channels is weak,[2)] and one is justified in employing the much simpler distorted-wave (DW)

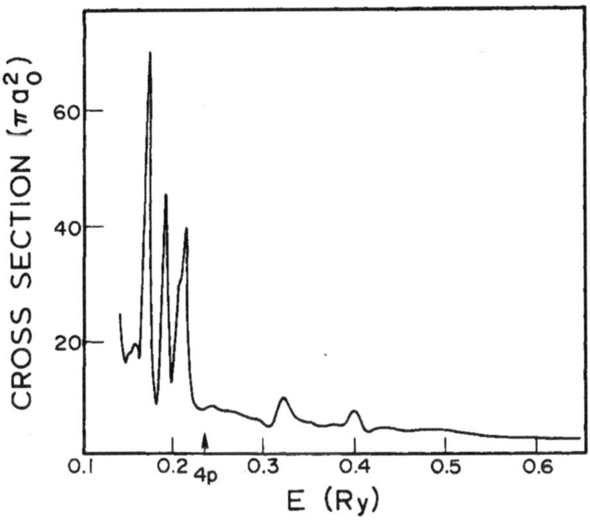

Fig. 3. The excitation cross section (in units of πa_0^2) for the 4s→3d transition in Ca^+ from a 6-state close coupling calculation.[1] The 3d threshold is at .125 rydbergs and the 4p threshold, at .231 rydbergs, is indicated by the small arrow along the energy axis.

approximation to determine the direct-excitation cross section. We drop all potential terms V_{ij} with $i \neq j$ in Eq. (5) and solve the DW differential equation

$$\left[-\frac{1}{2}\frac{d^2}{dr^2} + \frac{\ell_i(\ell_i+1)}{2r^2} - \frac{Z}{r} + V_{ii} - \frac{k_i^2}{2} \right] f_i(r) = 0 , \qquad (9)$$

where the asymptotic form of the radial wavefunction, $f_i(r)$, is

$$f_i(r) \underset{r\to\infty}{\sim} \sin(k_i x + \delta_i) , \qquad (10)$$

and where δ_i is the distorted-wave phase shift. The cross section for the transition $\alpha_i J_i \to \alpha_f J_f$ between two levels of total angular momentum J_i and J_f is given by

$$\sigma_{if} = \frac{4\pi}{k_i^2(2J_i+1)} \sum_{\substack{J\ell_i\ell_f \\ j_ij_f}} (2J+1) \mid \rho_{fi} \mid^2 , \qquad (11)$$

where J is the total angular momentum of the N-electron system, $j_i=\ell_i\pm 1/2$, $j_f=\ell_f\pm 1/2$, and ρ_{fi} is an element of the ρ matrix, with off-diagonal elements given by

$$\rho_{fi} = \frac{-2}{\sqrt{k_ik_f}} \int_0^\infty f_f(r)V_{fi}f_i(r)dr , \qquad (12)$$

and diagonal elements equal to zero.

Although, for highly ionized species, continuum coupling may have a small effect on the direct-excitation cross section, the contribution of resonances to the total-excitation cross section is still quite important. Thus it would be advantageous to include the effects of resonances within the distorted-wave formalism. If the coupling of the resonances to the adjacent continuum is relatively weak, we can impose the independent-processes approximation, and use perturbation theory to show that the total cross section from a level i to a level f is given by

$$\sigma_t(i\rightarrow f) = \sigma_d(i\rightarrow f) + \sum_j \sigma_r(i\rightarrow j)B^a(j\rightarrow f) , \qquad (13)$$

where $\sigma_d(i\rightarrow f)$ is the direct-excitation cross section from level i to level f, calculated from Eq. (11); $\sigma_r(i\rightarrow j)$ is the resonant-recombination cross section from level i of the (N-1)-electron ion to the doubly excited level j of the N-electron ion; and $B^a(j\rightarrow f)$ is the branching ratio for autoionization from level j to level f. $B^a(j\rightarrow f)$ is given by the equation

$$B^{a}(j\rightarrow f) = \frac{A_a(j\rightarrow f)}{\sum_k A_a(j\rightarrow k) + \sum_n A_r(j\rightarrow n)}\ , \tag{14}$$

where $A_a(j\rightarrow k)$ is the autoionizing rate from level j to a lower level k of the (N-1)-electron ion, and $A_r(j\rightarrow n)$ is the radiative rate from level j to a bound level n of the N-electron ion.

By the principle of detailed balance, the resonant-recombination cross section from level i to level j, averaged over a narrow energy bin, can be written in terms of the rate for the reverse process of autoionization from level j to level i as

$$\sigma_r(i\rightarrow j) = \frac{2\pi^2}{\Delta\varepsilon k_i^2}\,\frac{2J_j+1}{2(2J_i+1)}\,A_a(j\rightarrow i)\ , \tag{15}$$

where J_i and J_j are the total angular momenta of levels i and j, respectively, and $\Delta\varepsilon$ is an energy bin width larger than the largest resonance width. The autoionizing rates are calculated using distorted-wave continuum functions and the autoionizing and radiative rates necessary for the determination of resonant contributions from high Rydberg states are found by extrapolating calculated rates.

This formalism was first used by Cowan[3] to determine the resonant contributions to the $(2s^2)\,^1S \rightarrow (2s2p)\,^3P$ excitation in O^{4+}. He obtained good agreement with close-coupling calculations. Since then, it has been used to determine the resonant contribution to excitation for a number of highly ionized systems. For example, we employed this method[4] to estimate the resonant contribution from doubly excited states of the form $2p^5 3pn\ell$, $2p^5 3dn\ell$, and $2p^5 4\ell 4\ell'$ to the relatively weak transition $2p^6 \rightarrow 2p^5 3s$ in Ti^{12+}. The results are shown in Fig. 4, where the indirect cross sections, averaged over a narrow energy bin $\Delta\varepsilon$, were convoluted with a gaussian distribution function typical of experimental electron distributions. Obviously, the resonant contributions dominate the cross section.

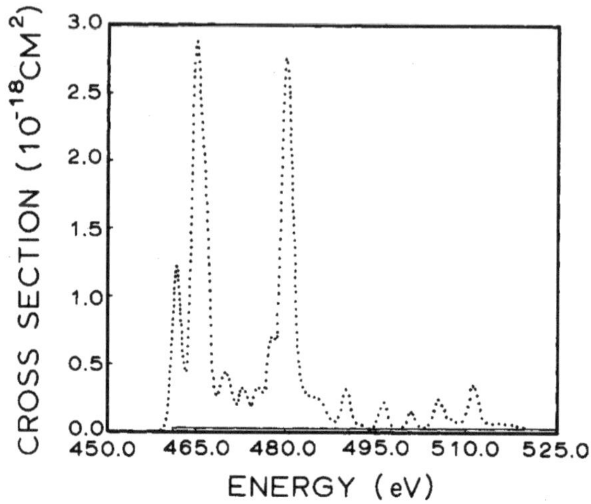

Fig. 4. Electron-impact $2p^6 \longrightarrow 2p^5 3s$ excitation cross section of Ti^{12+}. The solid curve, hardly visible on the scale of the graph, is the direct-excitation cross section, while the dotted curve is the sum of the direct and indirect cross sections.

3. ELECTRON-IMPACT IONIZATION

We now consider the electron-impact ionization of an N-electron ion X(Z,N). The direct-ionization process is

$$e^- + X(Z,N) \longrightarrow X(Z,N-1) + e^- + e^- \ . \tag{16}$$

However, in addition, we can have the indirect processes of inner-shell excitation followed by autoionization

$$\begin{array}{l} e^- + X(Z,N) \longrightarrow X(Z,N)^{**} + e^- \\ \qquad\qquad\qquad\qquad \hookrightarrow X(Z,N-1) + e^- \ ; \end{array} \tag{17}$$

resonant recombination to a doubly excited state of the (N+1)-electron ion followed by the sequential auto-ionization of two electrons

$$e^- + X(Z,N) \longrightarrow X(Z,N+1)^{**} \longrightarrow X(Z,N)^{**} + e^- \longrightarrow X(Z,N-1) + e^- \; ; \qquad (18)$$

and resonant recombination followed by the simultaneous autoionization of two electrons

$$e^- + X(Z,N) \longrightarrow X(Z,N+1)^{**} \longrightarrow X(Z,N-1) + e^- + e^- \, . \qquad (19)$$

The double asterisk in the above equations denotes a doubly excited autoionizing state.

For now we consider only the first two processes of direct ionization and excitation autoionization. If one employs an independent-processes approximation, the total ionization cross section from an initial configuration i is given by

$$\sigma_{tot}(i) = \sum_f \sigma_{ion}(i \rightarrow f) + \sum_j \sigma_{exc}(i \rightarrow j) B_j^a \, , \qquad (20)$$

where $\sigma_{ion}(i \rightarrow f)$ is the distorted-wave direct-ionization cross section from the initial configuration i to the final configuration f of the (N-1)-electron ion, and $\sigma_{exc}(i \rightarrow j)$ is the distorted-wave inner-shell excitation cross section from configuration i to a particular doubly excited level j. B_j^a is the branching ratio for autoionization to all lower levels of the (N-1)-electron ion, which is given by

$$B_j^a = \frac{\sum_k A_a(j \rightarrow k)}{\sum_k A_a(j \rightarrow k) + \sum_n A_r(j \rightarrow n)} \, . \tag{21}$$

There is a particular difficulty with the calculation of the direct-ionization cross section $\sigma_{ion}(i \rightarrow f)$ which is important to mention. It is associated with the fact that the final state involves two free electrons in the field of an ion, which is a problem in many-body theory that has not yet been solved in a manner suitable for practical calculations. In the distorted-wave approximation, it is assumed that each of two electrons experiences a spherical potential at infinity; thus, the phase between the two electrons is lost, and the relative phase between the direct and exchange terms within the cross-section formula is unknown. We employ the minimum-phase approximation of Peterkop[5]) which amounts to squaring the difference between the absolute values of the direct and exchange amplitudes; this leads to maximum destructive interference and seems to give the best agreement with experiment.

We have performed numerous calculations of total-ionization cross sections for highly ionized species using this approximation. However, in principle, one should include the possible quantum mechanical interference between direct ionization and excitation autoionization. Jakubowicz and Moores[6]) have developed a method of doing this by using either Coulomb or distorted waves to represent the scattered electron, but close-coupling wave-functions to describe the N-electron target states, before and after the collision. In this way, they include the interaction between the doubly excited autoionizing states and the adjacent continuum explicitly. However, the good agreement between calculations performed using this method[7]) with those performed using the independent-processes approximation[8] seems to indicate that interference between direct ionization and excitation autoionization has a small effect on the total-ionization cross section.

Excitation autoionization can dominate over direct ionization, even for highly ionized species. For example, consider ionization of Na-like ions from the $2p^6 3s$ ground state. In addition to direct ionization out of the 3s, 2p, and 2s subshells, we can have indirect contributions from inner-shell excitation to doubly excited configurations of the type $2s^2 2p^5 3sn\ell$ and $2s2p^6 3sn\ell$. Calculations for the ions Ti^{11+}, Cr^{13+}, Fe^{15+}, and Ni^{17+} using the independent-processes approximation[9] indicate that the excitation-autoionization contributions enhance the total cross section near threshold by nearly a factor of five, in agreement with cross-beam measurements for the first three of these ions.[10,11]

In Fig. 5, we show the calculated[9] and measured[11]

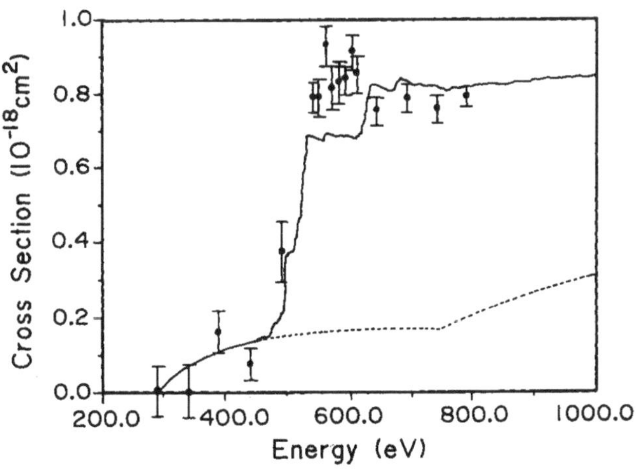

Fig. 5. Calculated ionization cross section for Ti^{11+}. Solid curve, total-ionization cross section; dashed curve, direct-ionization contributions only. Experimental points are from Ref. 11.

cross sections for Ti^{11+}. The overall agreement between experiment and theory is good; however, the data points are clearly above the calculated cross section in the energy range from about 550 eV to 610 eV. This may indicate the formation of recombination resonances of the type $2s^2 2p^5 3sn\ell n'\ell'$ and $2s2p^6 3sn\ell n'\ell'$ in Ti^{10+}, which can autoionize sequentially with the emission of two electrons to Ti^{12+} (see Eq. 18).

In the independent-processes approximation, the cross section for this process of resonant recombination double autoionization (RRDA) is given by

$$\sigma_{rrda} = \sum_j \sigma_r(i \rightarrow j) B_j^{da}, \tag{22}$$

where $\sigma_r(i \rightarrow j)$ is the resonant-recombination cross section from level i of the N-electron ion to a particular doubly excited level j of the (N+1)-electron ion (Eq. 15). B_j^{da} is the branching ratio for double autoionization, which is given by

$$B_j^{da} = \sum_d \left[\frac{A_a(j \rightarrow d)}{\sum_k A_a(j \rightarrow k) + \sum_n A_r(j \rightarrow n)} \times \frac{\sum_f A_a(d \rightarrow f)}{\sum_f A_a(d \rightarrow f) + \sum_m A_r(d \rightarrow m)} \right], \tag{23}$$

where the first term inside the square brackets is the branching ratio for autoionization to a particular doubly excited autoionizing level d of the N-electron ion; the second term is the branching ratio for autoionization from the level d to any lower level of the (N-1)-electron ion; and finally, the sum over k in the first term includes autoionization to all lower levels of the N-electron ion, both autoionizing and bound.

Because of the extremely large number of levels involved in the determination of the double autoionization branching ratios, such a calculation for a case like that of Ti^{11+} becomes very difficult to perform, even at this level of approximation. LaGattuta and Hahn[12] employed average rates between configurations, rather than between individual levels, to obtain an estimate for RRDA in Na-like Fe^{15+}, but comparison with experiment[10] indicates that this method far over estimates the contribution of the resonances. Chen and Reed[13] have just completed a very large calculation of the RRDA process for Fe^{15+} between the individual levels, and they find that the size of the resonance contribution to the ionization cross section agrees well with the experimental measurement.[10]

Now consider the resonant-recombination process followed by the simultaneous emission of two electrons (Eq. 19). We shall refer to this as resonant recombination auto-double ionization (RRAD). The cross section for RRAD is given by

$$\sigma_{rrad} = \sum_j \sigma_r(i\rightarrow j)B_j^{ad}, \tag{24}$$

where B_j^{ad} is the branching ratio for auto-double ionization

$$B_j^{ad} = \frac{\sum_f A_{ad}(j\rightarrow f)}{\sum_f A_{ad}(j\rightarrow f) + \sum_k A_a(j\rightarrow k) + \sum_n A_r(j\rightarrow n)}, \tag{25}$$

and where $A_{ad}(j\rightarrow f)$ is the auto-double ionization rate from the doubly excited level j the (N+1)-electron ion to level f of the (N-1)-electron ion.

Auto-double ionization is strictly an electron-electron correlation effect, since it involves de-excitation of one electron and the simultaneous ejection of two electrons. We have used many-body perturbation theory to estimate the contribution of RRAD to the ionization cross section of Li-like ions.[14] In particular we considered the the RRAD process

$$e^- + 1s^2 2s \longrightarrow (1s2s^2 2p)\,^3P \longrightarrow 1s^2 + e^- + e^- . \qquad (26)$$

The $(1s2s^2 2p)\,^3P$ term of the corresponding Be-like ion was chosen because there exist no open channels for sequential double autoionization, and because of its relative simplicity. The term $(1s2s^2 2p)\,^1P$ was not included because it has a very small resonant-recombination cross section.

The auto-double ionization rate was calculated in lowest-order perturbation theory by summing up over a large number of continuum and bound intermediate states. As in the case of direct ionization, the final state involves two continuum electrons, and the phase between the direct and exchange terms is unknown. In order to obtain a range of possible rates, we used a set of different phase choices. The minimum rate was obtained by using the minimum-phase approximation, similar to that used in direct-ionization calculations, in which one squares the difference between the absolute values of the direct and exchange amplitudes. The maximum rate was obtained by employing the maximum-phase approximation, in which one squares the sum of the absolutes values of the direct and exchange amplitudes.

Although our calculations indicated that the contribution from this RRDA process should be a maximum of only 0.64% of the direct ionization cross section in C^{3+}, the resonant contributions from the terms $(1s2s^2 2p)\,^3P$ and $(1s2s2p^2)\,^3D$ to the ionization cross section of this ion were recently measured by Müller et al. at Giessen.[15] Their measured cross section for the $(1s2s^2 2p)\,^3P$ resonance was 1.1×10^{-20} cm^2, while our calculated cross section was 0.96×10^{-20} cm^2 using the minimum-phase approximation and

1.7×10^{-20} cm^2 using the maximum-phase approximation.

Müller et al.[16] have just completed an experiment on the ionization of Li^+, in which they observed the indirect processes of double excitation followed by autoioniation

$$e^- + Li^+(1s^2) \longrightarrow Li^+(2\ell 2\ell') + e^- \qquad \llcorner\!\!\longrightarrow Li^{2+}(1s) + e^- , \qquad (27)$$

and double excitation with resonant recombination followed by auto-double ionization

$$e^- + Li^+(1s^2) \longrightarrow Li(2\ell^2 2\ell') \longrightarrow Li^{2+}(1s) + e^- + e^- . \qquad (28)$$

Shown in Fig. 6 is the total-ionization cross section, minus the background direct-ionization cross section, from

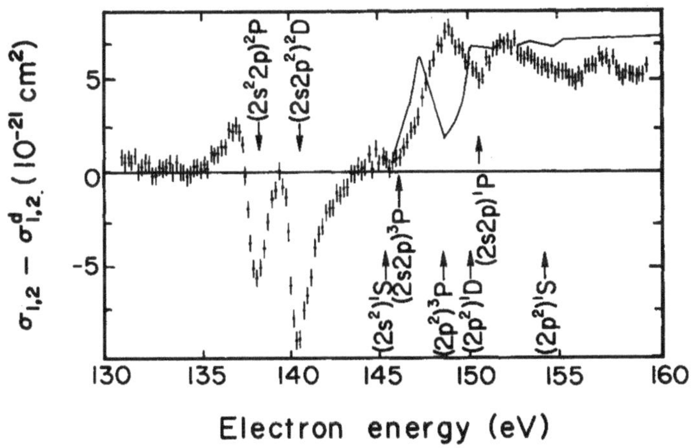

Fig. 6. The data points from Müller et al.[16] show the total-ionization cross section minus the direct-ionization cross section for Li^+ in the energy range where indirect processes contribute. The solid curve is the cross section for the $1s^2 \rightarrow 2\ell 2\ell'$ double excitation from an 11-state close-coupling calculation. The positions of the $(2s^2 2p)^2P$ and $(2s2p^2)^2D$ terms of Li are from Ref. 16, while the positions of the doubly excited states of Li^+ were obtained from Hartree-Fock calculations with configuration interaction.

this experiment in the energy range where these indirect processes contribute. The shape of the curve where the recombination resonances $(2s^2 2p)^2P$ and $(2s2p^2)^2D$ can be populated indicates strong interference with the direct-ionization background. A theoretical description of the interference between these processes involving a double-electron continuum has not yet been worked out; however, the shape of the curve closely resembles the familiar Fano profile, even though Fano theory[17] applies only to a doubly excited state interfering with a single-electron continuum.

In order to investigate the process of double excitation followed by autoionization (Eq. 27), we performed an 11-state close-coupling calculation $[(1s^2)^1S$, $(1s2s)^3S$, $(1s2s)^1S$, $(1s2p)^3P$, $(1s2p)^1P$, $(2s^2)^1S$, $(2s2p)^3P$, $(2s2p)^1P$, $(2p^2)^3P$, $(2p^2)^1D$, and $(2p^2)^1S]$ of the cross section for the $1s^2 \longrightarrow 2\ell 2\ell'$ double excitation. The 1s orbital, was obtained from a Hartree-Fock calculation of the $1s^2$ configuration, while the 2s and 2p orbitals were from Hartree-Fock calculations for the 1s2s and 1s2p configurations, respectively. Although, these particular 2s and 2p orbitals were used in performing the close-coupling calculations, they give poor energies for the doubly excited terms; therefore the positions of these terms were determined from a configuration-interaction calculation using 2s and 2p orbitals obtained from Hartree-Fock calculations for the $2s^2$ and $2p^2$ configurations. The close-coupling calculations were carried out using the program IMPACT.[18] The branching ratios for autoionization to the $1sk\ell$ continuum for these doubly excited states are equal to one.

The results of this calculation are shown by the solid curve in Fig. 6. The magnitude of the calculated cross section is in reasonable agreement with the experiment above the last doubly excited state. However, the shape of the calculated cross section, near the excitation threshold, is quite different from the measurement. In addition the calculation includes contributions from recombination resonances of the type $2\ell 2\ell' n\ell''$, which may, or may not, contribute to the ionization cross section, depending on the branching ratios for double autoionization (Eq. 23). Thus, a great deal more work will be required to properly treat this process. Obviously, these new ionization experiments are severly testing the current state of electron-ion collision theory.

4. ELECTRON-IMPACT RECOMBINATION

Finally we consider the recombination of free electrons with an (N-1)-electron ion. The first process is three-body recombination

$$e^- + e^- + X(Z,N-1) \longrightarrow X(Z,N) + e^- . \qquad (29)$$

It is just the reverse of electron-impact ionization, and due to its relatively low probability, it is only important at high electron densities. Next we have radiative recombination

$$e^- + X(Z,N-1) \longrightarrow X(Z,N) + h\nu . \qquad (30)$$

This is just the reverse of photoionization, and it dominates the recombination cross section at relatively low electron energies; however, its importance decreases rapidly with energy. Finally, we have the indirect process of dielectronic recombination (DR)

$$e^- + X(Z,N-1) \longrightarrow X(Z,N)^{**} \longrightarrow X(Z,N) + h\nu . \qquad (31)$$

This process involves resonant recombination to a doubly

excited state of the N-electron ion followed by radiative decay to a bound state. Of course, the doubly excited state can also autoionize and provide a resonance contribution to the elastic or excitation cross section of the (N-1)-electron ion. DR is the dominant recombination process at relatively high energies, and we shall now consider it in more detail.

In principle, radiative recombination and dielectronic recombination can interfere. A formalism to treat such interference has been developed;[19] however, model calculations[20] indicate that this will have a negligible effect on the total cross section. Furthermore, Bell and Seaton[21] have developed a coupled-channel formalism for DR, based on quantum-defect theory, which properly includes the interaction of the doubly excited states with the adjacent continuum, and with each other. However, the majority of calculations have been performed using an independent-processes approximation, and a comparison of results obtained with these two methods seems to indicate that the effects of overlapping, interfering resonances on the total-DR cross section are small.

In the isolated-resonance approximation, the DR cross section from an initial level i of the (N-1)-electron ion is given by

$$\sigma_{dr}(i) = \sum_j \sigma_r(i \longrightarrow j) B_j^r \tag{32}$$

where again, $\sigma_r(i \longrightarrow j)$ is the resonant-recombination cross section to a doubly excited level j of the N-electron ion. B_j^r is the branching ratio for radiative decay to a bound state of the N-electron ion. It is given by

$$B_j^r = \frac{\sum_n A_r(j \longrightarrow n)}{\sum_k A_a(j \longrightarrow k) + \sum_m A_r(j \longrightarrow m)} , \tag{33}$$

where the sum over n is over all lower bound levels of the N-electron ion, while the sum over m is over all lower levels of the N-electron ion, bound and autoionizing.

As an example let us consider dielectronic recombination associated with the 2s $\longrightarrow$ 2p excitation in Li-like C^{3+}

$$e^- + C^{3+}(1s^2 2s) \longrightarrow C^{2+}(1s^2 2pn\ell) \longrightarrow C^{2+}(1s^2 2sn\ell) + h\nu$$
$$\hookrightarrow C^{2+}(1s^2 2pn'\ell') + h\nu'. \quad (34)$$

The radiative transitions involving the Rydberg electron ($n\ell \longrightarrow n'\ell'$) can dominate for low Rydberg states, but their importance falls off rapidly with the principal quantum number n. On the other hand, the radiatve rates for the transitions involving the core electron (2p$\longrightarrow$2s) are nearly independent of the Rydberg electron. An energy-level diagram for this system is shown in Fig. 7.

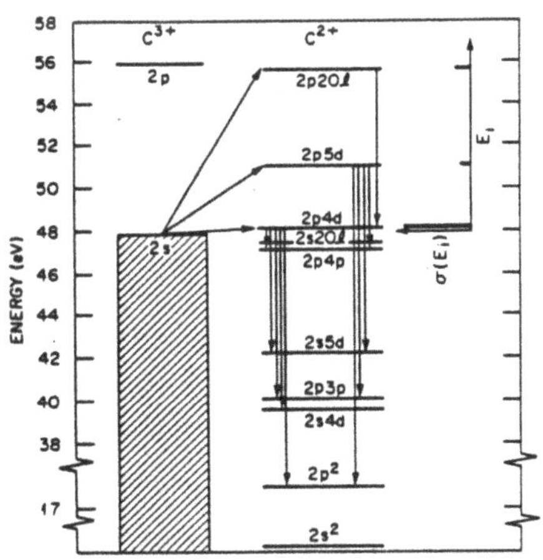

Fig. 7. An energy level diagram illustrating several DR transitions associated with the 2s$\longrightarrow$2p excitation in C^{3+}. A schematic diagram of the DR cross section resulting from transitions through the 2p4d, 2p5d, and 2p20ℓ configurations of C^{2+} is inset on the far right.

The low lying resonances such as 2p4d have relatively large cross sections because of the size of the radiative rates for transitions in which the Rydberg electron is the active electron; however, these resonances are rather widely spaced. On the other hand, the cross sections for the closely spaced high Rydberg states are smaller, but they continue to contribute to the cross section until we reach rather high values of the principal quantum number, n, where the resonant-recombination cross section (which falls of approximately as $1/n^3$) becomes so small that the states become essentially closed to recombination. Thus, the DR cross section associated with such $\Delta n = 0$ excitations are dominated by high Rydberg states.

External electric fields can have a pronounced effect on dielectronic recombination. First of all such fields can ionize electrons in high Rydberg states, and thereby decrease the DR cross section. Secondly, fields mix high Rydberg states with different values of the orbital angular momentum ℓ. The resonant recombination cross section is high for small values of ℓ and small for high values of ℓ, and in the absence of a field, the high angular momentum states are essentially closed to recombination. However external fields mix these states in such a way as to open up more channels for recombination and thereby enhance the cross section.

In DR experiments, the fields in the interaction region are relatively small, while the fields in the analyzing region are very large, typically of the order of 10 kV/cm. Thus the smaller fields in the interaction region enhance the DR cross section, and then after recombination has occurred, the fields in the interaction region strip off the high Rydberg states.

The amount of field ionization can be determined by employing hydrogenic field-ionization formulas such as the one developed by Damburg and Kolosov.[22] However, a much simpler method to estimate the effects of field ionization is to assume that states rapidly ionize for all values of $n>n_m$, where the cutoff, n_m, is given by the semiclassical formula

$$n_m = \left[\frac{6.2 \times 10^5}{E} q^3 \right]^{1/4} , \qquad (35)$$

where E is the electric field in V/cm and q is the charge of the ion before recombination. It has been found that the cutoff determined from this formula agrees closely with with the value of n for which the hydrogenic formulas predict a very rapid increase in the rate of field ionization.[23]

We determine the effects of field mixing in the interaction region by employing eigenvectors for the doubly excited states which are obtained by diagonalizing a Hamiltonian matrix that includes the stark matrix elements as well as the internal electrostatic and spin-orbit terms. Finally, the DR experiments do not measure the cross section, but rather the product of the electron velocity and the cross section convoluted with an electron velocity distribution function - a sort of rate coefficient which is a function of electron energy. Thus we combine our cross section with the experimental velocity distribution to make a theoretical prediction of the measurement.

The first DR measurements on a series of Li-like and Na-like ions were done by Dittner et al. at ORNL.[24,25] In this set of experiments, the field in the interaction region was expected to be about 30 V/cm. Although the agreement between experiment and our theoretical calculations[26] was quite good for the Na-like ions,[24] the measured rates were larger than the calculated rates for the Li-like ions (including the C^{3+} case considered above) even for the maximum field enhancement of the theoretical cross sections.[25] Furthermore, the velocity distribution in these experiments was sufficiently wide that a comparison between experiment and theory was impossible for low Rydberg states, which are not affected by electric fields.

Andersen and Bolko have now completed measurements of DR for C^{3+} and O^{5+} using the Aarhus EN-tandem accelerator and a beamline equipped with an electron cooler.[27] The electric field in the interaction region is expected to be less than 5V/cm and the electron distribution is very narrow, so that individual low Rydberg states are resolved.

Their measurements for C^{3+} in comparison with our most recent calculations for various small fields in the interaction region[28] are shown in Fig. 8. As can be seen, the agreement between experiment and theory in the region of the high Rydberg states for energies between 7.0 and 8.0 eV is excellent for a field of about 3 V/cm. However, the

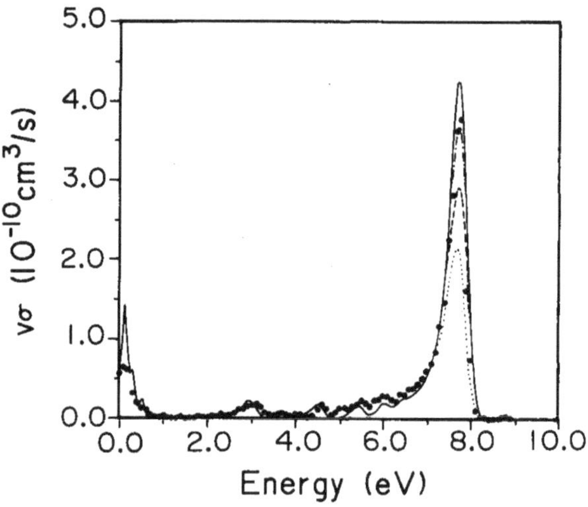

Fig. 8. Theoretical DR rate coefficients for C^{3+} (as measured in the Aarhus electron-ion merged-beam experiment) as a function of electron energy. ······ E = 0 V/cm; - - - E = 1 V/cm; -·-·-·- E = 3 V/cm; ——— E = 5 V/cm. The calculations include all resonances up to and including n_m = 35. The experimental points are from Ref. 27.

calculated rate for the peak between 0 and 0.6 eV, which is due to the 2p4d and 2p4f resonances, is nearly twice the measured rate. Since the cross section for resonant recombination varies as the inverse of the electron energy (see Eq. 15), and these resonances lie so close to zero, this discrepancy is most likely due to errors in the calculated positions of the doubly excited levels for these configurations. The agreement with experiment for O^{5+} is not quite as good, but still very reasonable.[27,28]

Additional experiments using electron coolers and ion storage rings are now being planned for more highly ionized species, and they will provide further tests of our understanding of electron-ion recombination.

5. CONCLUSIONS

Doubly excited autoionizing states play an extremely important role in electron-ion collisions. They are populated as intermediate states during electron-impact excitation, ionization and recombination. They are included in a natural way when one employs close-coupling wave functions, but in many cases, the independent-processes approximation provides a reliable means for determining total cross sections. However, as more sensitive electron-ion collision experiments are developed, more interference effects between channels will be detected, and theoretical methods which include the interaction between the doubly excited states and the adjacent continuum will be necessary.

6. ACKNOWLEDGMENTS

We wish to thank D. C. Gregory for providing us with his data on the ionization of Ti^{11+} prior to publication, A. Müller for providing us with his data on ionization of

Li^+ prior to publication, and L. Andersen and J. Bolko for furnishing us with their data on DR for C^{3+} and O^{5+} prior to publication. Conversations with M. H. Chen and K. J. Reed regarding the calculation of resonant recombination double autoionization in Fe^{15+} are also gratefully acknowledged. This work was supported by the Office of Fusion Energy, U. S. Department of Energy, under Contract No. DE-AC05-84OR21400 with Martin Marietta Energy Systems, Inc. and Contract No. DE-FG05-86ER53217 with Auburn University.

REFERENCES

1. Mitroy, J., Griffin, D. C., Norcross, D. W., and Pindzola, M. S., Phys. Rev. A38, 3339 (1988).
2. Pindzola, M. S. and Griffin, D. C., Phys. Rev. A39, 2385 (1989).
3. Cowan, R. D., J. Phys. B13, 1471 (1980).
4. Pindzola, M. S., Griffin, D. C., and Bottcher, C., Phys. Rev. A32, 822 (1985).
5. Peterkop, R. K., Zh. Eksp. Teor. Fiz. 41, 1938 (1961) [Sov. Phys. - JETP 14, 1377 (1962)].
6. Jakubowicz, H. and Moores, D. L., J. Phys. B14, 3733 (1981).
7. Moores, D. L. and Reed, K. J., Phys. Rev. A39, 1747 (1989).
8. Pindzola, M. S., Griffin, D. C., and Bottcher, C., submitted to Phys. Rev. A (1989).
9. Griffin, D. C., Pindzola, M. S., and Bottcher, C., Phys. Rev. A36, 3642 (1987).
10. Gregory, D. C., Wang, L. J., Rinn, K., and Meyer, F. W., Phys. Rev. A35, 3256 (1987).
11. Gregory, D. C., Wang, L. J., and S. Chantrenne, private communications.

12. LaGattuta, K. J. and Hahn, Y., Phys. Rev. A24, 2273 (1981).
13. Chen, M. H. and Reed, K. J., private communication.
14. Pindzola, M. S. and Griffin, D. C., Phys. Rev. A36, 2628 (1987).
15. Müller, A., Hoffmann, G., Tinschert, K., and Salzborn, E., Phys. Rev. Letts., 61, 1352 (1988).
16. Müller, A., Hofmann, G., Weissbecker, B., Stenke, M., Tinschert, K., Wagner, M., and Salzborn, E., private communication.
17. Fano, U., Phys. Rev. 124, 1866 (1961).
18. Cress, M. A., Seaton, M. J., and Wilson, P. M. H., Comput. Phys. Commun. 15, 23 (1978).
19. Alber, G., Cooper, J., and Rau, A. R. P., Phys. Rev. A30, 2845 (1984).
20. Griffin, D. C. and Pindzola, M. S., Phys. Rev. A35, 2821 (1987).
21. Bell, R. H. and Seaton, M. J., J. Phys. B18, 1589 (1985).
22. Damburg, R. J. and Kolosov, V. V., J. Phys. B12, 2637 (1979).
23. Bottcher, C., Griffin, D. C., and Pindzola, M. S., Phys. Rev. A34, 860 (1986).
24. Dittner, P. F., Datz, S., Miller, P. D., Pepmiller, P. L., and Fou, C. M., Phys. Rev. A33, 124 (1986).
25. Dittner, P. F., Datz, S., Miller, P. D., Pepmiller, P. L., and Fou, C. M., Phys. Rev. A35, 3668 (1987).
26. Griffin, D. C., Pindzola, M. S., and Bottcher, C., Phys. Rev. A33, 3124 (1986).
27. Andersen, L. H. and Bolko, J., submitted to Phys. Rev. A (1989).
28. Griffin, D. C., Pindzola, M. S., and Krylstedt, P., submitted to Phys. Rev. A (1989).

Shell Model Algorithms *

Michel Vallières
Department of Physics and Atmospheric Science
Drexel University, Philadelphia, PA 19104

Abstract

We will review in this lecture different algorithms used to perform nuclear shell-model calculations. In section 2 we will briefly mention the traditional shell-model approaches based on the m-scheme and j-j coupling; in particular in section 2.5 we will describe some of the successes of these approaches and their limitations. We then proceed to give our views about the possibilities to perform shell-model calculations beyond the *sd* shell. In section 4 and 5, we present two current schemes that allow calculations for much larger nuclei, and the consequences of these algorithms on the general philosophy of the shell-model approach.

1 Introduction

The shell-model constitutes a fundamental approach in nuclear, atomic and non-relativistic quark calculations. Over the years it has achieved considerable successes in these diverse fields; however it remains to this date a very difficult procedure to implement when the number of particles and/or the number of shells or the value of the quantum numbers specifying these shells become large. In this paper we will present first some of the "traditional" algorithms used to perform nuclear physics shell-model calculations as well as introduce two of the current schemes that extend considerably the range of these calculations. These extensions require a rethinking of the shell-model calculations in that the emphasis of the approach must shift from a description of single particle degrees of freedom to the collective ones.

In section 2, we will briefly review the traditional m-scheme, the j-j coupling scheme and a mixed scheme, as implemented in the OXBASH code. Then

* *This work was supported in part by the National Science Foundation, Grant PHY85-20634 and Grant PHY88-43235.*

section 3 will be devoted to a brief mention of the successes of the shell-model in the *sd* shell. In section 4, we will describe the FDU0 code, an implementation of a pseudo-LS scheme in terms of pairs. In section 5 we will describe the DUSM code, an entirely new algorithm for shell-model calculations, based on extensive use of the permutation group to efficiently build the multi-particle multi-shell configuration Hilbert space.

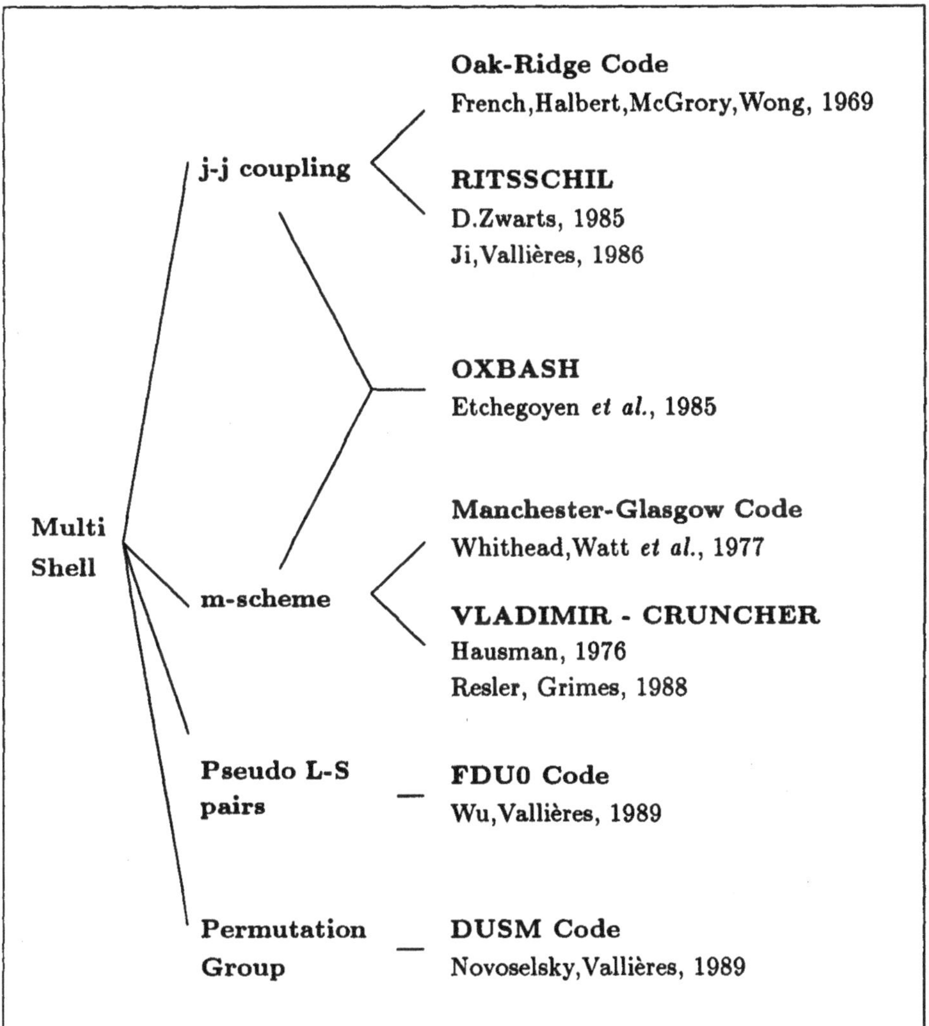

Figure 1: Different Shell-Model Codes

These different shell-model schemes can be viewed from a historical point of view as indicated in Fig. 1. The different sections of this paper will roughly correspond to the different historical events indicated in this figure.

2 Traditional Algorithms

The Oak-Ridge shell-model code [1] was the first implementation of a large scale shell-model algorithm in nuclear physics. It used a *j-j* coupled approach and became a very powerful tool to study nuclei in the *sd* shell. It is only recently that Zwarts [2] recoded this algorithm in a more flexible and presumably more powerful manner. Concurrently with the development of these codes, Whitehead *et al.* [3], the Lawrence Livermore group [4] and recently a Ohio University group [5] developed the *m*-scheme algorithm. This scheme aims at taking advantage of the arithmetic performed by computers using the binary system. It also had its successes, but mostly in the arena of "no-core" calculations for relatively small nuclei. In the following I will describe briefly the *m*-scheme algorithm, the *j-j* method, the Lanczos diagonalization procedure [6] and finally an hybrid approach implemented in the OXBASH code [7].

2.1 m-scheme

The *m*-scheme shell-model algorithm is based on the second quantization formulation of quantum mechanics and the similarity between the action of creation and annihilation operators on an N particle state and the logical AND and OR operations on the bits of computer words. A computer word is associated with each N-particle state in an uncoupled representation, namely a correspondence is established between

$$|N> \propto a^\dagger_{j_1 m_1} a^\dagger_{j_2 m_2} a^\dagger_{j_3 m_3} \ldots a^\dagger_{j_N m_N} |vac> \quad (1)$$

and

$$|N> \longrightarrow (001100110100) \qquad \textit{computer word} \quad (2)$$

A creation operator $a^\dagger_{j_k m_k}$ acting on such a state will then produce zero if the orbital is already occupied or a new state with the particular orbital occupied in the opposite case, times a phase which depends on the number of permutation necessary to let the operator $a^\dagger_{j_k m_k}$ act on the kth orbital. Therefore, the result of $a^\dagger_{j_k m_k}|N>$ can be obtained by a logical AND operation on the kth bit of the computer word corresponding to $|N>$ multiplied by a phase. A similar strategy applies to the annihilation operators. Therefore, calculating matrix elements of any operator written in second quantized form involves simple logical operations on computer words.

The obvious advantage of this scheme is that the bit manipulations involved in the scheme are very fast to perform on digital computers; this results in very efficient procedures to calculate the matrix representatives of any second quantized operators. Note also that none of the usual "angular momentum algebra" is necessary since the states are decoupled at all times. The symmetry of the hamiltonian on the other hand is preserved, i.e., a rotationally invariant hamiltonian will generate states with good J, but in a hidden way; however, this will not be apparent until the expectation values of relevant operators is taken, i.e., $< J^2 >$, in the resulting eigenstates.

A disadvantage of the approach is that the matrices generated are generally enormous. The current version of the Lawrence Livermore code can handle matrices up to 50,000 in size on a SUN computer [5]; however this may not be enough. For instance, ^{28}Si in the active *sd* shell requires a matrix of size 303,854 (M=0), while one of the largest matrix for the same nucleus in JT coupled scheme is for J=3, T=1 at size 6706. Another size limitation comes from the fact that the maximum number of orbitals that can be included is limited by the word length; multiple words can of course be combined to overcome this limitation, but this results in far less efficient coding. Another major disadvantage of the approach is that it is practically impossible to implement a truncation scheme of any type, short of diagonalizing the Casimir operators of some relevant group in this enormous space.

2.2 j-j coupling scheme

This was the very first scheme that was proposed to perform shell-model calculations [8]; excellent pedagogical references describing the details of this scheme can be found (i.e. in the book by DeShalit and Talmi [9] and in the more recent book by Brussard and Glaudemans [10]). In this approach, the antisymmetric N-particle states in each shell are first built recursively, with $N_i \leq N$; then the multi-shell Hilbert space is constructed from these single-shell states. The steps involved in this scheme can be listed very simply:

- Given the complete N-1 particle antisymmetrized states in a given shell, build the set of N-particle angular momentum coupled states (overcomplete and not fully antisymmetrized)

$$\left[|j^{N-1}\alpha' J' > |j>\right]^J \tag{3}$$

- Form antisymmetrized N particle states as linear combination of these coupled states

$$|j^N \alpha J >= \sum \left[j^{N-1}\alpha' J' j J|\}j^N \alpha J\right] \left[|j^{N-1}\alpha' J' > |j>\right]^J \tag{4}$$

  where $\left[j^{N-1}\alpha' J' j J|\}j^N \alpha J\right]$ are the Coefficients of Fractional Parentage (CFPs).

- Repeat these recursive steps to build antisymmetric states of arbitrary particle number $N_i \leq N$ in each shell.

- Form multi-shell antisymmetric states by the outer product of antisymmetric states in each shell.
- Compute the expectation value of the hamiltonian by insertion of complete sets of states for N-1 and N-2 systems (if the hamiltonian contains 1 and 2-body interaction terms) and using the identity

$$< j^N \alpha J || a^\dagger || j^{N-1} \alpha' J' > = \sqrt{N} \sqrt{(2j+1)} \, [j^{N-1} \alpha' J' j J |\} j^N \alpha J] \quad (5)$$

- Diagonalize the hamiltonian matrix (use the Lanczos algorithm [6] (see below) for large matrices).
- Compute the expectation values of the transition operators.

The most complicated and time consuming part of the algorithm indicated above is the calculation of the single-particle single-shell CFPs. The traditional algorithm to compute these CFPs dates back from Racah in 1949 [11]; he antisymmetrized the last particle wave function by force leading to a set of algebraic equations to solve for the CFPs. Redmond [12] found recurrence relations that solve Racah equations. However these yield an over-complete non-orthogonal set of basis states. A Schmidt orthogonalization procedure (the procedure used in the original RITSSCHIL code [2] to solve this problem) can lead to numerical instabilities when the dimensions of the space become large. Ji and Vallières [13] proposed to diagonalize the Grammian matrix (the matrix of overlaps) $G = < j^N \alpha' J | j^N \alpha J >$. Upon diagonalization, the large eigenvalues indicate states to be kept with norm equal to the eigenvalues and the small ones (at least 10^{-8} the large ones in VAX single precision) superfluous states to be discarded. But even this scheme has severe restrictions when the shells are large; the DUSM code which we will explain later in section 5 is the cure to this problem.

2.3 Oxford-Buenos Aires Shell Model (OXBASH) Code

This scheme resulted from an Oxford-Buenos Aires collaboration involving Rae, Etchegoyen, Godwin and Brown [7] . This shell-model code follows a hybrid algorithm between the m-scheme and the j-j scheme; it builds the basis states in m-scheme and then builds the hamiltonian matrix in a j-j coupled scheme. This approach was devised to avoid the difficulties inherent to each angular momentum schemes, namely to avoid angular momentum coupling algebra in setting up the j-j coupled basis and to avoid the very large matrices of the m-scheme. Of course, one has to pay the price for this apparent ease of the computation, namely the difficulty is in the transition from the m-scheme to the j-j coupled scheme.

2.4 The Lanczos Algorithm

The Lanczos algorithm [6] is a method to diagonalize large matrices when only few extreme (low or high) eigenvalues and eigenstates are needed. It is an algorithm that is used commonly in all shell-model codes when the dimensions of the matrices become large. It has strictly nothing to do with the m-scheme, eventhough the matrices in this scheme are almost always large enough to require the use of the Lanczos algorithm, a common misconception.

The principle of this algorithm is as follows:

- Start with ϕ_1, a vector in the Hilbert space which should contain components in the full space, but otherwise be arbitrary. Often a normalized vector $\frac{1}{\sqrt{\mathcal{N}}}(111...)$ is used.

- Build
$$H\phi_1 = \alpha_1\phi_1 + \beta_1\phi_2 \tag{6}$$

- Continue building
$$H\phi_2 = \beta_1\phi_1 + \alpha_2\phi_2 + \beta_2\phi_3 \tag{7}$$
where the hermiticity of the hamiltonian has been used

- Repeat the previous step $\mathcal{N}$ times to build an $\mathcal{N} \times \mathcal{N}$ tridiagonal matrix (because of the hermiticity of the hamiltonian) $$\begin{bmatrix} \alpha_1 & \beta_1 & & \\ \beta_1 & \alpha_2 & \beta_2 & \\ & \beta_2 & \alpha_3 & \beta_3 \\ & & . & . \end{bmatrix}$$

- Lanczos proved that the diagonalization of this matrix for $\mathcal{N} \leq N_{Hilbert}$ should provide a good approximation to the lowest (or highest) eigenvalues of the original matrix.

- The eigenvectors are then reconstructed using the eigenvectors of the small matrix and the basis states.

2.5 Results in the $0s$, $0p$ and sd shell

The greatest successes of the nuclear shell-model are undoubtedly in the *sd* shell. We refer here to nuclei from ^{16}O to ^{40}Ca for which the active or valence nucleons occupy the spherical *sd* major shell. The dimensions of the Hilbert space are not extremely large so that shell-model calculations are possible on moderate size computers using the traditional algorithms. In particular, Wildenthal [14] has performed systematics studies of all the *sd* shell nuclei whose purpose was to establish the best possible solution the shell-model could offer in this shell. He used a hamiltonian in second quantized form which included one and two body interaction terms. He fitted the energies of all the known states of a given parity for all the nuclei in the shell. The parameters of the fit include three single particle energies and 63 two body matrix elements. A systematic mass dependence, a $1/A^{0.3}$ factor in the two body matrix elements, was included. The results of these calculations are well documented in review articles; in particular refer to [14] for a recent account of the work. Suffice to say that the calculations reproduce all the systematic trends in the energy levels in a very impressive way; systematic deviations occur for nuclei at the beginning or at the end of the major shell where we expect to have to include adjacent shells in the model space. The transition rates and moments were also calculated. These require few extra parameters, the effective charges, that were fitted to the known transitions and moments. Again, the systematic trends are well reproduced.

The smaller nuclei, the *p* shell nuclei, can obviously be treated by the shell-model. An early systematic set of calculations were performed by Cohen and Kurath [15] using a simple interaction; other properties besides energies were fitted as well [16]. Also of interest for these nuclei were no-core calculations that were performed in the *m*-scheme. In particular, Jastrow functions were used with the bare Reid potential to treat these nuclei in a combination of variational calculation with shell-model approach [17]; one has also to be careful about center of mass spurious motion in these small nuclei [17].

3 Beyond the *sd* Shell

Beyond the *sd* shell lies a much more problematic region. Systematic studies of all states in these nuclei, similar to those in the *sd* shell, are no longer possible. The dimension of the Hilbert space becomes prohibitively large too quickly with the number of particles and/or the shell number. It is fair to say that so far only very preliminary calculations have been attempted: for instance, only to mention a few such attempts, are the studies of nuclei in the mass 90 region [18], Pb region [19], mass 50 isotones [20] and *of*-1*p* shell nuclei [21]. Also, Metsch [22] made some shell model studies of the mass 64 nuclei in his thesis.

It is our firm belief that shell-model calculations beyond the *sd* shell require a rethinking of the shell-model philosophy [23]. Attempts at reproducing all the available levels is no longer possible; the shell-model approach will only be aplicable in a systematic way if it aims at reproducing the collective states of nuclei in some large region of the periodic table. The task of the shell model theorist therefore becomes:

- Select a proper truncation of the model space; this truncation has to be drastic, to render the calculation possible, and systematic, to allow meaningful results.
- Derive an effective interaction to be used in the smaller model space.
- Develop efficient algorithms to perform these shell-model calculations, since the dimensions remain very large.

Concerning the problem of the truncation of the model space, it is evident that the first truncation is that of choosing an adequate set of valence orbitals. This choice is based on the well known existence of energy gaps between major shells. But this truncation is clearly not sufficient beyond the *sd* shell. This is when dynamical symmetries must be invoked in dictating a further truncation of the Hilbert space that will be adequate to describe collective degrees of freedom. In this arena, the only known truncation scheme in *j-j* coupling is the seniority scheme, which applies well for vibrational nuclei. In order to describe rotational nuclei, models that contain an $SU(3)$ group chain must be invoked. For nuclei in the *sd* shell, such models were proposed by Wigner [24], and then by Elliott [25]. These approaches are summarized in Fig. 2.

Truncation Schemes

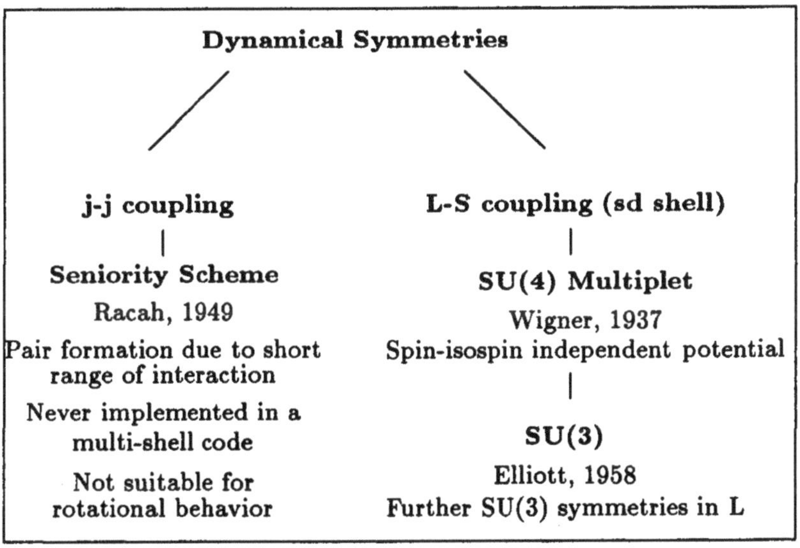

Figure 2: Truncation schemes in the *sd* shell.

For nuclei beyond the *sd* shell, Hecht [26], and independently Arima *et al.* [27], introduced the idea of an *L-S* coupling via the idea of pseudo angular momenta and spins. Fig. 3 shows the two schemes which implement this idea, the pseudo $SU(3)$ scheme which is a direct generalization of Elliott's $SU(3)$ scheme for the *sd* shell, and the Fermion Dynamical Symmetry Model (FDSM) [28]. The latter is a model based on the introduction of pair of particles coupled to angular momentum zero and two; we will say more about it in the next section. Either one of these models can be used as a model for the truncation of the Hilbert space for shell-model calculations. They define a subspace of the complete Hilbert space which is meant to be the relevant one to explain the collective degrees of freedom of the nuclei.

Beyond SD Shell

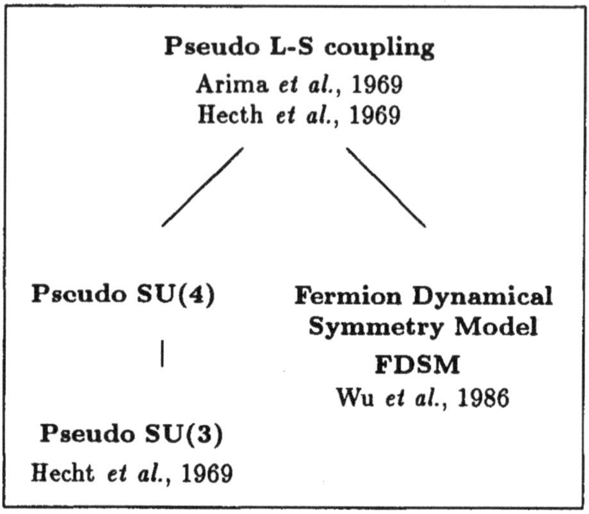

Figure 3: Truncation schemes beyond the *sd* shell.

4 FDU0

FDU0 is a shell-model code written by Hua Wu and the present author [29] which implements the truncation of the shell-model space as dictated by the Fermion Dynamical Symmetry Model (FDSM) [29]. The latter models the collective subspace of nuclei by using fermion pairs of angular momentum zero and two. Following Ginocchio [30] the spherical shells are mapped on a pseudo angular momentum, **k**, and pseudo spin, **i**, by a unitary transformation, **j** = **k** + **i**, yielding

$$b^\dagger_{km_k im_i} = \sum C^{jm_j}_{km_k im_i} a^\dagger_{jm_j}. \tag{8}$$

Pairs of fermions are then formed in which the pseudo angular momenta (spins) are coupled to angular momentum 0 or 2, the complementary spins (angular momenta) being coupled to zero, namely

$$A^{r\dagger}_m = \sum \sqrt{\Omega_{ki}/2}\left(b^\dagger_{ki} b^\dagger_{ki}\right)^{r0}_{m0} \tag{9}$$

or

$$A^{r\dagger}_m = \sum \sqrt{\Omega_{ki}/2}\left(b^\dagger_{ki} b^\dagger_{ki}\right)^{0r}_{0m}, \tag{10}$$

where r=0,2 and $\Omega_{ki} = \frac{1}{2}(2k+1)(2i+1)$. Ginocchio [30] found that the k and i values can only be 1 (k active scheme) or 3/2 (i active) if only S and D pairs are allowed. The FDSM [28] further requires that the L-S unitary transformation, i.e., **j**=**k**+**i**, maps the entire fermion space to the k-i basis when applied to one physical major shell under the condition that only one type of symmetry be used; this restricts the choice of k and i to unique values in each shell.

Besides the pair operators, multipole operators are also constructed in both the k and i active schemes:

$$p^r_u = \sum_i \sqrt{\Omega_{ki}/2}\left(b^\dagger_{1i}\tilde{b}_{1i}\right)^{r0}_{\mu 0} \tag{11}$$

and

$$p^r_u = \sum_k \sqrt{\Omega_{ki}/2}\left(b^\dagger_{k3/2}\tilde{b}_{k3/2}\right)^{0r}_{0\mu}. \tag{12}$$

In practice, the sum over k can be dropped since it is found that a unique value is generally sufficient to map the fermion space to the i-active scheme. In each schemes, the pair creation and annihilation operators and multipole operators form closed algebras: $Sp(6)$ is generated if k is active and $SO(8)$ if i is active. These groups have very little in common; their group chain decompositions for instance are very different. However, their canonical commutation relations differ in a relatively minor way, namely

$$[A^r_\mu, A^{s\dagger}_\nu] = \Omega\delta_{rs}\delta_{\mu\nu} - 2\sum_t K^{t\sigma}_{r-\mu,s\nu}(-)^\mu P^t_\sigma \tag{13}$$

$$[P^r_\mu, A^{s\dagger}_\nu] = \sum_t K^{t\sigma}_{r\mu,s\nu} A^{t\dagger}_\sigma \tag{14}$$

$$[P^r_\mu, P^s_\nu] = \frac{1}{2}\sum_t [(-)^t - (-)^{r+s}] K^{t\sigma}_{r\mu,s\nu} P^t_\sigma \tag{15}$$

where $\Omega = \frac{1}{2}\sum(2j+1)$, with the sum performed over the normal parity levels of the major shell, and

$$K^{t\sigma}_{r\mu,s\nu} = K^t_{r,s} C^{t\sigma}_{r\mu s\nu} \tag{16}$$

$$K^t_{r,s} = (-)^\alpha \sqrt{(2\alpha+1)(2r+1)(2s+1)} \left\{ \begin{array}{ccc} r & s & t \\ \alpha & \alpha & \alpha \end{array} \right\} \tag{17}$$

with $\alpha = 1$ for k active and $\alpha = \frac{3}{2}$ for i active. This suggests a possible numerical implementation of this scheme in a shell model language, where the multi-particle multi-shell wavefunctions are built out of pairs rather than single fermions. The approach then becomes a full shell-model technique performed in a systematically truncated fermion space according to the philosophy of the FDSM. The truncation of the fermion pair space to only pairs of angular momentum 0 and 2 is referred to as the u=0 subspace. The FDU0 program [29] implements this truncation scheme.

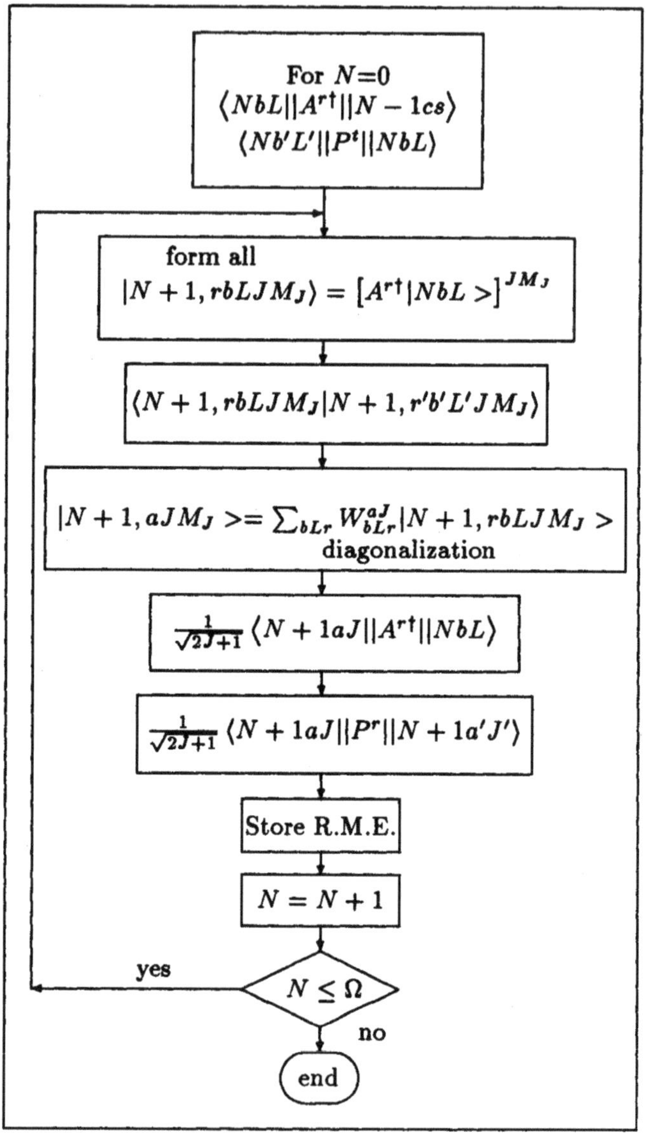

Figure 4: Iterative procedure to build the basis.

The building of the basis in a pair-coupled scheme parallels that of the j-j coupling scheme for single fermions and is best expressed in a flow chart as shown in Fig. 4. An overcomplete non-orthogonal basis for N+1 pair states is obtained by angular momentum coupling the last pair of fermions to the N-

pair wavefunctions. The procedure to compute the overlap matrix is to use the commutation relations to express the $N+1$ overlaps in terms of the N overlaps and multipole operators matrix elements. The procedure is thus recursive in N. The diagonalization of the Grammian matrix allows the selection and recursive build-up of an orthogonal basis. The calculation of matrix elements of any operator written in second quantized form by insertion of appropriate complete set of states is in principle straightforward, but tedious in practice. The details of this procedure can be found in Ref. [29].

The first use of this algorithm was to establish theoretical results: when the neutrons and protons are in shells which dictate different dynamical groups, the usual group decomposition techniques fail to provide a proof of the existence of rotational behavior. The code was then used to obtained such approximate rotational band using a generic "quadrupole-quadrupole" hamiltonian. The program was also used to obtain approximate vibrational spectra in neutron-proton systems of arbitrary group structure. These calculations are described in Ref. [29].

The procedure can also be used to "fit" energy spectra and transition rates. The FDU0 code is very efficient and can compute energies and transition rates for any even-even nucleus within the restricted FDSM model space.

5 DUSM

The Drexel University Shell Model (DUSM) [31,32,33,34] code implements a new approach to perform shell-model calculations based on permutation group concepts. It results from a collaboration of A. Novoselsky, R. Gilmore, J.Q. Chen, J. Katriel and the present author. In this approach multi-particle multi-shell wavefunctions are build in each angular momentum sector (i.e. in j and T (isospin space), L and S, or L, S, color and flavor for quarks) and then combined globally to form antisymmetric wavefunctions. This coupling order is opposite to the coupling order used in the traditional approaches. This is illustrated in Fig. 5. Note that this approach requires the construction of wavefunctions of arbitrary permutational symmetry in each subspace; the Young diagrams associated with each shell and each subspace in Fig. 5 indicate this arbitrary symmetry. Conjugate patterns (in the case of 2 subspaces) are combined in the last step to form final antisymmetric wavefunctions.

The calculation of the CFPs is performed by an efficient matrix diagonalization technique. In this approach the matrix representative of the quadratic Casimir operator of the subspace dictated unitary group is diagonalized in the angular momentum coupled basis of N-1 symmetrized state with the added last particle. The details of this method are explained in Ref. [31]. The same matrix diagonalization algorithm is used to compute the Outer Product and Inner product Isoscalar Factors (OISFs and IISFs) [32,33]; in this case the two-cycle permutation operator is used as the operator to diagonalize.

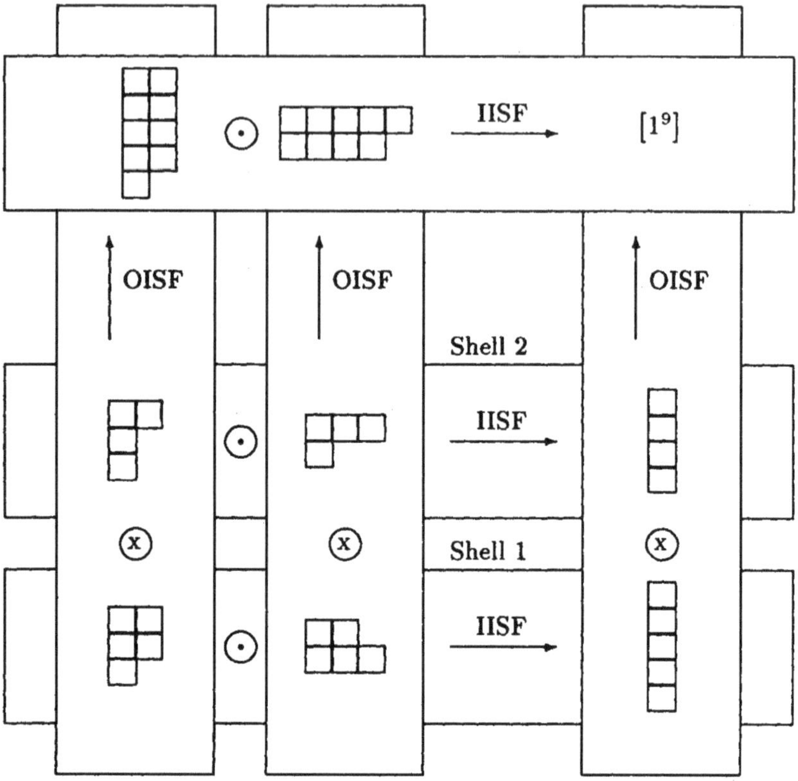

Figure 5: Coupling order in Shell-Model codes.

The matrix elements of one body and two body operators are easily computed once it is realized that the matrix elements of a creation and/or annihilation operator in any given shell is proportional to a CFPs for any permutational symmetry of the parent and daughter states. The notion of path in Young diagram space is introduced to help the bookkeeping involved in these calculations; this helps greatly in sorting out the Young diagrams in L subspace and the corresponding conjugate diagrams in S subspace. Matrix elements of one and two-body operators are then obtained by insertion of appropriate complete set of states in the N-1 and N-2 systems; this leads to the "summation over paths" technique. The details are to be found in Ref. [33].

The advantages of this approach are multifold:

- The calculation of the single particle CFP in each subspace have to deal with only one quantum number at a time; eventhough CFPs corresponding to all permutational symmetry are to be computed, this remains a very manageable task.

Table 1: CPU usage for the $(f_{7/2})^6$ case with isospin

Task	RITSSCHIL	DUSM
CFP	7 hours	54 min.
Reduced Matrix Elements	29 hours	30 min.
Hamiltonian Set-up and diagonalization	4 min.	90 sec.

- Typical calculations are multi-shell in only one of the angular momentum subspace: for instance, nuclear calculations are multi-shell in j but single shell in isospin, atomic calculations are multi shell in l but single shell in spin and non relativistic quark calculations are multi shell in l but single shell in spin, color and flavor. This results in a great economy in the calculation.

- The very efficient and general algorithm based on matrix diagonalization of Casimir operators of appropriate groups was adapted to compute all necessary coupling coefficients. This leads to the diagonalization of relatively small matrices due to the relatively few ways permutational symmetries can couple to each other in a recursive build-up of many particle systems. Furthermore, the construction and diagonalization of these matrices are easily vectorizable, making this an efficient algorithm for supercomputers.

- The permutational label is a very efficient quantum number in that it labels irreducible representation of relatively small dimensions. This explains the efficiency of the matrix diagonalization technique as quoted above.

- The Young diagram can be ordered in lexical order; this allows for the writing of efficient codes that do not require any searches.

The algorithm has proven very efficient so far. Six particles occupying the $f_{7/2}$ shell in j-j coupling with isospin can be used as a benchmark for the efficiency of the algorithm. The RITSSCHIL code is used to give a representative timing to compare with. The results are summarized Table 1. The same complete calculation can be completed in only 107 CPU seconds on the CRAY X/MP at the National Center for Supercomputer Applications (NCSA) in Illinois. The RITSSCHIL code does not adapt very well for supercomputer use in its current implementation because of too much I/O. Our current scheme does not suffer from this drawback.

The DUSM code is currently under development by it's authors; all of the necessary coupling coefficients codes are written and fully tested while the multi-shell code is near completion.

6 Conclusions

Shell-Model approach is a very exciting way to study nuclei; it is considered by most as a very fundamental approach that has had great successes in describing small nuclei. The new approaches to collective phenomena studies based on quasi-spin, quasi-angular momentum concept and dynamical symmetries, when coupled with the new shell-model algorithms, offer much for the future of the shell-model when applied to much larger nuclei.

Acknowledgments

It is a pleasure to acknowledge the many collaborators I have had recently in my shell-model research: J.-Q. Chen, R. Gilmore, X. Ji, J. Katriel, A. Novoselsky and H. Wu. Supercomputer time from the National Center for Supercomputer Applications (NCSA) in Illinois is acknowledged. This work was partially supported by the National Science Foundation, Grant PHY85-20634 and PHY88-43235.

References

[1] J.B. French, E.C. Halbert, J.B. McGrory and S.S.M. Wong, Adv. Nuc. Phys. **3** (1969) 193; E.C. Halbert, J.B. McGrory, B.H. Wildenthal and S.P Pandya, Adv. Nuc. Phys. **4** (1971) 375; see also J.B. McGrory, B.H. Wildenthal, Ann. Rev. Nucl. Part. Sci. **30** (1980) 383.

[2] D. Zwarts, Comp. Phys. Comm. **38** (1985) 565.

[3] R.R. Whitehead, A. Watt, B.J. Cole and I. Morrison, Adv. Nuc. Phys. **9** (1977) 123.

[4] VLADIMIR code, R.F. Hausman, Jr. Lawrence Livermore National Laboratory Report No UCRL-52178, Ph. D. Thesis, 1976.

[5] CRUNCHER code, D.A. Resler and S.M. Grimes, Computers in Physics, May-June 1988, 65.

[6] C. Lanczos, J. Res. Nat. Bur. Stand. **45** (1950) 55.

[7] A. Etchegoyen, W. D. M. Rae, N. S. Godwin, W. A. Richter, C. H. Zimmerman, B. A. Brown, W. E. Ormand and J. S. Winfield, MSU-NSCL report 524 (1985).

[8] M.G. Mayer, Phys. Rev. **75** (1949) 1968; O.J. Haxel, J.H.D. Jensen and H.E. Suess, Phys. Rev. **75** (1949) 1766; M.G. Mayer and J.H.D. Jensen, *Elementary Theory of Nuclear Shell Structure*, (John Wiley and Sons, (N.Y.) 1955).

[9] A. De-Shalit and I. Talmi, *Nuclear Shell Theory* (Academic Press, New York, 1963).

[10] P.J. Brussard and P.W.M. Glaudemans, *Shell Model Applications in Nuclear Spectroscopy* (North-Holland, Amsterdam, 1977).

[11] G. Racah, Phys. Rev. **63** (1943) 367.

[12] P.J. Redmond, Proc. Roy. Soc. **A222** (1954) 84.

[13] X. Ji and M. Vallières, Phys. Rev. **C35** (1987) 1583.

[14] B.H. Wildenthal, Prog. Part. Nuc. Phys., *ed.* D. Wilkenson (Pergamon, Oxford, 1984); B.H. Wildenthal in proceedings of "Nuclear Shell Models", *ed.* M. Vallières and B.H. Wildenthal, (World Scientific, 1984); B.A. Brown and B.H. Wildenthal, Ann. Rev. Nucl. Part. Sci. **38** (1988).

[15] S. Cohen and D. Kurath, Nuc. Phys., **73** (1965) 1.

[16] S. Cohen and D. Kurath, Nuc. Phys., **A141** (1970) 145; D. Kurath, Phys. Rev., **C7** (1973) 1390; D. Kurath, D.J. Millener, Nuc. Phys. **A238** (1975) 269.

[17] J.M. Irvine, G.S. Mani and M. Vallières, Czech. J. Phys. **B** (1974) 24; J.M. Irvine, G.S. Mani, V.F. Pucknell, M. Vallières and F. Yazici, Ann. Phys. (N.Y.) **102** (1976) 129.

[18] D.H. Gloeckner and F.J. Serduke, Nuc. Phys., **220** (1974) 477.

[19] J.B. McGrory aand T.T.S. Kuo, Nuc. Phys., **247** (1975) 283.

[20] X. Ji and B.H. Wildenthal, Phys. Rev., **C37** (1988) 1256; Phys. Rev., **C40** (1989) 389.

[21] J.B. McGrory, Phys. Rev. **C8** (1973) 693; T. Motoba aand K. Itomoga, Prog. Theor. Phys. Supp. **65** (1979) 136; S.T. Hsieh, X. Ji, R. Mooy and B.H. Wildenthal, AIP. Conf. Proc. (USA) **142** (1986) 357.

[22] B. Ch. Metsch, Ph.D. Thesis, U. of Utrech, 1981.

[23] X. Ji, B.H. Wildenthal and M. Vallières, Nuc. Phys. **A492** (1989) 815.

[24] E.P. Wigner, Phys. Rev. **51** (1937) 106.

[25] J.P. Elliott, Proc. Roy. Soc. **A245** (1958) 128.

[26] K. Hecht and A. Adler, Nuc. Phys. **A137** (1969) 129.

[27] A. Arima, M. Harvey and K. Shimuzu, Phys. Lett., **B30** (1969) 517.

[28] C.L. Wu, D.H. Feng, X.G. Chen, J.Q. Chen and M.W. Guidry, Phys. Lett. **168B** (1986) 313; Phys. Rev. **C36** (1987) 1157.

[29] Hua Wu and M. Vallières, Phys. Rev. **C39** (1989) 1066; FDU0 code, Hua Wu, unpublished.

[30] J.N. Ginocchio, Ann. Phys. (N.Y.) **126** (1980) 234.

[31] A. Novoselsky, J. Katriel and R. Gilmore, J. Math. Phys. **29** (1988) 1368.

[32] A. Novoselsky, M. Vallières and R. Gilmore, *Matrix Elements of Shell Model Hamiltonians in Multiple Angular Momentum Coupling Schemes*, **C38** (1988) 1440.

[33] J.-Q. Chen, A. Novoselsky, M. Vallières and R. Gilmore, *A New Approach to Multi-shell Calculations in Multiple Angular Momentum Coupling Schemes*, Phys. Rev. **C39**, 1088-1103, (1989).

[34] M. Vallières, A. Novoselsky and R. Gilmore, "New Shell-Model Algorithm" in *Proceedings of the International Conference on Computers in Physics*, L. Deyuan and D.H. Feng (Eds.), (World Scientific, Singapore, 1989).

ITERATIVE SOLUTION OF LARGE LINEAR SYSTEMS AND HEAVY PARTICLE COLLISIONS: ION-ION RECOMBINATION

E. J. Mansky
School of Physics
Georgia Institute of Technology
Atlanta, Georgia 30332

ABSTRACT

The solution of large sparse linear systems of algebraic equations arising from the discretization of coupled Boltzmann partial integro-differential equations, which model ion-ion recombination processes in dense gases, is discussed. The advantages and limitations of various representations of these equations is provided. A detailed analysis is given of the derivation and structure of the coefficient matrix $\mathcal{A}$ of the resultant algebraic equations. The need for preconditioning the algebraic equations through the calculation of the condition number of the matrix $\mathcal{A}$ is highlighted. Approximate methods of computing termolecular recombination rate coefficients via the Debye-Smoluchowski equation and diffusion models in energy space are also briefly discussed.

I. INTRODUCTION

In this lecture the numerical solution of large sets of linear algebraic equations by iterative methods will be discussed with particular application to problems in heavy particle collisions. The physical problem specifically addressed is that of ion-ion recombination at arbitrary gas densities. The determination of the rate of recombination is governed by the solution of a pair of coupled Boltzmann-like integro-differential equations (IDE's). The derivation of these coupled Boltzmann equations from a more basic perspective involving the BBGKY hierarchy of equations is reviewed in Section II. The solution of these coupled IDE's provides a general framework for discussing the problem of computing chemical reaction rates in dense plasmas. This is provided in

Section II along with a detailed discussion of the advantages and limitations of transforming the IDE's into a set composed solely of differential equations (DE's) or integral equations (IE's). In all three representations the problem of numerically solving the coupled Boltzmann equations reduces to one of solving a set of simultaneous linear algebraic equations composed of a large, sparse, real, positive definite, non-symmetric, ill-conditioned matrix. The solution of these algebraic equations by iterative techniques is highlighted in Section II.

Historically, until the advent of supercomputers, the direct solution of the coupled Boltzmann equations was generally avoided through the use of simplifying approximations because of the difficulty in solving large sets of algebraic equations. In Section III the link between the coupled Boltzmann equations and its approximations is given. In particular the formulation of the problem in terms of diffusion equations (in energy space) and Debye-Smoluchowski equations is accentuated in Section III. The conclusions of this lecture and a list of general references is given in Section IV.

II. BOLTZMANN EQUATION TREATMENT OF IONIC RECOMBINATION

The overall goal of the type of calculations described in this lecture is the prediction from a microscopic viewpoint, of the rate of chemical reactions in dense gases. The proto-type chemical reaction we are primarily interested in is that of ion-ion recombination at arbitrary gas densities,

$$X^+ + Y^- + Z \overset{\alpha}{\rightarrow} [XY]^* + Z \tag{1}$$

whereby free ions (X^+,Y^-) are converted into diatomic molecules XY (usually in some metastable state denoted by *). We will assume that the number density of third bodies Z is arbitrary, but that the free ion number densities is sufficiently low so that the interaction potential between ions is strictly Coulombic. This will necessarily exclude from discussion dynamic screening effects in dense plasmas. We will also not discuss the related problem of ion-atom association,

$$X^+ + Y + Z \rightarrow [XY^+]^* + Z$$

which is an important mechanism by which molecular ions are formed in interstellar media and in laboratory plasmas. To solve both problems from a microscopic standpoint will require a great deal of information on the full three-body sector of phase space which is beyond the scope of this lecture to provide. Anyway, before the latter two problems can be solved, a complete understanding of the solution of the termolecular recombination rate in the limit of low ionic density and arbitrary gas density will be needed.

We are interested in computing in this lecture microscopic reaction rates which the reader should take to mean that the reaction rates will be expressed in terms of the phase space distribution functions f_N for the N particles comprising the three component plasma (positively and negatively charged particles as well as neutral species) undergoing termolecular recombination. Our starting point is the BBGKY hierarchy of equations,

$$\frac{\partial f_s}{\partial t} = - \mathcal{H}_s f_s + n\mathcal{L}_s f_{s+1} \qquad s = 1,2,3,\ldots,N-1 \tag{2}$$

which is a set of coupled equations for the s-particle reduced distribution functions $f_s = f_s(x_1,x_2,\ldots,x_s;t) = V^{-(N-s)} \int dx_{s+1} \int dx_{s+2} \ldots \int dx_N \, f_N(x_1,\ldots,x_N;t)$ with V denoting the total volume of phase space and $x_i:(\vec{r}_i,\vec{p}_i)$ denotes the 6-dimensional phase space point for particle i. In equation (2), the Hamiltonian operator for s particles of equal mass m is defined,

$$\mathcal{H}_s = \sum_{i=1}^{s} \frac{\vec{p}_i}{m} \vec{\nabla}_{\vec{r}_i} + \frac{\vec{\mathcal{F}}_i}{m} \bullet \vec{\nabla}_{\vec{p}_i} - \sum_{1 \leq i < j \leq s} \theta_{ij} \tag{3}$$

where $\vec{\mathcal{F}}_i$ is the external force on particle i, and the interaction operator between particles i and j, is

$$\theta_{ij} = \frac{\partial\phi_{ij}}{\partial\vec{r}_i} \bullet \vec{\nabla}_{\vec{p}_i} + \frac{\partial\phi_{ij}}{\partial\vec{r}_j} \bullet \vec{\nabla}_{\vec{p}_j} \tag{4}$$

with the interaction potential between particles i and j denoted ϕ_{ij}. The phase-mixing operator $\mathcal{L}_s$ in the BBGKY equations (2) is defined,

$$\mathcal{L}_s = \sum_{i=1}^{s} \int dx_{s+1}\, \theta_{i,s+1} \tag{5}$$

and the number density n in (2) is n = N/V. For a detailed derivation of the BBGKY hierarchy (2) the reader is referred to the statistical mechanical literature (Akhiezer et al. [1], Balescu [3,4], Chapman and Cowling [13], Ferziger and Kaper [18] and Tolman [41]).

Since in ionic recombination we are interested in the formation of diatomic molecules, it is natural to assume that the most important reduced distribution functions in the three component plasma are those for one and two particles. Hence we will truncate the BBGKY hierarchy of equations at f_3 and concentrate on the equations for f_1, f_2. Also, since we are interested only in the recombination of positively and negatively charged particles to form neutral diatomic molecules, it follows that the main determining factor in computing α, the rate of reaction (1) will be the pair correlation function $g_2^{(+-)}$ between X^+ and Y^-. From this we conclude that a separate BBGKY hierarchy (2) will be required for each component of the plasma. These hierarchies for the three component plasma are,

$$\dot{f}_1^{(+)} = -\mathcal{H}_1^{(+)} f_1^{(+)} + \mathcal{L}_1^{(++)} f_2^{(++)} + \mathcal{L}_1^{(+-)} f_2^{(+-)} + \mathcal{L}_1^{(+n)} f_2^{(+n)} \tag{6a}$$
$$\dot{f}_1^{(-)} = -\mathcal{H}_1^{(-)} f_1^{(-)} + \mathcal{L}_1^{(-+)} f_2^{(-+)} + \mathcal{L}_1^{(--)} f_2^{(--)} + \mathcal{L}_1^{(-n)} f_2^{(-n)} \tag{6b}$$
$$\dot{f}_1^{(n)} = -\mathcal{H}_1^{(n)} f_1^{(n)} + \mathcal{L}_1^{(n+)} f_2^{(n+)} + \mathcal{L}_1^{(n-)} f_2^{(n-)} + \mathcal{L}_1^{(nn)} f_2^{(nn)} \tag{6c}$$

$$\dot{f}_2^{(++)} = -\mathcal{H}_2^{(++)} f_2^{(++)} + \mathcal{L}_2^{(+++)} f_3^{(+++)} + \mathcal{L}_2^{(++-)} f_3^{(++-)} + \mathcal{L}_2^{(++n)} f_3^{(++n)} \quad (7a)$$

$$\dot{f}_2^{(+-)} = -\mathcal{H}_2^{(+-)} f_2^{(+-)} + \mathcal{L}_2^{(+-+)} f_3^{(+-+)} + \mathcal{L}_2^{(+--)} f_3^{(+--)} + \mathcal{L}_2^{(+-n)} f_3^{(+-n)} \quad (7b)$$

$$\dot{f}_2^{(+n)} = -\mathcal{H}_2^{(+n)} f_2^{(+n)} + \mathcal{L}_2^{(+n+)} f_3^{(+n+)} + \mathcal{L}_2^{(+n-)} f_3^{(+n-)} + \mathcal{L}_2^{(+nn)} f_3^{(+nn)} \quad (7c)$$

$$\dot{f}_2^{(--)} = -\mathcal{H}_2^{(--)} f_2^{(--)} + \mathcal{L}_2^{(--+)} f_3^{(--+)} + \mathcal{L}_2^{(---)} f_3^{(---)} + \mathcal{L}_2^{(--n)} f_3^{(--n)} \quad (7d)$$

$$\dot{f}_2^{(-n)} = -\mathcal{H}_2^{(-n)} f_2^{(-n)} + \mathcal{L}_2^{(-n+)} f_3^{(-n+)} + \mathcal{L}_2^{(-n-)} f_3^{(-n-)} + \mathcal{L}_2^{(-nn)} f_3^{(-nn)} \quad (7e)$$

$$\dot{f}_2^{(nn)} = -\mathcal{H}_2^{(nn)} f_2^{(nn)} + \mathcal{L}_2^{(nn+)} f_3^{(nn+)} + \mathcal{L}_2^{(nn-)} f_3^{(nn-)} + \mathcal{L}_2^{(nnn)} f_3^{(nnn)} \quad (7f)$$

where the superscripts +, -, n indicates a positively or negatively charged particle or a neutral species, respectively, and the dots indicate differentiation w.r.t. time.

The set of coupled equations (6,7) are closed by use of the cluster expansion (Ferziger and Kaper [18]) wherein the 3 particle distributions f_3 are written as functionals of the 1 and 2-particle distribution functions. The latter functions are written in turn as functionals of f_1 thereby closing the set of equations (6,7). For a detailed derivation of these equations see Mansky [35] and Flannery and Mansky [28]. Since we are primarily interested in the numerical aspects of the problem of computing ionic recombination rates α in this lecture, we will omit the details of the subsequent reduction of the coupled equations (6,7) to the working equations (8), but refer the reader to the above two references as well as two earlier important papers of Flannery's [24,25].

Therefore, after reduction, the final steady state working equations are

$$\left[\frac{1}{r} - \lambda\right] \frac{\partial \rho^{+}(r,\lambda)}{\partial r} + \frac{1}{r}\left[\frac{1}{r} - 2\lambda\right][\rho^{+}(r,\lambda) - \rho^{-}(r,\lambda)] = \Gamma' \int_{-\infty}^{1/r} \rho^{+}(r,\mu)F(\lambda,\mu;r)d\mu - \Gamma' \rho^{+}(r,\lambda)\mathcal{F}(r,\lambda) \quad (8a)$$

$$-\left[\frac{1}{r} - \lambda\right] \frac{\partial \rho^{-}(r,\lambda)}{\partial r} = \Gamma' \int_{-\infty}^{1/r} \rho^{-}(r,\mu)F(\lambda,\mu;r)d\mu - \Gamma' \rho^{-}(r,\lambda)\mathcal{F}(r,\lambda) \quad (8b)$$

$$\left[\frac{1}{r} - \lambda\right] \frac{\partial \rho^{+}(r,\lambda)}{\partial r} = \Gamma' \int_{-\infty}^{1/r} \rho^{+}(r,\mu)F(\lambda,\mu;r)d\mu - \Gamma' \rho^{+}(r,\lambda)\mathcal{F}(r,\lambda) \quad (8c)$$

$$-\left[\frac{1}{r} - \lambda\right] \frac{\partial \rho^{-}(r,\lambda)}{\partial r} - \frac{1}{r}\left|\frac{1}{r} - 2\lambda\right| [\rho^{+}(r,\lambda)-\rho^{-}(r,\lambda)] = \Gamma' \int_{-\infty}^{1/r} \rho^{-}(r,\mu)F(\lambda,\mu;r)d\mu - \Gamma' \rho^{-}(r,\lambda)\mathcal{F}(r,\lambda) \quad (8d)$$

where equations (8a,b) are valid for region I: $-\infty < \lambda \le \frac{1}{2r}$ and (8c,d) are valid for region II: $\frac{1}{2r} \le \lambda \le \frac{1}{r}$ (see figure 1). The functions $\rho^{(\pm)}(r,\lambda)$ represented the number density of ion-pairs expanding (+) and contracting (-) at a given relative separation r and internal energy λ in phase space. The ρ's are just the ratio of the number density of ion-pairs undergoing recombination to the equilibrium number density (i.e., $\rho^{(\pm)}(r,\lambda) = n^{(\pm)}(r,\lambda)/n_{eq}$). The working equations (8) are written in terms of dimensionless natural variables which are defined,

$$r = r_{12}/R_e, \qquad \lambda = -E_i/kT, \qquad \mu = -E_f/kT$$

where $R_e = e^2/kT$ and $\Gamma' \equiv \Gamma\, e^{-\lambda}$, $\Gamma = \frac{\Lambda_{mfp}}{2\sqrt{\pi}}\Gamma_o$, $\Lambda_{mfp} = R_e/\lambda_{mfp}$, $\lambda_{mfp} = (N(Z)Q^D)^{-1}$. Hence the dependence on gas density in (8) is contained in the constant Γ, which also depends on the masses via,

$$\Gamma_o \equiv \begin{cases} [(1+c)/c]^{3/2} & \text{, charge transfer} \\ (1+a)^2/a^{3/2} & \text{, hard sphere} \\ \sqrt{3/2}\,(1+a)^{5/2}/a^{3/2} & \text{, polarization} \end{cases}$$

and,

$$\mathcal{F}(r,\lambda) \equiv \int_{-\infty}^{1/r} F(\lambda,\mu;r)d\mu$$

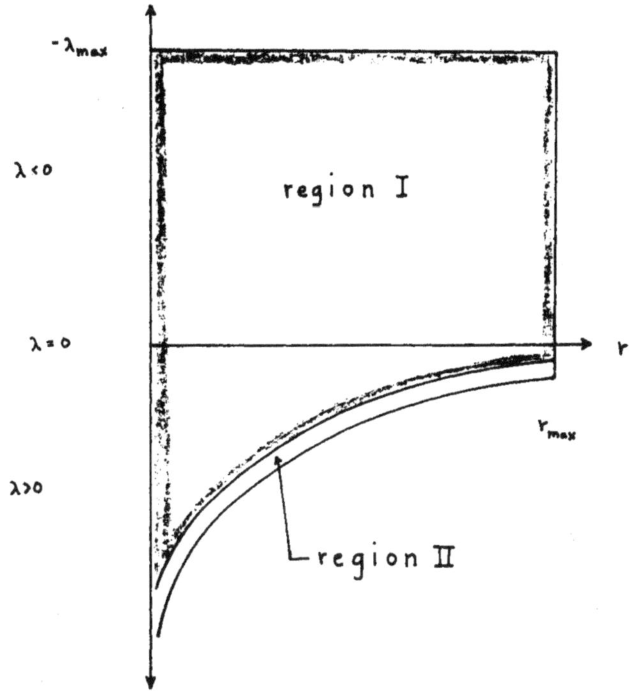

Figure 1: Illustration of λ, r phase space domain of equation (8).

The detailed formulae for the energy-change rate coefficients $F(\lambda,\mu;r)$ for the energy transfer mechanisms of charge-transfer, hard sphere and polarization collisions need not concern us here, but can be found in the original literature Flannery [21-23] and Flannery and Mansky [27]). In the definition of Γ_0 the mass ratio parameters are defined,

$$a = m_2 m_3 / m_1 (m_1 + m_2 + m_3)$$

$$c = m_1 / m_2$$

where m_2, m_2 and m_3 are the masses of X^+, Y^- and Z, respectively.

The functions $\rho^{\pm}(r,\lambda)$ are not completely determined by the working equations (8) until their associated boundary conditions are specified,

$$\rho^{\pm}(r,\lambda\to-\infty) = 1 \tag{9a}$$

$$\rho^{\pm}(r\to\infty,\lambda) = 1 \quad \text{for } \lambda \leq 0 \tag{9b}$$

$$\rho^{+}(r=o,\lambda) = \rho^{-}(r=o,\lambda) \tag{9c}$$

$$\rho^{+}(r,\frac{1}{r}) = \rho^{-}(r,\frac{1}{r}) \tag{9d}$$

Before we convert the working equations (8) into a set of practical, numerical equations we will discuss the relative computational merits of transforming the integro-differential equations (8) into equivalent differential equations or integral equations.

Technically, the working equations (8) are coupled Volterra partial integro-differential equations (PIDE's). They are 1st-order in r and 1-dimensional in λ meaning that the highest derivative in r appearing in (8) is the first, while only single integrals w.r.t. λ appear. To convert (8) into a set of PDE's define the functions $\tilde{\rho}^{\pm}(r,\lambda)$ by

$$\rho^{\pm}(r,\lambda) = \frac{1}{F(\mu,\lambda;r)}\frac{\partial\tilde{\rho}^{\pm}(r,\lambda)}{\partial\lambda} \tag{10}$$

yielding,

$$(\frac{1}{r}-\lambda)\frac{\partial}{\partial r}\left[\frac{1}{F(\mu,\lambda;r)}\frac{\partial\tilde{\rho}^{+}(r,\lambda)}{\partial\lambda}\right] + \frac{\frac{1}{r}\left[\frac{1}{r}-2\lambda\right]}{F(\mu,\lambda;r)}\left[\frac{\partial\tilde{\rho}^{+}(r,\lambda)}{\partial\lambda} - \frac{\partial\tilde{\rho}^{-}(r,\lambda)}{\partial\lambda}\right]$$

$$+ \Gamma'\frac{\mathcal{F}(r,\lambda)}{F(\mu,\lambda;r)}\frac{\partial\tilde{\rho}^{+}(r,\lambda)}{\partial\lambda} = \Gamma'[\tilde{\rho}^{+}(r,\frac{1}{r}) - \tilde{\rho}^{+}(r,-\infty)] \tag{11a}$$

$$-(\frac{1}{r}-\lambda)\frac{\partial}{\partial r}\left[\frac{1}{F(\mu,\lambda;r)}\frac{\partial\tilde{\rho}^{-}(r,\lambda)}{\partial\lambda}\right] + \Gamma'\frac{\mathcal{F}(r,\lambda)}{F(\mu,\lambda;r)}\frac{\partial\tilde{\rho}^{-}(r,\lambda)}{\partial\lambda} = \Gamma'[\tilde{\rho}^{-}(r,\frac{1}{r})-\tilde{\rho}^{-}(r,-\infty)] \tag{11b}$$

$$(\frac{1}{r}-\lambda)\frac{\partial}{\partial r}\left[\frac{1}{F(\mu,\lambda;r)}\frac{\partial\tilde{\rho}^{+}(r,\lambda)}{\partial\lambda}\right] + \Gamma'\frac{\mathcal{F}(r,\lambda)}{F(\mu,\lambda;r)}\frac{\partial\tilde{\rho}^{+}(r,\lambda)}{\partial\lambda} = \Gamma'[\tilde{\rho}^{+}(r,\frac{1}{r})-\tilde{\rho}^{-}(r,-\infty)] \tag{11c}$$

$$-\left(\frac{1}{r}-\lambda\right)\frac{\partial}{\partial r}\left[\frac{1}{F(\mu,\lambda;r)}\frac{\partial\tilde{\rho}^-(r,\lambda)}{\partial\lambda}\right]+\Gamma'\frac{\mathcal{F}(r,\lambda)}{F(\mu,\lambda;r)}\frac{\partial\tilde{\rho}^-(r,\lambda)}{\partial\lambda}-\frac{\frac{1}{r}\left|\frac{1}{r}-2\lambda\right|}{F(\mu,\lambda;r)}\left[\frac{\partial\tilde{\rho}^+}{\partial\lambda}-\frac{\partial\tilde{\rho}^-}{\partial\lambda}\right]$$

$$=\Gamma'[\tilde{\rho}^-(r,\tfrac{1}{r})-\tilde{\rho}^-(r,-\infty)] \qquad (11d)$$

Performing the indicated differentiation w.r.t. r and rearranging terms results in the following set of coupled hyperbolic 2nd-order partial differential equations,

$$\frac{\left[\frac{1}{r}-\lambda\right]}{F(\mu,\lambda;r)}\frac{\partial^2\tilde{\rho}^+(r,\lambda)}{\partial r\partial\lambda}+\left[\frac{\frac{1}{r}\left[\frac{1}{r}-2\lambda\right]}{F(\mu,\lambda;r)}-\frac{\left[\frac{1}{r}-\lambda\right]F'(\mu,\lambda;r)}{(F(\mu,\lambda,;r))^2}+\frac{\Gamma'\mathcal{F}(r,\lambda)}{F(\mu,\lambda;r)}\right]\frac{\partial\tilde{\rho}^+(r,\lambda)}{\partial\lambda}$$

$$-\frac{\frac{1}{r}\left[\frac{1}{r}-2\lambda\right]}{F(\mu,\lambda;r)}\frac{\partial\tilde{\rho}^-(r,\lambda)}{\partial\lambda}=\Gamma'[\tilde{\rho}^+(r,\tfrac{1}{r})-\tilde{\rho}^+(r,-\infty)] \qquad (12a)$$

$$\frac{-\left[\frac{1}{r}-\lambda\right]}{F(\mu,\lambda;r)}\frac{\partial^2\tilde{\rho}^-(r,\lambda)}{\partial r\partial\lambda}+\left[\frac{\Gamma'\mathcal{F}(r,\lambda)}{F(\mu,\lambda;r)}+\frac{\left[\frac{1}{r}-\lambda\right]F'(\mu,\lambda;r)}{(F(\mu,\lambda,;r))^2}\right]\frac{\partial\tilde{\rho}^-(r,\lambda)}{\partial\lambda}$$

$$=\Gamma'[\tilde{\rho}^-(r,\tfrac{1}{r})-\tilde{\rho}^-(r,-\infty)] \qquad (12b)$$

$$\frac{\left[\frac{1}{r}-\lambda\right]}{F(\mu,\lambda;r)}\frac{\partial^2\tilde{\rho}^+(r,\lambda)}{\partial r\partial\lambda}+\left[\frac{\Gamma'\mathcal{F}(r,\lambda)}{F(\mu,\lambda;r)}-\frac{\left[\frac{1}{r}-\lambda\right]F'(\mu,\lambda;r)}{(F(\mu,\lambda,;r))^2}\right]\frac{\partial\tilde{\rho}^+(r,\lambda)}{\partial\lambda}$$

$$=\Gamma'[\tilde{\rho}^+(r,\tfrac{1}{r})-\tilde{\rho}^+(r,-\infty)] \qquad (12c)$$

$$\frac{-\left[\frac{1}{r}-\lambda\right]}{F(\mu,\lambda;r)}\frac{\partial^2\tilde{\rho}^-(r,\lambda)}{\partial r\partial\lambda}+\left[\frac{\frac{1}{r}\left|\frac{1}{r}-2\lambda\right|}{F(\mu,\lambda;r)}+\frac{\left[\frac{1}{r}-\lambda\right]F'(\mu,\lambda;r)}{(F(\mu,\lambda,;r))^2}+\frac{\Gamma'\mathcal{F}(r,\lambda)}{F(\mu,\lambda;r)}\right]\frac{\partial\tilde{\rho}^-(r,\lambda)}{\partial\lambda}$$

$$-\frac{\frac{1}{r}\left|\frac{1}{r}-2\lambda\right|}{F(\mu,\lambda;r)}\frac{\partial\widetilde{\rho}^{+}(r,\lambda)}{\partial\lambda} = \Gamma'[\widetilde{\rho}^{-}(r,\frac{1}{r}) - \widetilde{\rho}^{-}(r,-\infty)] \qquad (12d)$$

with the boundary conditions at the turning point ($\lambda = 1/r$) and in the continuum ($\lambda \to -\infty$) incorporated into the RHS of (12). Equations (12a,b) are valid only in region I, while (12c,d) are valid only in region II. Note that primes on $F(\mu,\lambda;r)$ in (12) denote differentiation w.r.t. r, and that $F(\mu,\lambda;r) = F(\lambda,\mu;r)$.

To convert (8) into an analogous set of coupled integral equations, define the functions $\overline{\rho}^{\pm}(r,\lambda)$ by,

$$\rho^{\pm}(r,\lambda) = \int_{0}^{r} \overline{\rho}^{\pm}(r',\lambda)dr' \qquad (13)$$

yielding,

$$(\frac{1}{r} - \lambda)[\overline{\rho}^{+}(r,\lambda) - \overline{\rho}^{+}(o,\lambda)] + \frac{1}{r}(\frac{1}{r} - 2\lambda)\int_{0}^{r}[\overline{\rho}^{+}(r',\lambda) - \overline{\rho}^{-}(r',\lambda)]dr'$$

$$= \Gamma'\int_{-\infty}^{1/r} d\mu\, F(\lambda,\mu;r)\int_{0}^{r}\overline{\rho}^{+}(r',\mu)dr' - \Gamma'\,\mathcal{F}(r,\lambda)\int_{0}^{r}\overline{\rho}^{+}(r',\lambda)dr' \qquad (13a)$$

$$-(\frac{1}{r} - \lambda)[\overline{\rho}^{-}(r,\lambda) - \overline{\rho}^{-}(o,\lambda)] = \Gamma'\int_{-\infty}^{1/r} d\mu F(\lambda,\mu;r)\int_{0}^{r}\overline{\rho}^{-}(r',\mu)dr'$$

$$- \Gamma'\mathcal{F}(r,\lambda)\int_{0}^{r}\overline{\rho}^{-}(r',\lambda)dr' \qquad (13b)$$

$$(\frac{1}{r} - \lambda)[\overline{\rho}^{+}(r,\lambda) - \overline{\rho}^{+}(o,\lambda)] = \Gamma'\int_{-\infty}^{1/r} d\mu\, F(\lambda,\mu;r)\int_{0}^{r}\overline{\rho}^{+}(r',\mu)dr'$$

$$- \Gamma'\,\mathcal{F}(r,\lambda)\int_{0}^{r}\overline{\rho}^{+}(r',\lambda)dr' \qquad (13c)$$

$$-(\frac{1}{r} - \lambda)[\bar{\rho}^-(r,\lambda) - \bar{\rho}^-(o,\lambda)] - \frac{1}{r}\left|\frac{1}{r} - 2\lambda\right| \int_o^r [\bar{\rho}^+(r',\lambda) - \bar{\rho}^-(r',\lambda)]dr'$$

$$= \Gamma' \int_{-\infty}^{1/r} d\mu\, F(\lambda,\mu;r) \int_o^r \bar{\rho}^-(r',\mu)dr' - \Gamma'\, \mathcal{F}(r,\lambda) \int_o^r \bar{\rho}^-(r',\lambda)dr' \tag{13d}$$

Rearranging terms in (13) to show the couplings present between the integral equations gives,

$$(\frac{1}{r} - \lambda)\, \bar{\rho}^+(r,\lambda) + [\, \frac{1}{r}(\, \frac{1}{r} - 2\lambda) + \Gamma'\, \mathcal{F}(r,\lambda)] \int_o^r \bar{\rho}^+(r',\lambda)dr'$$

$$-\Gamma' \int_{-\infty}^{1/r} d\mu\, F(\lambda,\mu;r) \int_o^r \bar{\rho}^+(r',\mu)dr' - \frac{1}{r}(\frac{1}{r} - 2\lambda) \int_o^r \bar{\rho}^-(r',\lambda)dr'$$

$$= (\frac{1}{r} - \lambda)\bar{\rho}^+(o,\lambda) \tag{14a}$$

$$-(\frac{1}{r} - \lambda)\, \bar{\rho}^-(r,\lambda) + \Gamma'\, \mathcal{F}(r,\lambda) \int_o^r \bar{\rho}^-(r',\lambda)dr'$$

$$-\Gamma' \int_{-\infty}^{1/r} d\mu\, F(\lambda,\mu;r) \int_o^r \bar{\rho}^-(r',\mu)dr' = -(\frac{1}{r} - \lambda)\, \bar{\rho}^-(o,\lambda) \tag{14b}$$

$$(\frac{1}{r} - \lambda)\, \bar{\rho}^+(r,\lambda) + \Gamma'\, \mathcal{F}(r,\lambda) \int_o^r \bar{\rho}^+(r',\lambda)dr'$$

$$-\Gamma' \int_{-\infty}^{1/r} d\mu\, F(\lambda,\mu;r) \int_o^r \bar{\rho}^+(r',\mu)dr' = (\frac{1}{r} - \lambda)\, \bar{\rho}^+(o,\lambda) \tag{14c}$$

$$-(\frac{1}{r} - \lambda)\, \bar{\rho}^-(r,\lambda) + [\, \frac{1}{r}\left|\frac{1}{r} - 2\lambda\right| + \Gamma'\, \mathcal{F}(r,\lambda)] \int_o^r \bar{\rho}^-(r',\lambda)dr'$$

$$-\Gamma' \int_{-\infty}^{1/r} d\mu\, F(\lambda,\mu;r) \int_o^r \bar{\rho}^-(r',\mu)dr' - \frac{1}{r}\left|\frac{1}{r} - 2\lambda\right| \int_o^r \bar{\rho}^+(r',\lambda)dr'$$

$$= -(\frac{1}{r} - \lambda)\, \bar{\rho}^-(o,\lambda) \tag{14d}$$

where the boundary condition at r = o has been incorporated into the RHS of (14). The three representations of the coupled Boltzmann equations (equations (8), (12) and (14)) all require the same number of quadratures to obtain a solution - namely two each for $\rho^+(r,\lambda)$ and $\rho^-(r,\lambda)$. However, our reason for giving the details of the transformation between representations (cf. equations (10), (13)) is to highlight the different types of boundary conditions required in each case. In the case of the PDE's (12), it is clear from (10) that the required boundary conditions on $\tilde{\rho}^{\pm}(r,\lambda)$ should be global in energy and local in r, while in the case of (14) the boundary conditions on $\bar{\rho}^{\pm}(r,\lambda)$ should be local in energy and global in r. We use the word global to indicate that the integrand of an integral w.r.t. the specified degree of freedom is required as a boundary condition. Otherwise it is called a local boundary condition (e.g., equation (9) is local in both r and λ). Therefore, from (10) and (13) it is clear that the boundary conditions for the PDE and IE representations are of a mixed nature, and will be difficult to implement numerically. It should be clear however that in all three cases (eqs. (8), (12) and (14)), after discretization, the basic problem numerically is the same - namely one of solving a set of simulataneous algebraic equations for the PIDE representation (8) (these are the practical equations mentioned earlier), and leave it to the reader to write down the corresponding sets of equations for the other representations (12), (14).

Numerical Solution of PIDE's

To convert the coupled PIDE's (8) into algebraic equations, four steps need to be taken:

(i) Replace all derivatives with finite differences. If the PIDE is part of an initial value problem, the choice of either forward or backward differences will depend on the boundary conditions.

(ii) Replace all integrals with quadrature sums. The choice of quadrature rule is crucial in determining the overall stability and convergence rate of the resulting algorithm. The type of quadrature rule chosen in turn depends on the global behavior of the integrand over all of phase space. Hence, for a multidimensional kernal, this step can easily be the most time consuming one in preparing for the full solution.

(iii) Impose all boundary conditions on the algebraic equations resulting from steps (i) and (ii). Make sure that the boundary conditions used lead to a well-posed problem with a non-singular coefficient matrix.

(iv) Finally, choose a technique for solving the resultant set of algebraic equations which takes advantage as much as possible of the structure of the coefficient matrix. Compute the condition number of the coefficient matrix and determine whether the algebraic equations need preconditioning.

Therefore, discretizing $r \to r_i = \{o, r_1, r_2, \ldots, r_{max} \equiv r_{N_r}\}$ with $N_r + 1$ equally spaced points (step size h), and replacing integrals with quadrature sums (with weight functions ϕ_k) yields the following for (8),

$$T_{ij} \frac{[\rho^+_{i+1,j} - \rho^+_{i-1,j}]}{2h} + \mathcal{G}_{ij}(\rho^+_{ij} - \rho^-_{ij}) + \Gamma'_j \mathcal{F}_{ij}\rho^+_{ij} - \Gamma'_j \sum_k \phi_k F_{jki}\rho^+_{ik} = 0 \tag{15a}$$

$$-T_{ij} \frac{[\rho^-_{i+1,j} - \rho^-_{i-1,j}]}{2h} + \Gamma'_j \mathcal{F}_{ij}\rho^-_{ij} - \Gamma'_j \sum_k \phi_k F_{jki}\rho^-_{ik} = 0 \tag{15b}$$

$$T_{ij} \frac{[\rho^+_{i+1,j} - \rho^+_{i-1,j}]}{2h} + \Gamma'_j \mathcal{F}_{ij}\rho^+_{ij} - \Gamma'_j \sum_k \phi_k F_{jki}\rho^+_{ik} = 0 \tag{15c}$$

$$-T_{ij} \frac{[\rho^-_{i+1,j} - \rho^-_{i-1,j}]}{2h} - |\mathcal{G}_{ij}|(\rho^+_{ij} - \rho^-_{ij}) + \Gamma' \mathcal{F}_{ij}\rho^-_{ij} - \Gamma'_j \sum_k \phi_k F_{jki}\rho^-_{ik} = 0 \tag{15d}$$

where $\Gamma_j \equiv \Gamma e^{-\lambda_j}$, $F_{jki} = F(\lambda_j, \mu_k; r_i)$, $\mathcal{F}_{ij} = \mathcal{F}(r_i, \lambda_j)$, $\mathcal{G}_{ij} = \frac{1}{r_i}(\frac{1}{r_i} - 2\lambda_j)$,

$|\mathcal{G}_{ij}| = \frac{1}{r_i}\,\left|\frac{1}{r_i} - 2\lambda_j\right|$, $T_{ij} = \frac{1}{r_i} - \lambda_j$ and $\rho^{\pm}_{ij}$ denotes $\rho^{\pm}(r_i,\lambda_j)$.

It will prove convenient to rearrange equations (15) into the order (15a,c), followed by (15b,d), so that the resultant coefficient matrix is positive definite (Golub and van Loan [30]). Then applying the boundary conditions on r yields,

$$-(1-\delta_{io})T_{ij_1}\rho^{+}_{i-1,j_1} - 2h\sum_{k_1}\left[\Gamma'_{j_1}\phi_{k_1}F_{j_1k_1i} - \delta_{k_1j_1}(\mathcal{G}_{ij_1} + \Gamma'_{j_1}\mathcal{F}_{ij_1})\right]\rho^{+}_{ik_1}$$

$$- 2h\,\Gamma'_{j_1}\sum_{k_2}\phi_{k_2}F_{j_1k_2i}\,\rho^{+}_{ik_2} - 2h\,\mathcal{G}_{ij_1}\,\rho^{-}_{ij_1} + (1-\delta_{N_r})\rho^{+}_{i+1,j_1}$$

$$= \delta_{io}\rho^{+}(r=o,\lambda_{j_1}) + \delta_{iN_r}\rho^{+}(r_{max},\lambda_{j_1}) \qquad (16a)$$

$$-(1-\delta_{io})T_{ij_2}\rho^{+}_{i-1,j_2} - 2h\,\Gamma'_{j_2}\sum_{k_1}\phi_{k_1}F_{j_2k_1i}\,\rho^{+}_{ik_1}$$

$$- 2h\sum_{k_2}\left[\Gamma'_{j_2}\phi_{k_2}F_{j_2k_2i} - \delta_{k_2j_2}\,\Gamma'_{j_2}\mathcal{F}_{ij_2}\right]\rho^{+}_{ik_2} + (1-\delta_{iN_r})\rho^{+}_{i+1,j_2}$$

$$= \delta_{io}\rho^{+}(r=o,\lambda_{j_2}) + \delta_{iN_r}\rho^{+}(r_{max},\lambda_{j_2}) \qquad (16b)$$

$$(1-\delta_{io})T_{ij_1}\rho^{-}_{i-1,j_1} - 2h\sum_{k_1}\left[\Gamma'_{j_1}\phi_{k_1}F_{j_1k_1i} - \delta_{k_1j_1}\,\Gamma'_{j_1}\mathcal{F}_{ij_1}\right]\rho^{-}_{ik_1}$$

$$- 2h\,\Gamma'_{j_1}\sum_{k_2}\phi_{k_2}F_{j_1k_2i}\,\rho^{-}_{ik_2} - (1-\delta_{N_r})\rho^{-}_{i+1,j_1}$$

$$= \delta_{io}\rho^{-}(r=o,\lambda_{j_1}) + \delta_{iN_r}\rho^{-}(r_{max},\lambda_{j_1}) \qquad (16c)$$

$$(1-\delta_{io})T_{ij_2}\rho^-_{ij_2} - 2h\,|\mathcal{G}_{ij_2}|\rho^+_{ij_2} - 2h\,\Gamma'_{j_2}\sum_{k_1}\phi_{k_1}F_{j_2k_1i}\,\rho^-_{ik_1}$$

$$- 2h\sum_{k_2}\left[\Gamma'_{j_2}\phi_{k_2}F_{j_2k_2i} - \delta_{k_2j_2}(|\mathcal{G}_{ij_2}| + \Gamma'_{j_2}\mathcal{F}_{ij_2})\right]\rho^-_{ik_2} - (1-\delta_{iN_r})\rho^-_{i+1j_2}$$

$$= \delta_{io}\rho^-(r{=}o,\lambda_{j_2}) + \delta_{iN_r}\rho^-(r_{max},\lambda_{j_2}) \qquad (16d)$$

and where the indices are defined: $1 \leq i \leq N_r-1$, $-N_c \leq j_1 \leq N_{b_1}$, $N_{b_1} \leq j_2 \leq N_{b_2}$ and $-N_c \leq k_1 \leq N_{b_1}$, $N_{b_1} \leq k_2 \leq N_{b_2}$. The energy range $[-\infty, \frac{1}{r}]$ has been discretized into three grids: $[-\lambda_{max}, o]$, $[o, \frac{1}{2r_i}]$, $[\frac{1}{2r_i}, \frac{1}{r_i}]$ composed of $N_c + 1$, $N_{b_1} + 1$ and $N_{b_2} - N_{b_1}$ points, respectively. The total number of points in the energy grid is $N_c + N_{b_2}$. Hence the subscripts j_1, k_1 indicate that the energy is restricted to region I: $[-\lambda_{max}, \frac{1}{2r_i}]$, while j_2, k_2 indicate that the energy is restricted to region II: $[\frac{1}{2r_i}, \frac{1}{r_i}]$. The parameter λ_{max} represents the largest free ion energy considered. More details on the energy grid are given below.

Now, before applying the boundary conditions on λ (equations 9a,d)), a word is needed on continuity conditions. In equation (8) the λ,r phase space is divided into 2 regions. Since we assume that the unknowns $\rho^{\pm}(r,\lambda)$ are smooth functions of λ and r, we must insure that the computed solutions of (16) are continuous across the boundary between regions I and II. We do this by recognizing that (16a,b) are 2 equations for the *same* unknown when $j_1 = j_2 = N_{b_1}$. Therefore, we can add the two equations together and divide by 2. This insures that we will not have an overdetermined system of equations. In the case where j_1 and j_2 don't equal N_{b_1}, the appropriate terms from the k_1 and k_2 summations (i.e., the last in k_1 and the first in k_2) must be added together in (16a,b). These two steps will insure that $\rho^+(r,\lambda)$ is continuous at $\lambda = 1/2r$. Analogous additions in equations (16c,d) will likewise insure that $\rho^-(r,\lambda)$ will be continuous at $\lambda = 1/2r$. Therefore, applying the steps above to insure continuity in the solution, and the boundary conditions on λ yields the following set of practical algebraic equations,

$$-(1-\delta_{io})T_{ij_1}\rho^+_{i-1,j_1} - \sum_{k_1}{}' \mathcal{A}^{(i)+}_{j_1k_1}\rho^+_{ik_1} - \mathcal{C}^{(i)+}_{j_1N_{b_1}}\rho^+_{iN_{b_1}} - \sum_{k_2}{}' \mathcal{A}^{(i)+}_{j_1k_2}\rho^+_{ik_2}$$

$$- 2h\,\mathcal{G}_{ij_1}\rho^-_{ij_1} + (1-\delta_{iN_r})\rho^+_{i+1,j_1} = \mathcal{B}^{(i)+}_{j_1} \qquad (17a)$$

$$-(1-\delta_{io})T_{ij_2}\rho^+_{i-1,j_2} - \sum_{k_1}{}' \mathcal{A}^{(i)+}_{j_2k_1}\rho^+_{ik_1} - \mathcal{C}^{(i)+}_{j_2N_{b_1}}\rho^+_{iN_{b_1}} - \sum_{k_2}{}' \mathcal{A}^{(i)+}_{j_2k_2}\rho^+_{ik_2}$$

$$+ (1-\delta_{iN_r})\rho^+_{i+1,j_2} = \mathcal{B}^{(i)+}_{j_2} \qquad (17b)$$

$$(1-\delta_{io})T_{ij_1}\rho^-_{i-1,j_1} - \sum_{k_1}{}' \mathcal{A}^{(i)-}_{j_1k_1}\rho^-_{ik_1} - \mathcal{C}^{(i)-}_{j_1N_{b_1}}\rho^-_{iN_{b_1}} - \sum_{k_2}{}' \mathcal{A}^{(i)-}_{j_1k_2}\rho^-_{ik_2}$$

$$- (1-\delta_{iN_r})\rho^-_{i+1,j_1} = \mathcal{B}^{(i)-}_{j_1} \qquad (17c)$$

$$(1-\delta_{io})T_{ij_2}\rho^-_{i-1,j_2} - 2h\,|\mathcal{G}_{ij_2}|\,\rho^+_{ij_2} - \sum_{k_1}{}' \mathcal{A}^{(i)-}_{j_2k_1}\rho^-_{ik_1} - \mathcal{C}^{(i)-}_{j_2N_{b_1}}\rho^-_{iN_{b_1}}$$

$$- \sum_{k_2}{}' \mathcal{A}^{(i)-}_{j_2k_2}\rho^-_{ik_2} - (1-\delta_{iN_r})\rho^-_{i+1,j_2} = \mathcal{B}^{(i)-}_{j_2} \qquad (17d)$$

where the coefficient matrix $\mathcal{A}$ and column vector $\mathcal{B}$ are defined as,

$$\mathcal{A}^{(i)+}_{j_1k_1} = 2h\left[\Gamma'_{j_1}\ \phi_{k_1}\ F_{j_1k_1i} - \delta_{k_1j_1}(\mathcal{G}_{ij_1} + \Gamma'_{j_1}\ \mathcal{F}_{ij_1})\right] \tag{18a}$$

$$\mathcal{A}^{(i)+}_{j_1k_2} = 2h\ \Gamma'_{j_1}\ \phi_{k_2}\ F_{j_1k_2i} \tag{18b}$$

$$\mathcal{A}^{(i)+}_{j_2k_1} = 2h\ \Gamma'_{j_2}\ \phi_{k_1}\ F_{j_2k_1i} \tag{18c}$$

$$\mathcal{A}^{(i)+}_{j_2k_2} = 2h\left[\Gamma'_{j_2}\ \phi_{k_2}\ F_{j_2k_2i} - \delta_{k_2j_2}\ \Gamma'_{j_2}\ \mathcal{F}_{ij_2}\right] \tag{18d}$$

$$\mathcal{A}^{(i)-}_{j_1k_1} = 2h\left[\Gamma'_{j_1}\ \phi_{k_1}\ F_{j_1k_1i} - \delta_{k_1j_1}\ \Gamma'_{j_1}\ \mathcal{F}_{ij_1}\right] \tag{18e}$$

$$\mathcal{A}^{(i)-}_{j_1k_2} = 2h\ \Gamma'_{j_1}\ \phi_{k_2}\ F_{j_1k_2i} \tag{18f}$$

$$\mathcal{A}^{(i)-}_{j_2k_1} = 2h\ \Gamma'_{j_2}\ \phi_{k_1}\ F_{j_2k_1i} \tag{18g}$$

$$\mathcal{A}^{(i)-}_{j_2k_2} = 2h\left[\Gamma'_{j_2}\ \phi_{k_2}\ F_{j_2k_2i} - \delta_{k_2j_2}(|\mathcal{G}_{ij_2}| + \Gamma'_{j_2}\ \mathcal{F}_{ij_2})\right] \tag{18h}$$

and,

$$\mathcal{B}^{(i)+}_{j_1} = \delta_{io}\rho^{+}(r{=}o,\lambda_{j_1}) + \delta_{iN_r}\rho^{+}(r_{max},\lambda_{j_1}) + \delta_{N_{b_2}k_2}\ 2h\Gamma'_{j_1}\phi_{N_{b_1}}F_{j_1N_{b_2}i}\rho^{+}_{iN_{b_2}}$$

$$+\ \delta_{k_1\text{-}N_c}\rho^{+}_{i,-N_c}\ 2h\left[\Gamma'_{j_1}\ \phi_{-N_c}\ F_{j_1\text{-}N_ci} - \delta_{-N_cj_1}(\mathcal{G}_{ij_1} + \Gamma'_{j_1}\ \mathcal{F}_{ij_1})\right] \tag{19a}$$

$$\mathcal{B}^{(i)+}_{j_2} = \delta_{io}\rho^{+}(r{=}o,\lambda_{j_2}) + \delta_{iN_r}\rho^{+}(r_{max},\lambda_{j_2}) + \delta_{k_1\text{-}N_c}\ 2h\Gamma'_{j_2}\phi_{-N_c}F_{j_2\text{-}N_ci}\rho^{+}_{i\text{-}N_c}$$

$$+\ \delta_{k_2N_{b_2}}\rho^{+}_{iN_{b_2}}\left[2h\ \Gamma'_{j_2}\ \phi_{N_{b_2}}F_{j_2N_{b_2}i} - \delta_{k_2j_2}\ \Gamma'_{j_2}\ \mathcal{F}_{ij_2}\right] \tag{19b}$$

$$\mathcal{B}^{(i)-}_{j_1} = \delta_{io}\rho^-(r{=}o,\lambda_{j_1}) + \delta_{iN_r}\rho^-(r_{max},\lambda_{j_1}) + \delta_{N_{b_2}k_2}\, 2h\Gamma'_{j_1}\phi_{N_{b_2}} F_{j_1N_{b_2}i}\rho^-_{iN_{b_2}}$$

$$+ \delta_{k_1-N_c}\rho^-_{i,-N_c}\, 2h\left[\Gamma'_{j_1}\, \phi_{-N_c}\, F_{j_1-N_ci} - \delta_{k_1j_1}\, \Gamma'_{j_1}\, \mathcal{F}_{ij_1}\right] \tag{19c}$$

$$\mathcal{B}^{(i)-}_{j_2} = \delta_{io}\rho^-(r{=}o,\lambda_{j_2}) + \delta_{iN_r}\rho^-(r_{max},\lambda_{j_2}) + \delta_{k_1-N_c}\, 2h\Gamma'_{j_2}\phi_{-N_c}F_{j_2-N_ci}\rho^-_{i,-N_c}$$

$$+ \delta_{j_2N_{b_2}}\, 2h|\mathcal{G}_{ij_2}|(\rho^+_{i,N_{b_2}} - \rho^-_{i,N_{b_2}})$$

$$+ \delta_{k_2N_{b_2}}\, 2h\left[\Gamma'_{j_2}\, \phi_{N_{b_2}}\, F_{j_2N_{b_2}i} - \delta_{k_2j_2}\, \Gamma'_{j_2}\, \mathcal{F}_{ij_2}\right]\rho^-_{iN_{b_2}} \tag{19d}$$

The elements of the coefficient matrix $\mathcal{A}$ which govern the continuity of the solutions $\rho^{\pm}$ at the boundary between regions I and II are denoted by $\mathcal{C}$ in (17) and are defined,

$$\mathcal{C}^{(i)+}_{j_1N_{b_1}} = 4h\ \Gamma'_{j_1}\phi_{N_{b_1}} F_{j_1N_{b_1}i} - \delta_{j_1N_{b_1}}\, 2h\ \Gamma'_{j_1}\mathcal{F}_{ij_1} \tag{20a}$$

$$\mathcal{C}^{(i)+}_{j_2N_{b_1}} = 4h\ \Gamma'_{j_2}\phi_{N_{b_1}} F_{j_2N_{b_1}i} \tag{20b}$$

$$\mathcal{C}^{(i)-}_{j_1N_{b_1}} = 4h\ \Gamma'_{j_1}\phi_{N_{b_1}} F_{j_1N_{b_1}i} - \delta_{j_1N_{b_1}}\, 2h\ \Gamma'_{j_1}\mathcal{F}_{ij_1} \tag{20c}$$

$$\mathcal{C}^{(i)-}_{j_2N_{b_1}} = 4h\ \Gamma'_{j_2}\phi_{N_{b_1}} F_{j_2N_{b_1}i} \tag{20d}$$

The index i in (17) is defined as before, while in (17a,c) j_1 is defined: $-N_c+1 \le j_1 \le N_{b_1}$, and in (17b,d) j_2 is defined: $N_{b_1}+1 \le j_2 \le N_{b_2}-1$. The primes on the k_1 and k_2 summations in (17) denote that the λ boundary terms are omitted (i.e., $k_1 = -N_c$ and $k_2 = N_{b_2}$). The primes also indicate that the terms with $k_1 = N_{b_1}$ and $k_2 = N_{b_2}$ have been factored-out of the summation (and are represented by the $\mathscr{C}$ terms).

The algebraic equations (17) can be written in the familiar matrix notation,

$$\mathscr{A}\ \rho = \mathscr{B} \tag{21}$$

where $\mathscr{A}$ and $\mathscr{B}$ are given by (18,19), respectively, and ρ is the unknown column vector composed of the discretized elements of ρ^+ first, then ρ^-. The known column vector $\mathscr{B}$, composed of boundary conditions on $\rho^\pm$, has been written in full detail in (19) in order to show exactly how each boundary condition contributes to the problem. In practice, after application of (9), (19) will simplify considerably. From (18) it is clear that the coefficient matrix $\mathscr{A}$ is non-symmetric due to the presence of the $\mathscr{G}_{ij}$ and $|\mathscr{G}_{ij}|$ terms. The overall structure of the coefficient matrix is shown in figure 2.

The energy grid chosen was nonlinear due to the skew discontinuity present in the kernals $F(\lambda,\mu;r)$ (Flannery [22,23]). In particular, we use the 3tan(u) prescription of Bates and Mendaš [8] for the pivot points used in the Simpson's rule quadratures in (17). That is, the weights and pivots used in the energy quadratures are defined for the three grids as,

$$\lambda_k = 3\tan(k\xi), \qquad \phi_k = r_k\ \xi\ \sec^2(k\xi)$$

with,

(i) continuum part of region I: $-\lambda_{max} \leq \lambda \leq 0 \qquad -N_c \leq k \leq 0$

$$\xi = \tan^{-1}(\lambda_{max}/3)/N_c$$

$$r_k = \begin{cases} 1 & , k = -N_c,\ 0 \\ 4 & , k = \text{even integer in } (-N_c,0) \\ 2 & , k = \text{odd integer in } (-N_c,0) \end{cases}$$

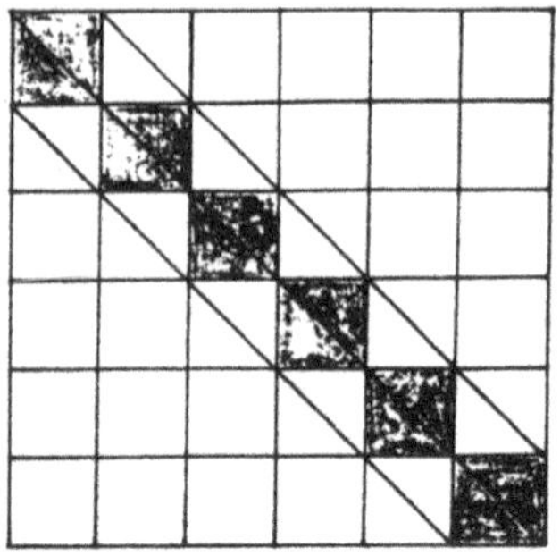

a

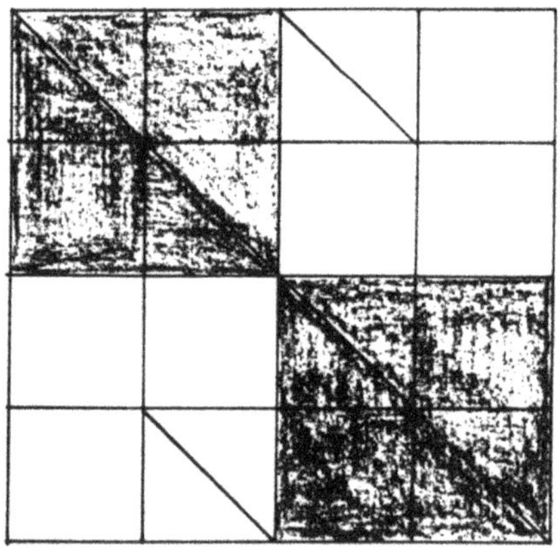

b

Figure 2: Structure of coefficient matrix $\mathcal{A}$ in equation (21). a) block tri-diagonal in r, b) non-symmetric, positive-definite in λ.

(ii) bound state part of region I: $0 \le \lambda \le \dfrac{1}{2r_i}$ $\quad 0 \le k \le N_{b_1}$

$$\xi = \tan^{-1}(1/6r_i)/N_{b_1}$$

$$r_k = \begin{cases} 1 & , \; k = 0,\; N_{b_1} \\ 4 & , \; k = \text{even integer in } (0, N_{b_1}) \\ 2 & , \; k = \text{odd integer in } (0, N_{b_1}) \end{cases}$$

and,

(iii) region II: $\frac{1}{2r_i} \leq \lambda \leq \frac{1}{r_i}$ $\quad N_{b_1} \leq k \leq N_{b_2}$

$$\lambda_k = 3\tan[\xi_{min} + (k-N_{b_1})\xi] \qquad \phi_k = r_k\, \xi\, \sec^2[\xi_{min} + (k-N_{b_1})\xi]$$

$$\xi_{min} = \tan^{-1}(1/6r_i) \qquad \xi = [\tan^{-1}(1/3r_i) - \tan^{-1}(1/6r_i)]/(N_{b_2}-N_{b_1})$$

$$r_k = \begin{cases} 1 & , k = N_{b_1}, N_{b_2} \\ 4 & , k = \text{even integer in } (N_{b_1}, N_{b_2}) \\ 2 & , k = \text{odd integer in } (N_{b_1}, N_{b_2}) \end{cases}$$

Denoting the total number of points in the energy quadrature by $N_\lambda \equiv N_c + N_{b_2} - 2$, and the total number of simultaneous equations in (21) by $N \equiv (N_r-1)\ 2\ N_\lambda = 2(N_r-1)(N_c+N_{b_2}-2)$. A typical value for N is 20,988 for $N_r = 100$, $N_c = 36$ and $N_{b_2} = 72$. Hence, due to the very large sparse nature of the coefficient matrix $\mathcal{A}$, the use of direct techniques like Gaussian elimination to solve (21) will be totally out of the question because of the time and memory requirements involved. Therefore iterative techniques like Lanczos algorithms (Cullum and Willoughby [15], Golub and van Loan [30], accelerated successive overrelaxation methods (Young [42], Hageman and Young [32]) and Tchebychev iteration (Manteuffel [36,37]) should be used to solve (21).

Consequently, before continuing our discussion of the solution of (21), we will review some of the iterative techniques used to deal with large linear systems. For readers interested in a complete treatment consult the books listed in section IV - especially those by Young and Hageman and Young.

Iterative Methods for Large Linear Systems

Solve the matrix equation $Ax = B$ where A is an $N \times N$ real, symmetric positive definite matrix, and x, B are column vectors of length N. Decompose A into three parts: (a) its diagonal elements (D), (b) all elements below the diagonal (C_L) and (c) all elements above the diagonal

(C_u). That is, we can write $A = D - C_L - C_u$. Now, scale the problem so the diagonal elements of the new coefficient matrix are 1. This can be done by multiplying by $D^{1/2}$,

$$(D^{-1/2}\ A\ D^{-1/2})(D^{1/2}\ x) = D^{-1/2}\ b \tag{22}$$

$$\rightarrow D^{-1/2}(D - C_L - C_u)D^{-1/2} = D^{-1/2}\ D\ D^{-1/2} - D^{-1/2}C_L D^{-1/2} - D^{-1/2}\ C_u\ D^{-1/2}$$

$$\begin{aligned} &= I - L - L^t \qquad (23)\\ &\equiv I - B \end{aligned}$$

where I is the identity matrix and $L \equiv D^{-1/2}\ C_L\ D^{-1/2}$, $B \equiv L + L^t$. Then, from (22), defining the column vectors $u \equiv D^{1/2}\ x$, $c \equiv D^{-1/2}\ b$ allows the original problem to be case into the suggestive form,

$$u = B\ u + c \tag{24}$$

which is then solved iteratively.

Two of the most widely used iterative methods are the Jacobi method,

$$u^{(n+1)} = B\ u^{(n)} + c \tag{25a}$$

and the Gauss-Seidel method,

$$\begin{aligned} u^{(n+1)} &= L\ u^{(n+1)} + L^t\ u^{(n)} + c\\ &= \mathcal{L}\ u^{(n)} + k \qquad (25b) \end{aligned}$$

where $k = (I - L)^{-1}\ c$, $\mathcal{L} = (I-L)^{-1}\ L^t$ and in (25) $u^{(n)}$ denotes the nth iteration of u. More recently variants on the Gauss-Seidel method have been developed to speed the convergence of the iteration procedure. One of the variants is the successive overrelaxation method (SOR),

$$u^{(n+1)} = \omega\,(L\,u^{(n+1)} + L^t\,u^{(n)} + c) + (1-\omega)\,u^{(n)}$$
$$= \mathcal{L}_\omega\,u^{(n)} + k_\omega \qquad (25c)$$

with $\mathcal{L}_\omega = (I - \omega L)^{-1}\,(\omega\,L^t + (1-\omega)I)$, $k_\omega = \omega(I-\omega L)^{-1}\,c$. The parameter ω is known as the acceleration parameter. When $\omega = 1$ (25c) reduces to the Gauss-Siedel method (25b). A generalization of the SOR method involves 2 SOR sweeps for each iteration of u, and is called the symmetric successive overrelaxation method (SSOR),

$$u^{(n+1/2)} = \mathcal{L}_\omega\,u^{(n)} + k_\omega^{(for.)} \qquad (25d)$$

$$u^{(n+1)} = \mathcal{U}_\omega\,u^{(n+1/2)} + k_\omega^{(back.)} \qquad (25e)$$

where the operator $\mathcal{L}_\omega$ is defined above, and $\mathcal{U}_\omega = (I - \omega L^t)^{-1}(\omega L + (1-\omega)I)$ and $k_\omega^{(for.)} = \omega(I - \omega L)^{-1}\,c$, $k_\omega^{(back.)} = \omega(I - \omega L^t)^{-1}\,c$. Equations (25d,e) can be combined into one iteration step by defining the operator $\mathcal{S}_\omega = \mathcal{U}_\omega\,\mathcal{L}_\omega$,

$$u^{(n+1)} = \mathcal{S}_\omega\,u^{(n)} + k_\omega \qquad (25f)$$

where $k_\omega = \omega\,(2-\omega)(I-\omega L^t)^{-1}(I-\omega L)^{-1}\,c$.

In the SOR and SSOR methods one must choose an acceleration parameter ω which will be optimal for a given coefficient matrix A. Two widely used methods of acceleration are Chebyshev acceleration and conjugate gradient acceleration. Writing the basic iteration procedure (24) as $u^{(n+1)} = \mathcal{G}\,u^{(n)} + k$, Chebyskev acceleration is defined,

$$u^{(n+1)} = \rho_{n+1}\{(\gamma\,\mathcal{G}\,u^{(n)} + k) + (1-\gamma)u^{(n)}\} + (1-\rho_{n+1})u^{(n-1)} \qquad (26a)$$

where $\gamma = 2/[2-M(\mathcal{G})-m(\mathcal{G})]$ and,

$$\rho_{n+1} = \begin{cases} 1 & , n = s \\ (1 - \sigma^2/2)^{-1} & , n = s+1 \\ [1 - (\sigma/2)^2 \rho_n]^{-1} & , n \geq s+2 \end{cases}$$

with $\sigma = [M(\mathcal{G})-m(\mathcal{G})]/[2 - M(\mathcal{G})-m(\mathcal{G})]$, $M(\mathcal{G})$ = maximum eigenvalue of matrix $\mathcal{G}$, $m(\mathcal{G})$ = minimum eigenvalue of $\mathcal{G}$. The integer s is initially zero, then increased as the adaptive procedure proceeds (Grimes et al. [31]). That is, one can reassign the overrelaxation parameter ρ several times during the iteration process. A disadvantage of Chebyshev acceleration is that it requires estimates of the smallest and largest eigenvalues of $\mathcal{G}$ to be made.

Conjugate gradient acceleration is defined,

$$u^{(n+1)} = \rho_{n+1} (\gamma_{n+1} \delta^{(n)} + u^{(n)}) + (1 - \rho_{n+1})u^{(n-1)} \tag{26b}$$

$$\delta^{(n+1)} = \rho_{n+1} (\gamma_{n+1} \mathcal{G} \delta^{(n)} + (1-\gamma_{n+1}) \delta^{(n)}) + (1 - \rho_{n+1})\delta^{(n-1)} \tag{26c}$$

where $\delta^{(n)}$ is a pseudo-residual vector given by: $\delta^{(n)} = \mathcal{G} u^{(n)} + k - u^{(n)}$. The acceleration parameters ρ and γ are defined,

$$\rho_{n+1} = \begin{cases} 1 & , n = 0 \\ \dfrac{1}{1 - \left[\dfrac{\gamma_{n+1}}{\gamma_n \rho_n}\right] \dfrac{\delta^{(n)t} W^t W \delta^{(n)}}{\delta^{(n-1)t} W^t W \delta^{(n-1)}}} & , n \geq 1 \end{cases}$$

$$\gamma_{n+1} = \left[1 - \frac{\delta^{(n)t} W^t W \mathcal{G} \delta^{(n)}}{\delta^{(n)t} W^t W \delta^{(n)}}\right]^{-1}$$

where W is a nonsingular symmetrization matrix. For the Jacobi method $W = D^{1/2}$, while for the SSOR method $W = (1/\omega) D^{-1/2}(D - \omega C_L^t)$. While the conjugate gradient acceleration method doesn't require estimation of the eigenvalues of $\mathcal{G}$, the number of arithmetic operations required for its implementation is greater.

We have necessarily been selective and brief in our discussion of iterative methods due to the vastness of this area of linear algebra. The discussion of this subsection is based upon appendix A in the technical report of Grimes et al. [31]. Two excellent books which provide good introductions to this area are Young [42] and Hageman and Young [32] (others can be found in the list of general references in section IV). As for software which implements the iterative techniques discussed above (as well as others not discussed here), excellent packages are the ITPACK library (Grimes et al. [31]) and the package of Lanczos algorithms of Cullum and Willoughby [15].

So far we have only discussed iterative techniques for symmetric coefficient matrices $\mathcal{A}$. However, the central numerical problem of this lecture is to solve equation (21) for a large, sparse non-symmetric matrix $\mathcal{A}$. Unfortunately there is much less known in linear algebra about iterative techniques for non-symmetric matrices. One way to handle non-symmetric matrices $\mathcal{A}$ is to consider (instead of (21) for example) the associated equation $\mathcal{A}^t\mathcal{A}\rho = \mathcal{A}^t\mathcal{B}$, where $\mathcal{A}^t\mathcal{A}$ is a symmetric coefficient matrix. However, in many practical applications the condition number of $\mathcal{A}^t\mathcal{A}$ is much greater than that of $\mathcal{A}$ - thereby indicating that this technique will not necessarily yield a problem which will converge rapidly using one of the iterative techniques discussed above. Another way to deal with non-symmetric problems is to develop the appropriate generalizations of (25,26) directly (Young and Jea [43], Manteuffel [36,37]). However, these generalizations require knowledge of the eigenvalue spectrum of $\mathcal{G}$ which is difficult to obtain in practice. In summary then, while some progress has been made in linear algebra towards handling the non-symmetric case, much more work needs to be done in devising criteria by which one can select acceleration parameters which will be optimal for a given general coefficient matrix $\mathcal{G}$.

Returning now to our discussion of the numerical solution of (21) for non-symmetric $\mathcal{A}$, we will need to know the condition number $\kappa(\mathcal{A})$ for the coefficient matrix $\mathscr{A}$ before using one of the iterative techniques above. The condition number of a matrix $\mathscr{A}$ is defined: $\kappa(\mathscr{A}) = ||\mathscr{A}|| \bullet ||\mathscr{A}^{-1}||$. Computing the condition number of the coefficient matrix in a given problem (e.g., equation (21)) is necessary because the iterative techniques discussed above (like SSOR) work best for well-conditioned matrices (i.e., those with small κ). Otherwise, the number of iterations required for convergence will increase greatly, and since κ provides a measure of the sensitivity of a given problem to round-off errors and perturbations, if κ is too large for a given $\mathscr{A}$, then iterative techniques will not work due to accumulation of round-off errors. Physically, the condition number of a matrix $\mathscr{A}$ provides a measure of the distance between $\mathscr{A}$ and the set of angular matrices. Hence, when $\kappa(\mathscr{A})$ is very large the matrix $\mathscr{A}$ is considered ill-conditioned, which means that the numerical solution u, of (24) say, will be very sensitive to round-off error which are unavoidably accumulated during the iteration process. More details about condition numbers in linear algebra can be found in Golub and van Loan [30].

Using the routine LFCRG in the IMSL library [33] to estimate the condition number $\kappa_1(\mathscr{A})$, via the algorithm of Cline et al. [14], of the coefficient matrix of equation (17) we find that $\kappa_1(\mathscr{A}) \simeq 4.20\ 10^{19}$. In contrast $\kappa_1(\mathscr{A}) = 1751.1$ for the coefficient matrix $\mathscr{A}$ resulting from the discretization of the quasi-equilibrium integral equation (29). This indicates that the condition number for the coefficient matrix of (17) needs to be reduced approximately 16 orders of magnitude before the iterative techniques (25,26) will become effective. This can be accomplished either by row scaling (Golob and van Loan [30]) or pre-conditioning (Faber and Manteuffel [17]) the coefficient matrix. Unfortunately, these calculations have not been completed at the time of writing but will be reported in a forthcoming paper (Flannery and Mansky [28]). Hence, it is still an open question whether the iterative techniques discussed above are effective in solving (17). However our discussion of the condition number $\kappa_1(\mathscr{A})$ of $\mathscr{A}$ given by (18) has revealed the underlying reason why the earlier work of [8], failed to converge quickly as a function of gas density N(Z). Also, by illustrating the structure of the coefficient matrix in figure 2 and equation (18), a deeper insight into the role the gas density plays in the nonlinear pressure regime is obtained. In fact once the problem of the ill-condition of $\mathcal{A}$ is solved, a number of other problems in chemical physics should become amenable to the iterative techniques discussed in this lecture including the prediction of microscopic three-body ion-neutral association rates (Bates and McKibben [7]), and the inclusion of non-thermal effects into ion-ion recombination (Bates et al. [6]).

Before ending section II we wish to discuss the numerical solution of integral equations briefly. This is necessary because much of the numerical analysis of PIDE's relies heavily upon the expertise gained in solving related one-dimensional integral equations. The IE we will use as an example will be the quasi-equilibrium integral equation arising in ion-ion recombination. We will necessarily be brief since the numerical analysis of integral equations (even one-dimensional ones) is a vast field and we only wish to highlight points about the numerical treatment of IE's which are related to our earlier discussion of PIDE's. For a complete discussion of the numerical treatment of IE's see Baker [2].

In discussing the quasi-equilibrium theory of ion-ion recombination it will prove useful to define the functions $\rho^S(r,\lambda)$ and $\rho^D(r,\lambda)$,

$$\rho^D(r,\lambda) = \frac{1}{2}\,[\rho^+(r,\lambda) - \rho^-(r,\lambda)] \tag{27a}$$

$$\rho^S(r,\lambda) = \frac{1}{2}\,[\rho^+(r,\lambda) - \rho^-(r,\lambda)] \tag{27b}$$

which describe physically the net and total numbers of ion pairs undergoing recombination at a given relative separation r and internal energy λ. Substituting (27) into (8) yields two coupled PIDE's valid for $-\infty < \lambda \leq \frac{1}{r}$,

$$(\frac{1}{r} - \lambda)\,\frac{\partial\rho^D(r,\lambda)}{\partial r} + \frac{1}{r}\,(\frac{1}{r} - 2\lambda)\rho^D(r,\lambda) = \Gamma'\int_{-\infty}^{1/r}\rho^S(r,\mu)F(\lambda,\mu,;r)d\mu$$
$$- \Gamma'\,\mathcal{F}(r,\lambda)\rho^S(r,\lambda) \tag{28a}$$

$$\left(\frac{1}{r} - \lambda\right) \frac{\partial \rho^S(r,\lambda)}{\partial r} + \frac{1}{r}\left|\frac{1}{r} - 2\lambda\right|\rho^D(r,\lambda) = \Gamma' \int_{-\infty}^{1/r} \rho^D(r,\mu)F(\lambda,\mu,;r)d\mu$$
$$- \Gamma' \mathcal{F}(r,\lambda)\rho^D(r,\lambda) \qquad (28b)$$

In the quasi-equilibrium theory the motion of the center of mass of the ion pairs is taken to be in thermodynamic equilibrium with the third bodies, while a quasi-equilibrium distribution in highly excited internal energy states of the ion pairs is established effectively instantaneously due to collisions with the much more numerous third bodies. In the establishment of this distribution it is assumed that the distribution of separations of ion pairs does not effect the quasi-equilibrium distribution in internal energy. That is, the r dependence of the distributions $\rho^+(r,\lambda)$, $\rho^-(r,\lambda)$ is not influenced by the recombination proceeding in the plasma. Hence we can assume that the r-distributions of contracting and expanding ion pairs is in thermodynamic equilibrium, thereby implying that $\rho^D(r,\lambda) = 0$. This results in (28b) indicating that $\rho^S(r,\lambda)$ is a constant w.r.t. r. Multiplying (28a) by r^2 and integrating w.r.t. r yields the quasi-equilibrium integral equation for the distribution over internal energy states $\rho^S_{QE}(\lambda)$,

$$\int_{-\infty}^{\epsilon} F(\lambda,\mu)\rho^S_{QE}(\mu)d\mu = \rho^S_{QE}(\lambda) \int_{-\infty}^{\omega} F(\lambda,\mu)d\mu \qquad (29)$$

where ω is the maximum binding energy of an ion pair, and ϵ is the stabilization energy of an ion pair. We refer the reader to the original literature (Bates and Moffett [12], Bates and Flannery [5], Bates and Mendaš [11], and Flannery [22,23]) for the details.

Equation (29) is valid only in the low gas density limit where the flow of contracting and expanding ion pairs balance, at higher gas densities however a net contraction of ion pairs occurs so that the full PIDE (8) must be solved. We quote the expression for the recombination rate coefficient α for the low density limit from the original literature cited above,

$$\alpha/\alpha_T = \Gamma_o \int_{-\infty}^{\nu} d\lambda \int_{\nu}^{\omega} d\mu \; F(\lambda,\mu)[\rho_{QE}^{S}(\lambda) - \rho_{QE}^{S}(\mu)] \tag{30}$$

where $\nu = -E/kT$ is an arbitrary energy level, and α_T is the Thomson rate coefficient (Thomson [40]). From (30) it is clear that once IE (29) is solved for the $\rho_{QE}^{S}(\lambda)$, a bi-cubic spline quadrature will yield α.

To solve the quasi-equilibrium integral equation (29) we impose the boundary conditions: $\rho_{QE}^{S}(\lambda \leq o) = 1$, $\rho_{QE}^{S}(\lambda \geq \epsilon) = 0$ yielding,

$$\int_{o}^{\epsilon} \rho_{QE}^{S}(\mu)\; F(\lambda,\mu)d\mu - \rho_{QE}^{S}(\lambda) \int_{-\infty}^{\omega} F(\lambda,\mu)d\mu = \int_{-\infty}^{o} F(\lambda,\mu)d\mu \tag{31}$$

When (31) is discretized the result is a system of algebraic equations similar to (21) which can be solved either by iterative methods (SSOR) or direct techniques (Gaussian elimination), due to the much smaller size coefficient matrix $\mathcal{A}$ in the quasi-equilibrium case. As an example of the type of results obtained, we show in figures 3 and 4 the quasi-equilibrium distribution $\rho_{QE}^{S}(\lambda)$ and recombination rate α/α_T, respectively for the energy-transfer mechanisms of charge-transfer, hard-sphere and polarization collisions. We should also mention that, in addition to the smaller size, the quasi-equilibrium coefficient matrix is also symmetric - a fact which greatly helps in the numerical solution of (29). We have not discussed the numerical solution of the PIDE (28) in the same detail as that of (8), even though they are equivalent, because it results in a system of algebraic equations with a non-symmetric, nonpositive-definite coefficient matrix $\mathcal{A}$ - a problem much more difficult than (17). Finally, the quadrature rule used to determine the weights and pivots used in solving the quasi-equilibrium integral equation (31) were the same nonlinear Simpson's rule weights and pivots of Bates and Mendaš [8] discussed earlier. We conclude section II with a summary of the types of IE's found in the literature.

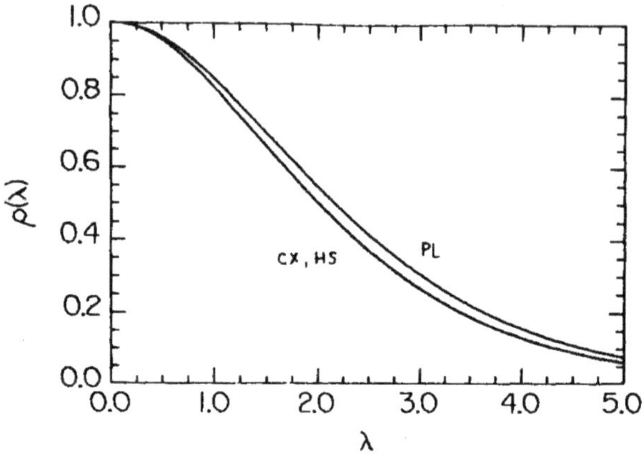

Figure 3: Quasi-equilibrium distribution function $\rho_{QE}(\lambda)$ for the case of a = 1/3 ($m_1=m_2=m_3$), and energy-change mechanisms of charge-transfer (CX), hard-sphere (HS) and polarization (PL) collisions.

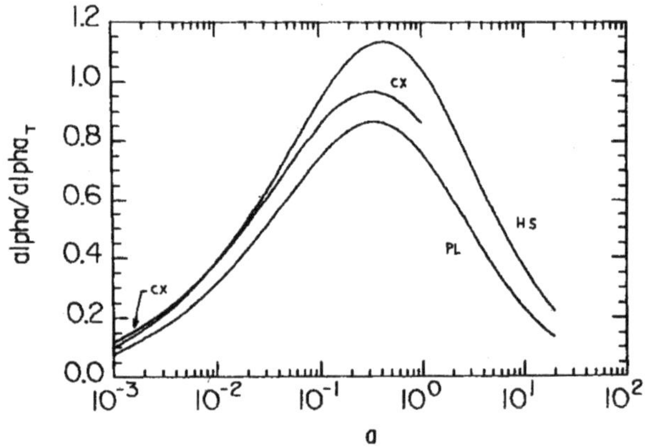

Figure 4: Quasi-Equilibrium recombination rate coefficient (α/α_T) versus mass ratio parameter (a) for energy-change mechanisms of charge-transfer (CX), hard-sphere (HS) and polarization (PL) collisions.

Numerical Solution of IE's

The three basic types of integral equations are Fredholm equations,

$$\int_a^b K(x,y)f(y)dy = g(x)f(x)$$

and Volterra equations of the first kind,

$$\int_0^t K(t,s)f(s)ds = g(t)$$

and second kind,

$$f(t) - \int_0^t K(t,s)f(s)ds = g(t)$$

where g is a known function, f is the unknown function and K is the kernal of the integral equation. We have already encountered Fredholm equations and Volterra equations of the 2nd-kind in (29) and (8), respectively. We will not encounter Volterra integral equations of the 1st-kind in this lecture. However, upon discretization, all three types of IE's above reduce to a problem of solving a system of linear algebraic equations. The particular technique used to solve the algebraic equations depends upon the structure of the coefficient matrix $\mathcal{A}$, which in turn depends on the behavior of the kernal K. Recalling our steps in the numerical solution of PIDE's, we find that steps (ii) - (iv) also provide a good prescription for the numerical solution of one-dimensional IE's. We have been brief in our summary of the numerical treatment of integral equations due again to the breadth of the area. For a complete introduction to the numerical solution of IE's see Baker [2] and Delves and Mohamed [16]. For an excellent account of Volterra equations see Linz [34].

III. APPROXIMATE TREATMENTS OF IONIC RECOMBINATION

As stated in the introduction, until the advent of super-computers, the direct solution of the PIDE's arising from the Boltzmann equation treatment of ionic recombination, was generally avoided due to the difficulty in solving systems of algebraic equations composed of 1000 or more equations and unknowns. The paper (Bates and Mendaš [9]) which originally derived the coupled PIDE's (8) solved them by a power series expansion in $\Lambda_{mfp.}$ which converged slowly with gas density and whose coefficients were difficult to compute in general. As discussed in section II, the slow convergence rate of the power series solution of (8) is directly related to the ill-conditioned nature of the coefficient matrix $\mathcal{A}$ in (21). Hence other methods of solving for the recombination rate α are needed. One such method which has proven quite successful is the Monte Carlo simulation of ion-ion recombination processes (Bates and Mendaš [10], and Morgan et al. [38]). We will not cover this type of calculation in this lecture since our main interest is in discussing techniques which lead to PDE's or PIDE's.

In this section we will discuss the Debye-Smoluchowski and diffusion equation approaches to ionic recombination. The starting point for the Debye-Smoluchowski equation is the macroscopic continuity equation for the number density of ion pairs, undergoing recombination of time t and separation R,

$$\frac{dn(\vec{R},t)}{dt} = \frac{\partial n(\vec{R},t)}{\partial t} + \vec{\nabla} \bullet \vec{J}(\vec{R},t) = 0 \tag{32}$$

for $R \geq S \equiv$ sink radius, and is solved subject to the asymptotic boundary condition $n(R \to \infty,t) = \tilde{N}(X^+)\ \tilde{N}(Y^-)$ where $\tilde{N}$ is the equilibrium number density. The net current $\vec{J}(\vec{R},t)$ of ion pairs expanding at time t is,

$$\vec{J}(\vec{R},t) = -D\ \vec{\nabla}\ n(\vec{R},t) + (K/e)(\vec{\nabla}\ V)\ n(\vec{R},t) \tag{33}$$

where V(R) is the interaction potential between X^+ and Y^-, and D, K are the relative diffusion and mobility coefficients of X^+ and Y^- in a background gas Z. The introduction of a sink, with an assigned local three-body reaction rate α_3 at the surface, allows one to avoid dealing with the complicated collision kernels $F(\lambda,\mu;r)$ and full PIDE nature of (8) by replacing the problem with a phenomenological model. After substituting the current (33) into (32), and discretizing, the problem reduces to a boundary value problem involving a time-dependent diffusion equation in R. An example of the resultant solution is given in figure 5 which shows the time-dependent number density of ion pairs n(R,t) versus R for a specified sink radius. For further details on the Debye-Smoluchowski equation and ion-ion recombination see Flannery and Mansky [29].

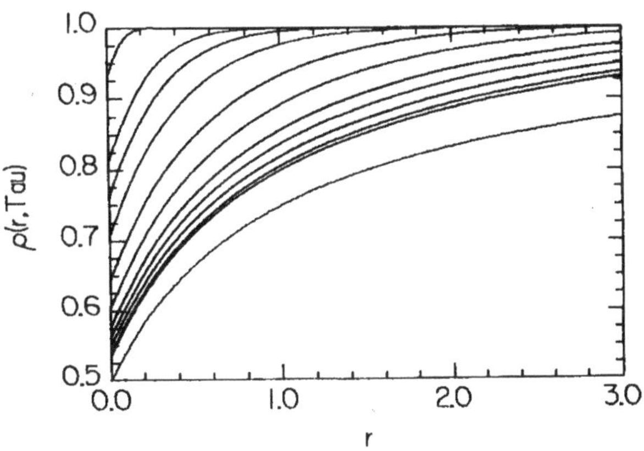

Figure 5: Solution of Debye-Smoluchowski equation for $\rho = n(r,\tau)/N_o \exp(-V/kT)$ $r \equiv R/S-1$, $\tau \equiv Dt/S^2$. τ ranges (from the top curve down) from 0.05, 0.5, 1, 2, 5, 10, 20, 30 to 100. The lowest curve is the steady state (equilibrium) distribution. Assigned parameters are S = 0.5, $\alpha_\infty/\alpha_{TR}$ = 0.5 (see [29] for details).

Diffusion equations in energy have also been used to model ionic recombination since the work of Pitaevskii [39]. In fact, the quasi-equilibrium theory of ion-ion recombination discussed in section II can be considered a Markov process (Flannery [19]). Writing the quasi-equilibrium integral equation (29) in terms of energies E, E_i (with the time-dependence reinserted - see Flannery [30], p. 17),

$$\frac{\partial n(E,t)}{\partial t} = N(Z) \int_{-E_s}^{\infty} n(E_i,t)\, K(E_i,E) dE_i - n(E,t) \int_{-D}^{\infty} K(E,E_i) dE_i \qquad (34)$$

Combining the integrals in (34) and Taylor series expanding the resultant integrand results in the Fokker-Planck equation,

$$\frac{\partial n(E,t)}{\partial t} = - \frac{\partial}{\partial E} [\Delta_1 n(E,t)] + \frac{1}{2} \frac{\partial^2}{\partial E^2} [\Delta_2\, n(E,t)] \qquad (35)$$

where,

$$\Delta_n = N(Z) \int_{-D}^{\infty} (E_f - E_i)^n\, K(E,E_f) dE_f$$

and one assumes that the energy-transfer between the ion-pairs and third bodies is small so that the Taylor series expansion of (34) converges. This necessarily limits the diffusion model to electron-ion recombination processes. As an example of the type of results obtained by the energy space diffusion equation, we show in figure 6 the steady-state distribution $\rho(\lambda)$ obtained by Pitaevskii compared with the corresponding results of the quasi-equilibrium theory. While Pitaevskii's treatment only becomes accurate in the limit of electron-ion recombination, it's similarity with the quasi-equilibrium results in figure 6, for the case of equal mass constitutents, is striking. For a more complete discussion of energy space diffusional theories of termolecular recombination see Flannery [26].

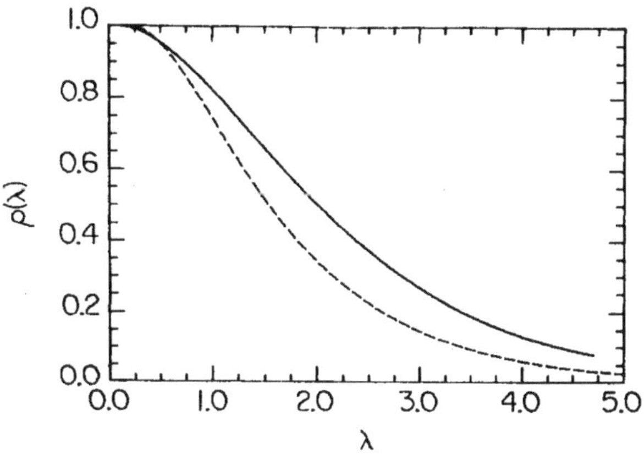

Figure 6: Comparison of quasi-equilibrium distribution (___) and Pitaevskii's distribution (---_ for the case of a = 1/3 and charge-transfer collisions.

IV. CONCLUSIONS AND GENERAL REFERENCES

In this lecture we have provided a detailed prescription for handling numerically the coupled partial integro-differential equations which arise from the Boltzmann equation treatment of ionic recombination. We have also given a brief summary of some of the approximate methods of treating ionic recombination. The reason for our detailed treatment of PIDE's is that there is little in either the physics or mathematics literature on how to tackle the problem of solving numerically a system of multi-dimensional PIDEs (i.e., systems with more than 1 independent variable). In this lecture we have tried to fill this gap.

Our main conclusion is that iterative techniques are numerically the most efficient way of solving the large systems of algebraic equations which result from PIDE's like (8). While this is not entirely unexpected, it is the first time techniques like the SSOR have been applied to problems in ionic recombination. With resolution of the problem of the ill-conditioning of $\mathscr{A}$ in (21), a number of long-standing problems in chemical physics will be able to be solved. We end this lecture with a list of general references which we have found useful on the subject of solving numerically IE's and PIDE's.

- Baker, C. T. H., 1977, The Numerical Treatment of Integral Equations, Oxford University Press.
- Cullum, J. K., and Willoughby, R. A., 1985, Lanczos Algorithms for Large Symmetric Eigenvalue Computations, Vol. I, Theory, Vol. II, Programs, Birkhäuser (Boston).
- Delves, L. M. and Mohammed, J. L., 1985, Computational Methods for Integral Equations, Cambridge University Press.
- Feldstein, A., and Sopka, J. R., 1974, SIAM J. Numer. Analy. 11, 826-46.
- Golub, G. H., and van Loan, C. F., 1983, Matrix Computations, Johns Hopkins University Press.
- Hageman, L. A., and Young, D. M., 1981, Applied Iterative Methods, Academic Press.
- Linz, P., 1985, Analytical and Numerical Methods for Volterra Equations SIAM Press, (Philadelphia).
- Young, D. M., 1971, Iterative Solution of Large Linear Systems, Academic Press.
- Young, D. M., and Jea, K. C., 1980, Lin. Algebra Appl. 34, 159-94.
- Wilkinson, J., 1965, The Algebraic Eigenvalue Problem, Oxford Univ. Press.

Acknowledgements

This work is supported by AFOSR under grant No. 84-0233. The invitation to lecture, and the partial support received from Oak Ridge National Laboratory are also gratefully acknowledged.

References

1. Akhiezer, A. I., Akhiezer, I. A., Polovin, R. V., Sitenko, A. G., and Stepanov, K. N., 1975, Plasma Electrodynamics, Vol. 1, Linear Theory, Pergamon Press.
2. Baker, C. T. H., 1977, The Numerical Treatment of Integral Equations, Oxford University Press.

3. Balescu, R., 1963, Statistical Mechanics of Charged Particles, Interscience - Wiley.
4. ibid., 1975, Equilibrium and Nonequilibrium Statistical Mechanics, Wiley.
5. Bates, D. R., and Flannery, M. R., 1968, Proc. Roy. Soc. (Lon.) A 302, 367-83.
6. Bates, D. R., Hays, P. B., and Spevak, D., 1971, J. Phys. B: At. Mol. Phys. 4, 962-8.
7. Bates, D. R., and McKibbin, 1974, Proc. Roy. Soc. (Lon.) A 339, 13-28.
8. Bates, D. R., and Mendaš, I., 1975, J. Phys. B: At. Mol. Phys. 8, 1770-5.
9. ibid., 1978, Proc. Roy. Soc. (Lon.) A 359, 275-85.
10. ibid., 1978, Proc. Roy. Soc. (Lon.) A 359, 287-301.
11. ibid., 1982, J. Phys. B: At. Mol. Phys. 15, 1949-56.
12. Bates, D. R., and Moffett, 1966, Proc. Roy. Soc. (Lon.) A 291, 1-8.
13. Chapman, S. and Cowling, T. G., 1970, The Mathematical Theory of Non-Uniform Gases, Cambridge University Press, 3rd. ed.
14. Cline, A. K., Moler, C. B., Stewart, G. W., and Wilkinson, J. H., 1979, SIAM J. Numer. Analy. 16, 368-75.
15. Cullum, J. K., and Willoughby, R. A., 1985, Lanczos Algorithms for Large Symmetric Eigenvalue Computations, Vol. I, Theory, Vol. II, Programs, Birkhäuser (Boston).
16. Delves, L. M., and Mohamed, J. L., 1985, Computational Methods for Integral Equations, Cambridge University Press.
17. Faber, V., and Manteuffel, T. A., 1989, p. 37-61, in Transport Theory, Invariant Imbedding and Integral Equations, (P. Nelson, V. Faber, T. A. Manteuffel, D. A. Seth and A. B. White, Jr., eds.) Marcel Dekker.
18. Ferziger, J. H., and Kaper, H. G., 1972, Mathematical Theory of Transport Processes in Gases, North-Holland.
19. Flannery, M. R., 1971, Ann. Phys. (N.Y.) 67, 376-88.
20. ibid., 1972, Chap. 1, p. 1-90, in Case Studies in Atomic Collision Physics, Vol. 2, (E. W. McDaniel and M. R. C. McDowell, eds.), North-Holland.

21. ibid., 1973, Ann. Phys. (N.Y.) 79, 480-517.
22. ibid., 1980, J. Phys. B: At. Mol. Phys. 13, 3649-64.
23. ibid., 1981, J. Phys. B: At. Mol. Phys. 14, 915-34.
24. ibid., 1982, Phil. Trans. Roy. Soc. (Lon.) A 304, 447-97.
25. ibid., 1987, J. Phys. B: At. Mol. Phys. 20, 4929-38.
26. ibid., 1987, J. Chem. Phys. 87, 6947-56.
27. Flannery, M. R., and Mansky E. J., 1988, J. Chem. Phys. 88, 4228-41.
28. ibid., 1989, "Kinetic Theory Foundations of Ion-Ion Recombination", J. Chem. Phys., to be submitted.
29. ibid., 1989, Chem. Phys. 132, 115-36.
30. Golub, G. H., and van Loan, C. F., 1983, Matrix Computations, Johns Hopkins University Press.
31. Grimes, R. G., Kincaid, D. R., and Young, D. M., 1979, ITPACK 2.0 User's Guide, CNA-150, Center for Numerical Analysis, Univ. Texas (Austin).
32. Hageman, L. A., and Young, D. M., 1981, Applied Iterative Methods, Academic Press.
33. IMSL Math/Library User's Manual, Vol. 1, 1989.
34. Linz, P., 1985, Analytical and Numerical Methods for Volterra Equations, SIAM Press (Philadelphia).
35. Mansky, E. J., 1985, Ph.D. thesis, Georgia Institute of Technology.
36. Manteuffel, T. A., 1977, Numer. Math. 28, 307-27.
37. ibid., 1978, Numer. Math. 31, 183-208.
38. Morgan, W. L., Whitten, B. L., and Bardsley, J. N., 1980, Phys. Rev. Lett. 45, 2021-4.
39. Pitaevskii, L. P., 1962, Soviet Phys. JETP 15, 919-21.
40. Thomson, J. J., 1924, Phil. Mag. 47, 337-78.
41. Tolman, R. C., 1938, The Principles of Statistical Mechanics, Oxford University Press.
42. Young, D. M., 1971, Iterative Solution of Large Linear Systems, Academic Press.
43. Young, D. M., and Jea, K. C., 1980, Lin. Algebra Appl. 34, 159-94.

THE NUMERICAL SOLUTION OF PARTIAL DIFFERENTIAL EQUATIONS IN ATOMIC SCATTERING THEORY

E. J. Mansky
School of Physics
Georgia Institute of Technology
Atlanta, Georgia 30332

ABSTRACT

The numerical solution of coupled partial and ordinary differential equations in electron-atom scattering theory are compared. In particular, a case study is made of the transition H(1s → 2s,2p) excited by electrons in the intermediate energy region. The results of the multichannel eikonal theory (MET) and the close coupling theory (CC) for this transition are compared and contrasted with experiment and each other. The principle conclusion is that the configuration space and angular momentum representations employed by the two theories provides information about the excitation process which is complementary. Specifically, the contrasting differences between the MET and CC results at small and large scattering angles for the modulus and phase angle of the complex scattering amplitudes $f_{ni}(\theta)$ sheds new light on the computational problems that need to be solved in order for the λ, R and I problem to be resolved.

I. INTRODUCTION

In this lecture the numerical solution of partial differential equations in electron-atom scattering theory will be discussed and contrasted with the problem of solving ordinary differential equations in scattering theory. In particular, the results obtained by the multichannel eikonal and close coupling theories for the electron impact excitation of hydrogen will be examined in detail. In Section II the partial differential equations of the multichannel eikonal theory are presented together with the ordinary differential equations of the close coupling theory. The advantages and limitations inherent in the representations employed by both theories is also discussed. In addition, the parallelizability of the algorithms used in the numerical solution of the PDE's and the ODE's in the two theories is discussed in Section II. Section III contains a discussion of the results of the two theories for the integral and differential cross sections and the complex scattering amplitudes for the electron impact excitation of hydrogen. The conclusions are presented in Section IV together with a list of general references.

II. THEORY

Here we are concerned with the scattering of a structureless projectile B at a distance $\vec{R}$ from a target atom A with electronic coordinate $\vec{r}$. In this case, the time-independent Schrödinger equation is,

$$\mathcal{H}_i \Psi_i = E_i \Psi_i \tag{1}$$

where the Hamiltonian operator $\mathcal{H}_i$ is given by,

$$\mathcal{H}_i = \frac{-\hbar^2}{2m_A} \nabla_A^2 - \frac{\hbar^2}{2m_B} \nabla_B^2 - \frac{\hbar^2}{2m_e} \nabla_e^2 + V_{AB} + V_{Ae} + V_{Be} \tag{2a}$$

$$= \frac{-\hbar^2}{2m_T} \nabla_{CM}^2 - \frac{\hbar^2}{2\mu} \nabla_{AB}^2 - \frac{\hbar^2}{2m} \nabla_r^2 + V_{AB} + V_{Ae} + V_{Be} \tag{2b}$$

$$\equiv \mathcal{H}_{CM} + \mathcal{H}_i' \tag{2c}$$

In equation (2a) the first three terms on the right-hand side of the equation are the kinetic energy operators for the indicated particles, while in equations (2b,c) the separation into center-of-mass (CM) and relative motion terms are shown. As is well known, the separability of the CM and relative motion terms in this case allows one to write the system wavefunction Ψ_i as a product of a plane wave with a wavefunction ψ_i for the relative motion of projectile B in the field of force of target A. The target atom A in the present case is assumed to be hydrogenic, generalization to other cases is straightforward. The masses in (2b) are defined as $m = m_e(m_A+m_B)/m_T$, $\mu = m_A m_B/(m_A+m_B)$ and $m_T = m_A + m_B + m_e$.

Therefore, Schrödinger's equation (1) now becomes,

$$\mathcal{H}_i' \Psi_i = E' \Psi_i \tag{3}$$

To solve (3) one generally expands the wavefunction ψ_i in a sum over eigenstates χ_n of the target A,

$$\psi_i = \sum_n F_n(\vec{r}_{AB})\, \chi_n(\vec{r}_{AB}, \vec{r}) \tag{4}$$

where F_n is the (unknown) wavefunction for the relative motion of projectile B a distance $\vec{r}_{AB}$ from target A (in channel n). In this lecture we are primarily concerned with contrasting the numerical solution of PDE's and ODE's in atomic scattering theory and hence the expansion given by (4) will be sufficient. However, if one is interested in resolving the spin structure of the target atom A, or the resonances in the cross section near threshold, additional terms (antisymmetrization, correlation, etc.) need to be added to (4). These additional terms will ultimately result in a larger set of coupled equations to be solved, but will not change in a material way our basic discussion of the numerical solution of coupled partial and ordinary differential equations in electron-atom scattering. References to calculations which do include the above effects are given in Section III.

Substitution of (4) into (3) results in the following set of coupled ordinary differential equations,

$$- \frac{\hbar^2}{2\mu} \nabla^2_{AB} F_n + \frac{\hbar^2 k_n^2}{2\mu} F_n - V_{nn} F_n = \sum_m{}' V_{nm}(\vec{r}_{AB}) F_m$$

$$- \sum_m \frac{\hbar^2}{2\mu} [F_m(\chi_n, \nabla^2_{AB} \chi_m) + 2 \vec{\nabla}_{AB} F_m (\chi_n, \vec{\nabla}_{AB} \chi_m)] \tag{5}$$

where the prime on the summation sign indicates omission of the term n=m, and the inner product (f,g) is defined.

$$(f,g) = \int f(\vec{r})\ g(\vec{r})d\vec{r}$$

To convert (5) into a set of PDE's, write $F_n(\vec{r}_{AB}) = A_n(\vec{r}_{AB})\ \exp[iS_n(\vec{r}_{AB})]$ and note that $r_{AB}^2 = \rho^2 + z^2$ with ρ, z, ϕ the usual cylindrical coordinates centered at A. This yields the following set of equations,

$$\Lambda_n \nabla_{AB}^2 A_n + 2\vec{\nabla}_{AB} \Lambda_n \bullet \vec{\nabla}_{AB} A_n + \left\{\nabla_{AB}^2 \Lambda_n + (k_n^2 - \frac{2\mu}{\hbar^2} V_{nn})\ \Lambda_n\right\} A_n$$

$$= \sum_m{}' \frac{2\mu}{\hbar^2} V_{nm}\ \Lambda_m\ A_m - \sum_m \frac{2\mu}{\hbar^2} \vec{\beta}_{nm} \bullet \vec{\nabla}_{AB}(A_m\ \Lambda_m) \tag{6}$$

with $\Lambda_n \equiv \exp[i\ S_n(\vec{r}_{AB})]$ and $\vec{\beta}_{nm} \equiv (\chi_n, \vec{\nabla}_{AB}\ \chi_m)$. After writing the gradient and Laplacian operators in (6) in terms of cylindrical coordinates, the coupled Hamilton-Jacobi partial differential equations are solved for the amplitude functions $A_n(\vec{r}_{AB})$. In these equations the eikonal phase $S_n(\vec{r}_{AB})$ is assumed to be known exactly. To obtain an equation for S_n set the term inside the curly brackets in (6) equal to zero,

$$\nabla_{AB}^2\ \Lambda_n(\vec{r}_{AB}) + (k_n^2 - \frac{2\mu}{\hbar^2} V_{nn})\ \Lambda_n(\vec{r}_{AB}) = 0 \tag{7}$$

where the V_{nm} in (6) are the instantaneous electrostatic interaction between the projectile and target i.e., $V_{nm} = V_{nm}(\vec{r}_{AB}) = (\chi_n, V(\vec{r}_{AB},\vec{r})\chi_m)$. Defining the local wavenumber $\kappa_n(\vec{r}_{AB}) \equiv k_n^2 - (2\mu/\hbar^2)\ V_{nn}$ and writing the Laplacian in (7) in spherical coordinates yields,

$$\frac{1}{r_{AB}^2} \frac{\partial}{\partial r_{AB}} \left[r_{AB}^2 \frac{\partial}{\partial r_{AB}}\ e^{iS_n(r_{AB})}\right] + \left[\frac{L_{AB}^2}{r_{AB}^2} + \kappa_n^2\right] e^{iS_n(r_{AB})} = 0$$

$$\rightarrow i\, S_n'' + i\, \frac{2}{r}\, S_n' - (S_n')^2 + \omega_n^2 = 0 \tag{8}$$

where the primes denote differentiation with respect to r_{AB} and L_{AB}^2 is the eigenvalue of the relative angular momentum operator $\vec{L}_{AB} = (\hbar/i)\, \vec{r}_{AB} \times \vec{\nabla}_{AB}$. The 'frequency' ω_n is defined, $\omega_n^2 \equiv (L_{AB}^2/r_{AB}^2) + \kappa_n^2$. Here we are interested in electron-atom collisions where the relative motion of the electron to a good approximation is a straight line. That is, we assume that the eikonal phase factor $S_n(\vec{r}_{AB})$ is a slowly varying function of $\vec{r}_{AB}$; or, equivalently, that the density of the classical ensemble of particles varies sufficiently slowly along the classical trajectory such that $S_n'' \sim 0$, and the L_{AB}^2 term in (8) can be ignored. In this case, the real part of (8) is integrated to give,

$$S_n(r_{AB}) \equiv S_n(\rho,z) = k_n z + \int_{-\infty}^{z} \left[\left[k_n^2 - \frac{2\mu}{\hbar^2} V_{nn}(\rho,z')\right]^{1/2} - k_n \right] dz' \tag{9}$$

In the case of heavy particle collisions the curvature of the trajectory of the projectile must be included in the eikonal phase. An example of this type of calculation, for ion-molecule collisions, is the work of McCann and Flannery [26,27]. With the choice of (9) for the eikonal phase the coupled equations (6) are independent of A_n on the left-hand-side. In order to further simplify these 2nd-order partial differential equations for A_n, we in addition assume that the term $\nabla_{AB}^2 A_n$ is small and that the eigenstates χ_n are independent of $\vec{r}_{AB}$. The latter condition insures that the second summation on the RHS of (6) vanishes and is consistent with our omission of electron correlation effects in the wave-function expansion. These terms become important when there is significant configuration mixing in the target atom.

With the above approximations, the coupled equations (6) reduce to a set of first-order partial differential equations,

$$i \frac{\hbar^2}{\mu} \kappa_n \frac{\partial C_n(\rho,z)}{\partial z} + [\frac{\hbar^2}{\mu} \kappa_n(\kappa_n-k_n) + V_{nn}]C_n = \sum_j V_{nj} C_j \exp[i(k_j-k_n)z] \quad (10)$$

where $A_n(\rho,z) = C_n(\rho,z)\, e^{i\Delta_{in}\phi} \exp\{-i \int_{-\infty}^{z} (\kappa_n-k_n)dz'\}$, and $\Delta_{in} \equiv m_i-m_n$.

These equations are solved subject to the boundary condition $C_n(\rho,-\infty) = \delta_{ni}$. The coupled equations (10) are the basis of the multichannel eikonal theory (MET) of Flannery and McCann [11-16]. The three principal advantages of the semiclassical equations (10) are:

(i) The equations are first-order in z, hence the numerical techniques used for ordinary differential equations can be used to solve (10). This also means that no matrix diagonalization needs to be done in the numerical solution of (10), as is the case with 2nd-order ODE's.

(ii) The second variable ρ (the projectile's impact parameter) appears in (10) only as a parameter. This indicates that the coupled PDE's (10) will be readily parallelizable. While no calculations have yet been performed with (10) solved on a parallelizable machine, when this is done, a great deal of time should be saved. This is important since the numerical solution of (10) is the principle bottleneck in the MET calculations.

(iii) The memory and time required to solve (10) is a linear function of the number of eigenstates χ_n used in the basis set (4). This is in contrast to the case of 2nd-order ODE's where the time required for the matrix diagonalization is a *cubic* function of the number of elements in the matrix to be diagonalized. This in turn is a result of the direct methods used for the matrix diagonalization, and hence represents a major hurdle to the use of large basis sets in the solution of 2nd-order ODE's. The ultimate reason behind the difficulty in using large basis sets in solving 2nd-order ODE's by matrix techniques lies in the use of an angular momentum representation for the wavefunctions F_n rather than a coordinate (i.e., configuration space) representation.

The main disadvantage of solving the 1st-order partial differential equations (10) is the fact that they must be solved over a 2-dimensional grid rather than a one-dimensional grid as is necessary in the solution of 2nd-order ODE's. A consequence of this is that the memory requirements are an order of magnitude larger for the former calculation as compared to the latter.

The close coupling 2nd-order ODE's which arise from using an angular momentum representation for $F_n(\vec{r}_{AB})$ in (4), are,

$$\left[\frac{d^2}{dr_{AB}^2} - \frac{\ell_i(\ell_i+1)}{r_{AB}^2} + \frac{2Z}{r_{AB}} + k_i^2\right] F_i(r_{AB})$$

$$= 2\sum_j V_{ij}(r_{AB})\, F_j(r_{AB}) + \int_0^\infty W_{ij}(r_{AB},r')\, F_j(r')dr'$$

$$+ \sum_{n\ell} \lambda_{n\ell}^{(i)}\, P_{n\ell}(r_{AB})\delta_{\ell\ell i} \qquad (11)$$

where ℓ_i is the orbital angular momentum quantum number of the projectile electron in state i, and the $\lambda_{n\ell}^{(i)}$ are Lagrange multipliers chosen such that the target orbitals $P_{n\ell}$ are orthogonal to the F_i. The matrix elements V_{ij} are the same as those defined previously, while the W_{ij} are the electrostatic matrix elements arising from inclusion of electron exchange (i.e., antisymmetrization) terms and correlation terms in the wavefunction expansion (4). The close coupling equations (11) are well known in the literature, hence their derivation need not be repeated here. However, attention is drawn to the following papers and reviews for those interested in further details [2,5-7,29].

Technically, the close coupling equations (11) are Fredholm ordinary integro-differential equations which, using the technique of Marriott [25], can be cast in the form of a larger set of purely ordinary differential equations. Then, after discretization of the Laplacian in (11), the problem is converted into one of matrix diagonalization.

The advantage of using the close coupling equations (11) is that very accurate inelastic cross sections can be obtained close to threshold - especially the resonance structure between the inelastic threshold and the threshold for ionization. However, as the energy crosses the ionization threshold (where the number of open channels becomes infinite), or in the case of transitions between excited states, the close coupling equations (11) become increasingly difficult to solve via matrix techniques due to the large number of basis states and partial waves ℓ_i required for convergence. This problem can partly be alleviated through the use of pseudo-states. Another way around the bottleneck of basis set size in the solution of (11) is through the use of multi-tasking on the CRAY-XMP. Important recent work in this regard is that of Sawey et al. [30]. Clearly further work on the numerical solution of (11), both by matrix diagonalization techniques and by solving the equivalent partial differential equations, is needed.

In this lecture we are interested in contrasting the numerical solutions of the coupled PDE's (10) with the ODE's (11). Hence we will only discuss the techniques used to solve numerically the PDE's (10), the techniques used to solve the close coupling equations (11) having been thoroughly described in [2,5,7]. In particular we will end Section II with a brief review of extrapolation methods used to solve 1st-order ordinary differential equations. A more complete discussion of the numerical solution of (10) including Runge-Kutta and predictor-corrector methods will be given in a forthcoming paper [22].

Extrapolation methods for ODE's

Consider the 1st-order ODE, $dy(t)/dt = F(t,y(t))$. When this is integrated for sufficiently small step sizes h, the solution of $y(t+h)$ can be written as a power series in h,

$$y(t+h) = y(t) + \sum_{i=1}^{m} \tau_i(t)h^i + \sigma(h^{m+1}) \tag{13}$$

The goal of extrapolation methods is to eliminate the power series in h in (13) above by integrating the differential equation for a sequence of step sizes $h_o, h_1, \ldots, h_m$, and then extrapolating the results to $h \to 0$. That is, the power series $\sum_i \tau_i(t)h^i$ is approximated by functions $R_m(t,h_i)$ which have m+1 unknowns. These unknowns are determined by the condition $R_m(t,h_j) = y(t+h_j)$, $j = 0,1,\ldots,m$. Hence the solution of the ODE y(t) is approximated by $R_m(t,0)$.

The two principle extrapolation methods are by polynomials and by rational functions. In polynomial extrapolation the function $R_m(t,h_i) \equiv R_m^{(i)}(h)$ is an mth degree polynomial in h and is computed recursively by,

$$R_m^{(i)} = R_m^{(i+1)} + \frac{R_{m-1}^{(i+1)} - R_{m-1}^{(i)}}{(h_i/h_{i+m})-1}$$

In rational function extrapolation the $R_m^{(i)} = P_m(h)/Q_m(h)$ where $P_m(h)$ and $Q_m(h)$ are polynomials in h of degree μ and υ, respectively. The $R_m^{(i)}(h)$ are computed in this case recursively via

$$R_{-1}^{(i)} = 0, \quad R_o^{(i)} = y(t+h_i)$$

$$R_m^{(i)} = R_{m-1}^{(i+1)} + \frac{R_{m-1}^{(i+1)} - R_{m-1}^{(i)}}{(h_i/h_{i+m})^2\left[1 - \dfrac{R_{m-1}^{(i+1)} - R_{m-1}^{(i)}}{R_{m-1}^{i+1} - R_{m-2}^{i+1}}\right] - 1}, \quad m \geq 1$$

The MET results discussed in Section III were obtained by solving the coupled 1st-order PDE's (10) using Bulirsch and Stoer's [4] method of rational extrapolation for ODE's. A full discussion of extrapolation techniques can be gound in Gear [17] and Dahlquist and Bjorck [10].

III. RESULTS

Before comparing the MET results obtained by solving the PDE's (10) with the close coupling equations (11) a short discussion on the practical numerical methods used to solve (10) are in order. To do this we quote the final expression used in the multichannel eikonal theory for the complex scattering amplitude for the transition $i \rightarrow n$,

$$f_{ni}(\theta) = -(i)^{\Lambda_{in}+1} \int_0^\infty J_{\Lambda_{in}}(q'\rho)\ [I_1(\rho,\gamma(\theta)) - i\ I_2(\rho,\gamma(\theta))]\rho d\rho \qquad (14)$$

where the integrals I_1 and I_2 are defined,

$$I_1(\rho,\gamma) = \int_{-\infty}^{\infty} dz\ \kappa_n(\rho,z) \frac{\partial C_n(\rho,z)}{\partial z} \exp[i\gamma(\theta)z] \qquad (15a)$$

$$I_2(\rho,\gamma) = \int_{-\infty}^{\infty} dz\ [\kappa_n(\kappa_n-k_n) + \frac{\mu}{h^2} V_{nn}]C_n(\rho,z) \exp[i\gamma(\theta)z] \qquad (15b)$$

and refer the reader to the original literature [12,13,16] for the details. In equations (14,15) $q' = k_n \sin\theta$ and $\gamma(\theta) = k_n(1-\cos\theta)$ and $\Lambda_{in} = m_i - m_n$. The 1st-order PDE's (10) are solved using Burlisch and Stoer's rational extrapolation technique for ODE's over a finite 2-dimensional grid: $o \leq \rho \leq \rho_{max}$, $-z_{max} \leq z \leq z_{max}$ for the amplitude functions $C_n(\rho,z)$. The values of ρ_{max} and z_{max} are varied until the cross section,

$$\int_0^{\rho_{max}} |C_n(\rho,z_{max})|^2 \rho d\rho \qquad (16)$$

is computed to within a tolerence ϵ (i.e., until subsequent evaluations change by less than an amount ϵ (%)). An additional criterion for the selection of optimal values of ρ_{max} and z_{max} is that the MET integral cross section, computed from the scattering amplitude (14) should converge to the 1st Born approximation at high incident energies. For this to be achieved it was found necessary to solve (20) using a non-linear grid in z in order that the rapid variation of C_n and $\partial C_n/\partial z$ near $z = 0$ be accurately represented. This was needed so that the subsequent evaluations of I_1, I_2 were accurate. The nonlinear grid in z used was:

$z_i = 3\tan(i\delta)$, $i = -N_z,\ldots,N_z$ with $\delta = \tan^{-1}(z_{max}/3)/N_z$. However, since I_1 and I_2 must be evaluated numerically from the tabulated solution of the coupled equations (10), the most efficient way of solving (10) is to make the grid points z_i used to solve (10) and the pivots used in the evaluation of I_1, I_2 *identical*. This avoids the need to interpolate w.r.t. z in the quadrature of I_1 and I_2 (interpolation w.r.t. ρ must still be done however). In the MET calculations discussed below, integrals I_1, I_2 were evaluated using Simpson's rule with the nonlinear pivots z_i chosen above and with weights $\omega_i = r_i\delta\sec^2(i\delta)$, where r_i are the usual Simpson's rule weights and N_z is the number of points used to discretize the z-range $[o,z_{max}]$. Hence while the number of points required for an accurate evaluation of I_1, I_2 is much larger using Simpson's rule as compared with using a higher-order quadrature method, the amount of time saved by eliminating the need to interpolate w.r.t. z more than makes up for the increased number of grid/ pivot points z_i required.

In figure 1 the real and imaginary parts of the amplitude function $C_n(\rho,z)$ for the 1s state of hydrogen are shown as an example of the type of behavior exhibited by the solutions of equation (10). For a more extensive exhibition of the solutions of the semiclassical equations (10) see [22].

In the remainder of Section III an overview of the MET results for e^-+H collisions will be given. This will include differential and integral cross sections as well as the complex scattering amplitudes. For a complete update and discussion of the present MET see Mansky and Flannery [23,24]. It should be clear that by comparing the results obtained for a wide range of physical observables, from the solution of equations (10) and (11), one not only gets an idea of the success or failure of a particular theory over a wider range of physical conditions, but also insight into the accuracy of the numerical solution of the coupled equations underlying a given theory. That is, by varying z_{max} and ρ_{max} in the semiclassical equations (10) until the integral cross sections computed from (14), for all states in the basis set, have

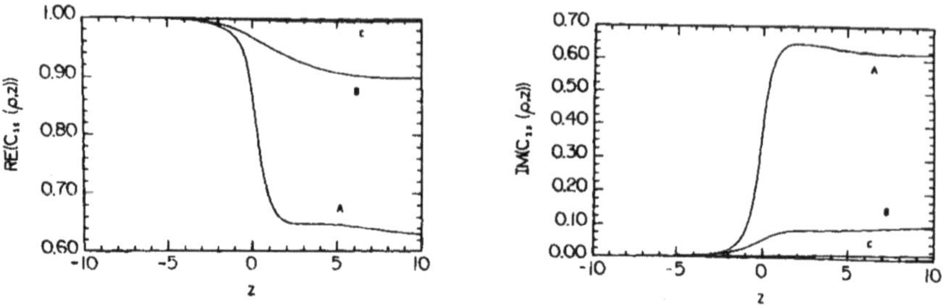

Figure 1: Real and imaginary parts of the MET amplitude function $C_{1s}(\rho,z)$ versus $z(a_o)$ for e^-+H collisions at E = 54.40 eV.

converged to their corresponding 1st Born approximation values in the limit of high energy, one obtains an idea of the minimum size 2-dimensional grid required to solve the coupled PDE's (10). These values of ρ_{max}, z_{max} can then be used to solve (10) for all other energies of interest. Note that a similar argument can be made about the numerical solution of the coupled ODE's (11) where the appropriate parameters are r_{max} and ℓ_{max} - the maximum value of the independent variable r_{AB} and the largest partial wave ℓ_i retained in the expansion.

The MET differential and integral cross sections are defined as,

$$\sigma_n(\theta) = \frac{k_n}{k_i} |f_{ni}(\theta)|^2 \tag{17a}$$

$$\sigma_n = \int_0^{2\pi} d\phi \int_0^{\pi} \sin\theta \, d\theta \, \sigma_n(\theta) = 2\pi \int_0^{\pi} \sigma_n(\theta) \sin(\theta) d\theta \tag{17b}$$

where the complex scattering amplitude for the transition i → n is given by (14). In figures 2,3 the integral and differential cross sections for

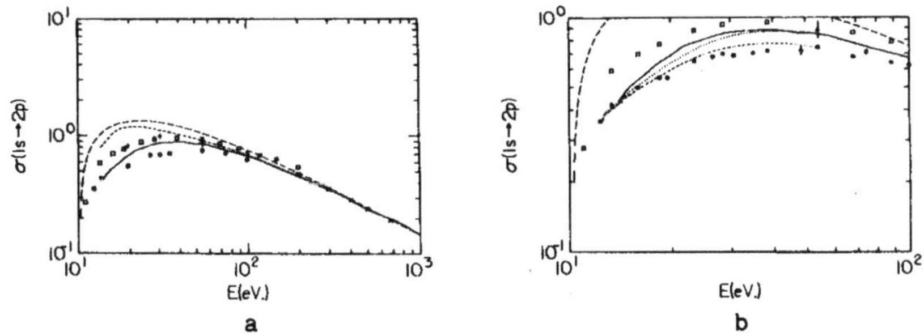

Figure 2a: Integral cross section σ_{2p} in units of πa_o^2 versus E (eV.). 1st Born (- - -), MET [23] (___), 3-state close coupling [18] (-----), DWSBA [20] (X), AVCC-18 state [9] (*), unitarized Born [31] (+), experimental data of Long et al. [21] as renormalized by Bransden and McDowell [3] (□), experimental data of Long et al. [21] as renormalized by van Wyngaarden and Walters [34] (0), experimental data of Williams [32] (Λ).

Figure 2b: Same as Fig. 2a except with AVCC-11 state (.....) [8].

e^-+H(1s → 2p) collisions are shown (results for other transitions in hydrogen are given in [23]). The MET results are in good overall agreement with experiment in figure 2, and clearly converge to the Born cross section at high energy. In particular, the agreement (cf. figure 2b) with the absolute measurement of Williams [32] at 54.40 eV is noteworthy. The differences between the original algebraic variational close coupling results of Callaway [8], and the same results as renormalized by van Wyngaarden and Walters [34], is a choice of normalization (i.e., normalization to experiment at 11 eV versus the pseudostate close coupling calculations of van Wyngaarden and Walters [35] at 350 eV). On the other hand, the differences between the 3-state close coupling results of Kingston, Fon and Burke [18] (cf. figure 2a) and the MET results is an indication of the lack of convergence w.r.t. basis set size in the former calculation. The importance of basis set size is evident in comparing the 3-state close coupling results of Kingston, Fon and Burke and the 18-state AVCC results of Callaway et al. [9] (cf. figure 2b). While the MET results are in good agreement with the results of Callaway et al. for energies E $\geq$ 70 eV, the differences observed in figure 2 at lower energies is due to the neglect of electron exchange terms in the MET. This is also evident in figure 3 by the rapid decrease of the MET differential cross sections $\sigma_{2s}(\theta)$, $\sigma_{2p}(\theta)$ at scattering angles $\theta > 40^o$ when compared to the close coupling results of Kingston, Fon and Burke [18] and van Wyngaarden and Walters [34]. The agreement between the MET and the experimental data of Williams [32] for $\theta \leq 20^o$ in figure 3 also indicates that the choices for z_{max} and ρ_{max} in the solution of (10) were correct.

From figure 3 one would conclude that electron exchange effects are only important at large scattering angles. This is incorrect. While a definitive calculation has not yet been done, the ongoing problem of theory to reproduce the experimental data for the λ, R and I parameters indicates that theory is still not handling adequately the numerical solution of the coupled equations (10) or (11). In figure 4 we show the λ, R and I parameters for e^-+H(1s → 2p) collisions at E = 54.40 eV and scattering angles $\theta \leq 50^o$. Clearly the MET results accurately reproduce the experimental data of Williams [32] only for $\theta \leq 20^o$, while at the level of $\sigma_{2p}(\theta)$ the corresponding angular range was $\theta \leq 40^o$. In contrast, the two close coupling results shown in figure 4 are in good agreement with experiment out to approximately 40°. However, at larger scattering angles ($\theta \geq 60^o$) the close coupling results fail to reproduce the second experimental minimum in the λ parameter observed around 100° and the magnitude of the R parameter in the range $70^o \leq \theta \leq 120^o$. These facts taken together indicate that while the values of z_{max}, ρ_{max} used to solve (10) in the MET are adequate at the level of differential and integral cross sections, the small z behavior of the amplitude functions still needs refinement for physical observables directly dependent on the complex scattering amplitude $f_{ni}(\theta)$. For completeness, the λ, R and I parameters are defined,

$$\lambda = |f_o|^2/[|f_o|^2 + 2|f_{+1}|^2] \quad (18a)$$

$$R = \sqrt{[\lambda(1-\lambda)/2]}\, \cos(\beta_1 - \beta_o) \quad (18b)$$

$$I = \sqrt{[\lambda(1-\lambda)/2]}\, \sin(\beta_1 - \beta_o) \quad (18c)$$

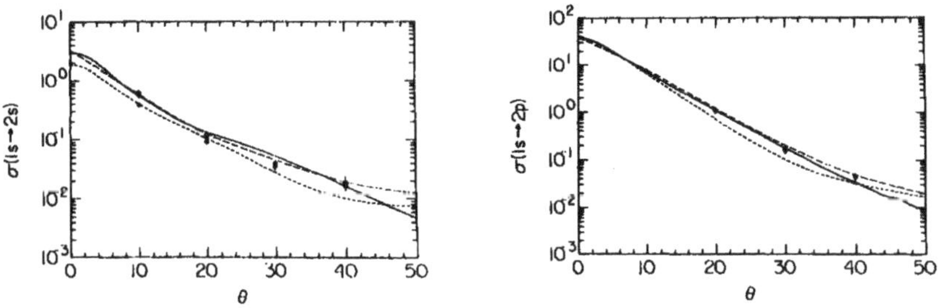

Figure 3: Differential cross sections $\sigma_{2s}(\theta)$, $\sigma_{2p}(\theta)$ in a_o^2/str. vs. θ for E = 54.40 eV. MET [23] (___), 3-state close coupling [18] (---), pseudostate close coupling [35] (----), experimental data of Williams [23] 0 (via sum), □ via ratio).

where $f_m(\theta) = |f_m(\theta)| \exp[i\beta_m(\theta)]$ and $f_m(\theta)$ denotes the complex scattering amplitude for $2p_m$ magnetic substate. For a complete discussion of the angular correlation and polarization correlation parameters in electron-atom scattering see Andersen et al. [1]. For more on the problem of the λ, R and I parameters in e^-+H collisions see Morgan [28].

To better understand what part of the solution of the PDE's (10) needs improvement, in regards to the λ, R and I problem discussed above, and where the electron exchange terms in (11) become important, we show the MET results for the scattering amplitudes $f_{2p_o}(\theta)$ and $f_{2p_1}(\theta)$ at E = 54.40 eV in figures 5 and 6, respectively. These are compared in figures 5,6 with the close coupling results of Kingston, Liew and Burke [19]. Two things are evident in these figures. First, that the electron exchange terms in the close coupling equations (11) manifest themselves quite differently in the modulus and phase angle of the scattering amplitude. For example, the phase angles for the singlet and triplet spin channels for the $f_{2p_o}(\theta)$ scattering amplitude differ from each other appreciably for $\theta \geq 30^o$, while for the $f_{2p_1}(\theta)$ they don't begin to differ greatly until $\theta \geq 40^o$. On the other hand, the modulii for the singlet and triplet spin channels of the $f_{2p_0}(\theta)$, $f_{2p_1}(\theta)$ scattering amplitudes only differ appreciably for $\theta \lesssim 20^o$, $\theta \lesssim 30^o$, respectively. That is, electron exchange terms are important at small scattering angles for the moduli of scattering amplitudes, while for the corresponding phase angles, they are important only at large scattering angles. This indicates that unraveling the relative contributions that direct and exchange terms make to a given scattering amplitude at a specific angle will be difficult.

The second point to note from figure 5 is that the MET results for the $f_{2p_0}(\theta)$ amplitude agrees quite closely at all angles with the triplet spin channel results of Kingston, Liew and Burke. On the other hand in figure 6 the MET results only agree with the singlet spin channel results over a limited angular range. In particular the MET results exceed both the singlet and triplet spin channel results for $|f_{2p_1}(\theta)|$

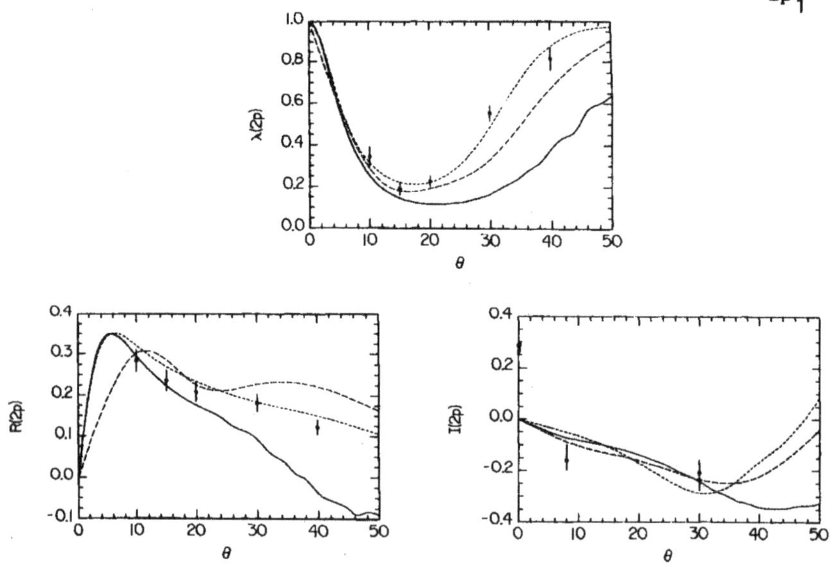

Figure 4: λ, R and I parameters for the 2p state at E = 54.40 eV. MET [23] (___), 3 state close coupling [18] (---), pseudostate close coupling [35] (-----), experimental data of Williams [32,33] O (λ, R from [32], I from [33]).

for $\theta \leq 20^{o}$. Note that we should not expect to see the MET results in figures 5,6 lying between the singlet and triplet spin channel results of Kingston, Liew and Burke. Rather, the observed behavior of the modulii and phase angles of the MET scattering amplitudes in figures 5,6 is a direct result of a complicated interplay between the z-behavior of the amplitude functions $C_n(\rho,z)$ and eikonal phases $S_n(\rho,z)$. A detailed discussion of these topics is beyond the scope of this lecture, but will be the subject of a forthcoming paper.

IV. CONCLUSIONS AND GENERAL REFERENCES

In this lecture we have contrasted solving coupled PDE's with ODE's in electron-atom collision theory. The principle conclusion of this lecture is that the solutions of (10) and (11) are complementary. That is, the configuration space representation employed by the MET, and the angular momentum representation employed by close coupling theories, complement one another, both in terms of information they provide about the scattering event, and in the energy ranges over which they are valid. This is important since it means that by solving (10) *and* (11) one gains additional insight into a particular excitation process that would not be obtained otherwise. This proved useful for example in the discussion of

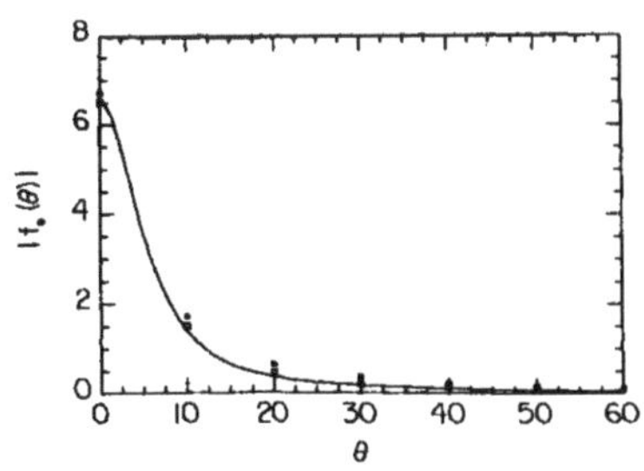

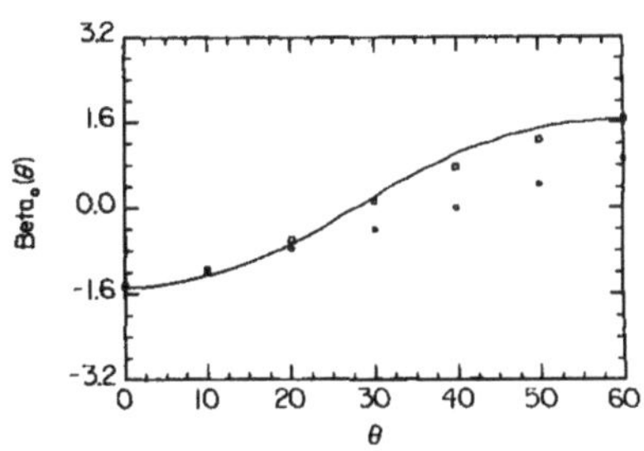

Figure 5: Modulus $|f_o|$ (in a_o) and phase angle β_o (in radians) for $f_{2p}(\theta)$ vs. θ for E = 54.40 eV. MET [23] (___), 3 state close coupling [19] O (singlet) □ (triplet).

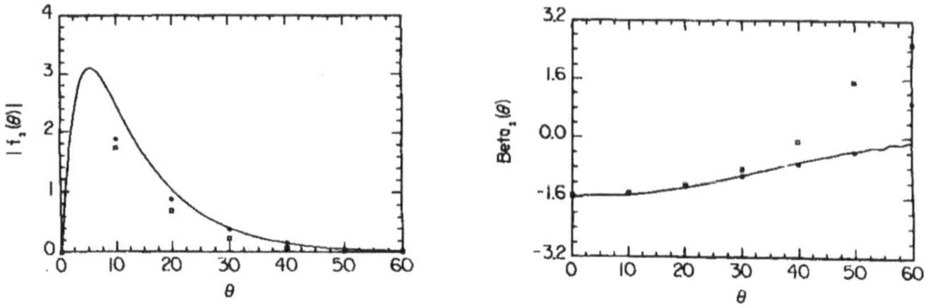

Figure 6: Same as Fig. 5 except for $f_{2p_1}(\theta)$.

the λ, R and I problem in Section III. We end this lecture with a short list of general references which we have found useful on the subject of coupled ordinary and partial differential equations.

- Collatz, L., 1960, The Numerical Treatment of Differential Equations 3rd ed., Springer-Verlag.
- Dahlquist, G., 1956, Math. Scandinavica 4 33-50, 1959, Trans. Roy. Inst. Tech., Stockholm, No. 130.
- Dahlquist, G., and Bjorck, A., 1974, Numerical Methods, Prentice Hall.
- Gear, C. W., 1971, Numerical Initial Value Problems in Ordering Differential Equations, Prentice-Hall.
- Henrici, P., 1962, Discrete Variable Methods for Ordinary Differential Equations, Wiley.
- Ince, E. L., 1956, Ordinary Differential Equations, Dover.
- Olver, P. J., 1986, Applications of Lie Groups to Differential Equations, Springer-Verlag.

Acknowledgements

This work is supported by AFOSR under grant no. 84-0233. The invitation to lecture, and the partial support received from Oak Ridge National Laboratory are also gratefully acknowledged.

References

1. Anderson, N., Gallagher, J. W., and Hertel, I. V., 1988, Phys. Rep. 165, 1-188.

2. Berrington, K. A., Burke, P. G., Le Dourneuf, M., Robb, W. D., Taylor, K. T. and Lan Vo Ky, 1978, Comp. Phys. Comm. 14, 367-412.

3. Bransden, B. H., and McDowell, M. R. C., 1978, Phys. Rep. 46, 249-394.

4. Bulirsch, R., and Stoer, J., 1966, Numer. Math. 8, 1-13.

5. Burke, P. G., and Eissner, W. B., 1983, Chapter 1 (p. 1-54) in Atoms in Astrophysics, Plenum Press (Burke, P. G., Eissner, W. B., Hummer, D. G. and Percival, I. C., eds.).

6. Burke, P. G., and Robb, W. D., 1971, Adv. Atom. Mol. Phys. 11, 143-214.

7. Burke, P. G., and Seaton, M. J., 1971, Methods in Comput. Phys. 10, 1-80.

8. Callaway, J., 1985, Phys. Rev. A 32, 775-83.

9. Callaway, J., Unnikrishnam, K., and Oza, D. H., 1987, Phys. Rev. A 36, 2576-84.

10. Dahlquist, G., and Bjorck, A., 1974, Numerical Methods, Prentice-Hall.

11. Flannery, M. R., and McCann, K. J., 1974, J. Phys. B: At. Mol. Phys. 7, L223-7.

12. ibid., J. Phys. B: At. Mol. Phys. 7, 2518-32.

13. ibid., J. Phys. B: At. Mol. Phys. 7, L522-7.

14. ibid., Phys. Rev. A 10, 2264-72.

15. ibid., 1975, J. Phys. B: At. Mol. Phys. 8, 1716-33.

16. ibid., Phys. Rev. A 12, 846-55.

17. Gear, C. W., 1971, Numerical Initial Value Problems in Ordinary Differential Equations, Prentice-Hall.

18. Kingston, A. E., Fon, W. C., and Burke, P. G., 1976, J. Phys. B: At. Mol. Phys. 9, 605-18.

19. Kingston, A. E., Liew, Y. C., and Burke, P. G., 1982, J. Phys. B: At. Mol. Phys. 15, 2755-66.

20. Kingston, A. E., and Walters, H. R. J., 1980, J. Phys. B: At. Mol. Phys. 13, 4633-62.

21. Long, R. L., Cox, D. M., and Smith, S. J., 1968, J. Res. NBS A 72, 521-35.

22. Mansky, E. J., 1989, J. Comput. Phys., to be submitted.

23. Mansky, E. J., and Flannery, M. R., 1989, "The Multichannel Eikonal Theory of Electron-Hydrogen Collisions I. Excitation of H(1s)", J. Phys. B, to be submitted.

24. ibid., "The Multichannel Eikonal Theory of Electron-Helium Collisions I. Excitation of He(1^1S)", J. Phys. B, to be submitted.

25. Marriott, R., 1958, Proc. Phys. Soc. (Lon.) 72, 121-9.

26. McCann, K. J., and Flannery, M. R., 1975, J. Chem. Phys. 63, 4695-4707.

27. ibid., 1978, J. Chem. Phys. 69, 5275-87.

28. Morgan, L. A., 1982, J. Phys. B: At. Mol. Phys. 15, 4247-57.

29. Percival, I. C., and Seaton, M. J., 1957, Proc. Camb. Phil. Soc. 53, 654-62.

30. Sawey, P. M. J., Berrington, K. A., Burke, P. G., and Kingston, A. E., 1989, 16th Int. Conf. on the Phys. of Electronic and Atom. Coll. (N.Y.) Abstracts of contrib. papers 214.

31. Somerville, W. B., 1963, Proc. Phys. Soc. (Lon.) 82, 446-55.

32. Williams, J. F., 1981, J. Phys. B: At. Mol. Phys. 14, 1197-1217.

33. ibid., 1986, Aust. J. Phys. 39, 621-32.

34. van Wyngaarden, W. L., and Walters, H. R. J., 1986, J. Phys. B: At. Mol. Phys. 19, L53-8.

35. ibid., J. Phys. B: At. Mol. Phys. 19, 929-68.

THREE-DIMENSIONAL HF AND TDHF CALCULATIONS WITH THE BASIS-SPLINE COLLOCATION TECHNIQUE

A. S. Umar
Department of Physics & Astronomy, Vanderbilt University
Nashville, TN 37235

ABSTRACT

We present the numerical details for performing static Hartree-Fock and time-dependent Hartree-Fock calculations using the Basis-Spline Collocation Method. Calculations are performed on a three-dimensional Cartesian lattice without any symmetry assumptions and quantal degeneracies. The three-dimensional Hartree-Fock calculations are compared with the solutions of the corresponding radial equations. The results demonstrate an unprecedented accuracy for relatively coarse meshes. We also present a zero impact parameter time-dependent Hartree-Fock calculation for the $^{16}O + {}^{16}O$ system.

1. INTRODUCTION

The mean-field formalism has been a succesful approximation to the nuclear many body problem for reproducing the principal properties of nuclei throughout the periodic table. The details of these calculations suggest that the Pauli principle plays an important role in simultaneously building up a mean-field and suppressing the strong N-N interaction terms. The time-dependent mean-field formalism has proved to be a valuable tool for studying a variety of nuclear phenomena; e.g. fusion, fission, deep-inelastic scattering, and nuclear molecular resonances. A comprehensive review of the previous applications can be found in Refs.[1,2]

On the other hand, due to the numerical complexity and demand of extensive computer time, most applications were limited to the study of simple systems or they involved physical symmetry simplifying assumptions which are not present in the basic theory. For example, most static Hartree-Fock (HF) calculations have been done for studying the ground state properties of even-even nuclei possesing spherical symmetry. Although there are well developed methods to extend these calculations to the study of highly excited deformed systems (e.g. energy surfaces and superdeformations), such applications have been few and restricted to light systems[3)]. Consequently, our knowledge of the effective interaction only incorporates the ground state properties of a few nuclei. Most time-dependent Hartree-Fock (TDHF) calculations have been performed with imposed spin-symmetry (no spin-orbit interaction) and using an axially symmetric geometry. Recently, we have shown that the relaxation of the spin-symmetry introduced considerable dissipation of the relative kinetic energy in low-energy heavy-ion collisions [4,5,6,7)].

From the numerical standpoint new techniques have been developed to handle the solution of the Hartree-Fock equations on a space-time lattice. In particular, equations of motion were obtained via the variation of the lattice representations of the constants of the motion, such as the total energy[8,4,9,10)] In this variation after discretization approach resulting equations exactly preserve the constants of the motion. Most numerical calculations employed low order finite-difference discretization techniques with accuracies appropriate for studying the gross features of the reaction. With the advent of computer technology it has become feasible to carry out more extensive nuclear structure and reaction studies without resorting to the symmetry assumptions employed in the earlier applications. In order to be able to perform such calculations, and to obtain a more detailed comparison with data, it is neccesary to exploit higher order interpolation techniques. Discretization of the energy functional on a Basis-Spline Collocation Lattice provides a highly accurate alternative to the finite-difference method. The details of the Basis-Spline Collocation Method (BSCM) will be published elsewhere[11)].

In this paper we present three-dimensional, unrestricted HF and TDHF calculations using the Basis-Spline Collocation Method. In section 2 we briefly introduce the basic HF and TDHF equations using the Skyrme interaction. In section 3 we employ the BSCM to obtain a collocation-lattice representation of the HF equations. This section also contains the methods for solving the resulting systems of equations. Section 4 presents the results for static and dynamic calculations and in section 5 we present a discussion of the results.

2. FORMALISM

In this section we will write down the static and dynamic Hartree-Fock equations using the Skyrme effective interaction.

2.1 Formal Equations

Given a many-body Hamiltonian containing two and three-body interactions

$$H = \sum_{i}^{N} t_i + \sum_{i<j}^{N} v_{ij} + \sum_{i<j<k}^{N} v_{ijk} \, , \tag{1}$$

the time-dependent action S can be constructed as[12)]

$$S = \int_{t_1}^{t_2} dt < \Phi_u(t)|H - i\hbar\partial_t|\Phi_u(t) > \, . \tag{2}$$

Here, $\Phi_u(t)$ denotes the time-dependent, uncorrolated many-body wavefunction which is commonly chosen to be a Slater determinant. Variation of the action gives us a most probable time-dependent path between the points t_1 and t_2 in the multi-dimensional space-time phase space

$$\delta S = 0 \rightarrow \Phi(t) = \Phi_0(t) \, . \tag{3}$$

In practice $\Phi_0(t)$ is chosen to be a Slater determinant

$$\Phi_0(t) = \frac{1}{\sqrt{N!}} det|\phi_\lambda(\mathbf{r}, t)| \, , \tag{4}$$

where $\phi_\lambda(\mathbf{r}, t)$ are the single-particle states with quantum numbers λ. If the variation in Eq.(3) is performed with respect to the single-particle states ϕ_λ^* we obtain a set of coupled, nonlinear, self-consistent initial value equations for the single-particle states

$$h\{\phi_\lambda\}\,\phi_\lambda = i\hbar\dot{\phi}_\lambda \qquad \lambda = 1, ..., N \, . \tag{5}$$

These are the fully microscopic time-dependent Hartree-Fock equations which preserve the major conservation laws such as the particle number, total energy etc.

Static equations can be obtained from Eq.(5) by taking out a trivial phase from the single-particle states

$$\begin{aligned} h(\{\chi_\lambda\})\chi_\lambda &= \epsilon_\lambda \chi_\lambda \\ \phi_\lambda(\mathbf{r}, t) &= e^{-i\epsilon_\lambda t/\hbar} \chi_\lambda(\mathbf{r}) \, . \end{aligned} \tag{6}$$

The initial time-dependent single-particle states can be constructed from the static solutions by multiplying them with an appropriate phase factor (or boost)

$$\phi_\lambda(t=0) = e^{i\mathbf{k}\cdot\mathbf{r}}\chi_\lambda(\mathbf{r})\,. \tag{7}$$

This corresponds to multiplying the static Slater determinant by a plane wave $exp(i\mathbf{k}\cdot\mathbf{R})$ where $\mathbf{R} = \sum_i \mathbf{r}_i$ and leads to a translating solution due to the Galilean invariance of the equations of motion.

2.2 Skyrme Potential

As the effective two-body interaction we employed the Skyrme potential with a generalized density-dependent term[13,14,15,16,17)]

$$\begin{aligned} v^{(2)}(\mathbf{r}_1,\mathbf{r}_2) = & \, t_0(1+x_0P_\sigma)\delta(\mathbf{r}_1-\mathbf{r}_2) + \frac{t_1}{2}[\delta(\mathbf{r}_1-\mathbf{r}_2)k^2 + k'^2\delta(\mathbf{r}_1-\mathbf{r}_2)] \\ & + t_2\mathbf{k}'\cdot\delta(\mathbf{r}_1-\mathbf{r}_2)\mathbf{k} + \frac{t_3}{6}(1+x_3P_\sigma)\rho^\alpha(\mathbf{R})\delta(\mathbf{r}_1-\mathbf{r}_2) \\ & + it_4(\boldsymbol{\sigma}_1+\boldsymbol{\sigma}_2)\cdot\mathbf{k}'\times\delta(\mathbf{r}_1-\mathbf{r}_2)\mathbf{k}\,, \end{aligned} \tag{8}$$

where

$$\begin{aligned} \mathbf{k} &= \frac{\nabla_1-\nabla_2}{2i} \qquad \mathbf{k}' = \mathbf{k}^\dagger \\ P_\sigma &= \frac{1+\boldsymbol{\sigma}_1\cdot\boldsymbol{\sigma}_2}{2}\,. \end{aligned} \tag{9}$$

In Eq.(8) the term proportional to t_3 is a density dependent term which has the same expectation value as a local three-body interaction for $\alpha = x_3 = 1$. The term with coefficient t_4 is the well known spin-orbit interaction used in earlier calculations.

The expectation value of H with the above interaction can be written in terms of the energy density $\mathcal{H}(\mathbf{r})$ as[13,14)]

$$E = <\Phi|H|\Phi> = \int d^3r\,\mathcal{H}(\mathbf{r})\,. \tag{10}$$

The energy density is commonly decomposed into various groups as given below

$$\mathcal{H} = \mathcal{H}_0 + \mathcal{H}_{LS} + \mathcal{H}_C + \mathcal{H}_Y \tag{11}$$

$$\begin{aligned} \mathcal{H}_0 = & \frac{\hbar^2}{2m}\tau + \frac{t_0}{2}(1+\frac{x_0}{2})\rho^2 - \frac{t_0}{2}(\frac{1}{2}+x_0)\sum_q \rho_q^2 \\ & + \frac{t_3}{12}(1+\frac{x_3}{2})\rho^{\alpha+2} - \frac{t_3}{12}(\frac{1}{2}+x_3)\rho^\alpha\sum_q\rho_q^2 \\ & + \frac{1}{4}(t_1+t_2)[\rho\tau - j^2] - \frac{1}{8}(t_1-t_2)\sum_q(\rho_q\tau_q - j_q^2) \end{aligned}$$

$$\mathcal{H}_{LS} = -\frac{t_4}{2}[\rho\nabla\cdot\mathbf{J} + \sum_q \rho_q(\nabla\cdot\mathbf{J}_q)]$$

$$\mathcal{H}_C = \frac{1}{2}e^2 \int d^3r' \rho_p(\mathbf{r})\frac{1}{|\mathbf{r}-\mathbf{r}'|}\rho_p(\mathbf{r}') - \frac{3}{4}e^2(\frac{3}{\pi})^{\frac{1}{3}}[\rho_p(\mathbf{r})]^{\frac{4}{3}} \tag{12}$$

and

$$\mathcal{H}_Y = \begin{cases} \int d^3r' \frac{exp(-|\mathbf{r}-\mathbf{r}'|/a)}{|\mathbf{r}-\mathbf{r}'|/a}[\frac{V_U}{2}\rho(\mathbf{r})\rho(\mathbf{r}') + \frac{V_L - V_U}{2}\sum_q \rho_q(\mathbf{r})\rho_q(\mathbf{r}')] \\ \text{or} \\ \frac{1}{16}(t_2 - 3t_1)\rho\nabla^2\rho + \frac{1}{32}(t_2 + 3t_1)\sum_q \rho_q\nabla^2\rho_q \end{cases} \tag{13}$$

The finite-range form of the $\mathcal{H}_Y$ term[9)] is usually preferred in dynamical calculations because of numerical stability and better surface properties. We also note that the Coulomb term $\mathcal{H}_C$ contains a Slater excange contribution in addition to the familiar direct term. Various quantities used in the energy density are written in terms of the spin-up and spin-down components of the single-particle states

$$\phi_\lambda = \begin{pmatrix} \psi_\lambda^+ \\ \psi_\lambda^- \end{pmatrix} , \tag{14}$$

as

$$\rho_q(\mathbf{r}) = \sum_{\substack{\lambda\in q \\ \mu}} n_\lambda\{|\psi_\lambda^+(\mathbf{r})|^2 + |\psi_\lambda^-(\mathbf{r})|^2\}$$

$$\tau_q(\mathbf{r}) = \sum_{\substack{\lambda\in q \\ \mu}} n_\lambda\{|\nabla\psi_\lambda^+(\mathbf{r})|^2 + |\nabla\psi_\lambda^-(\mathbf{r})|^2\}$$

$$\mathbf{j}_q(\mathbf{r}) = \sum_{\substack{\lambda\in q \\ \mu}} n_\lambda\{Im[\psi_\lambda^{+*}(\mathbf{r})\nabla\psi_\lambda^+(\mathbf{r}) + \psi_\lambda^{-*}(\mathbf{r})\nabla\psi_\lambda^-(\mathbf{r})\}$$

$$\mathbf{J}_q(\mathbf{r}) = -i \sum_{\substack{\lambda\in q \\ \mu\mu'=\pm}} n_\lambda\psi_\lambda^{\mu*}(\mathbf{r})(\nabla\times\sigma)\psi_\lambda^{\mu}(\mathbf{r}') .$$

Here, the subscript $q = n, p$ denotes the isospin quantum number and the total density is simply $\rho = \rho_n + \rho_p$ (same for all other quantities as well).

2.3 Hartree-Fock Hamiltonian

Using the Skyrme effective interaction and the one-body kinetic energy term, the Hartree-Fock Hamiltonian can be written as

$$h_q = -\nabla \cdot \frac{\hbar^2}{2m_q(\mathbf{r})}\nabla + U_q + U_Y + U_C + \frac{1}{2i}(\nabla \cdot \mathbf{I}_q + \mathbf{I}_q \cdot \nabla) - i\mathbf{B}_q \cdot (\nabla \times \sigma)\,. \tag{15}$$

Various terms in the above expression are given by

$$\begin{aligned}\frac{\hbar^2}{2m_q(\mathbf{r})} &= \frac{\hbar^2}{2m} + \frac{1}{4}(t_1+t_2)\rho + \frac{1}{8}(t_2-t_1)\rho_q \\ \mathbf{I}_q &= -\frac{1}{2}(t_1+t_2)\mathbf{j} - \frac{1}{4}(t_2-t_1)\mathbf{j}_q \\ \mathbf{B}_q &= -\frac{1}{8}(t_2-t_1)\mathbf{J}_q + \frac{t_4}{2}\nabla(\rho+\rho_q)\,,\end{aligned} \tag{16}$$

and

$$\begin{aligned}U_q = {} & t_0[(1+\frac{x_0}{2})\rho - (\frac{1}{2}+x_0)\rho_q] + \frac{1}{4}(t_1+t_2)\tau - \frac{1}{8}(t_1-t_2)\tau_q \\ & + \frac{t_3}{12}(1+\frac{x_3}{2})[\alpha+2]\rho^{\alpha+1} - \frac{t_3}{12}(\frac{1}{2}+x_3)[\alpha\rho^{\alpha-1}\sum_q \rho_q^2 + 2\rho^\alpha\rho_q] \\ & - \frac{t_4}{2}[\nabla\cdot\mathbf{J} + \nabla\cdot\mathbf{J}_q]\,.\end{aligned} \tag{17}$$

The Yukawa and Coulomb contributions are

$$U_Y = \begin{cases} \int d^3r' \, \frac{exp(-|\mathbf{r}-\mathbf{r}'|/a)}{|\mathbf{r}-\mathbf{r}'|/a}[V_U\rho(\mathbf{r}') + (V_L - V_U)\rho_q(\mathbf{r}')] \\ \text{or} \\ \frac{1}{8}(t_2-3t_1)\nabla^2\rho + \frac{1}{16}(t_2+3t_1)\nabla^2\rho_q \end{cases} \tag{18}$$

$$U_C = e^2\int d^3r' \frac{\rho_p(\mathbf{r}')}{|\mathbf{r}-\mathbf{r}'|} - e^2(\frac{3}{\pi})^{\frac{1}{3}}[\rho_p(\mathbf{r})]^{\frac{1}{3}}\,. \tag{19}$$

3. NUMERICAL IMPLEMENTATION - B-SPLINES

Currently, most HF and TDHF calculations are performed using low order finite-difference lattice techniques. It is desirable to investigate higher order interpolation methods which result in the improvement of the overall accuracy and reduction in the total number of lattice points. The lattice solution of differential equations on a discretized mesh of independent variables may be viewed to proceed in two steps: (1) Obtain a discrete representation of the functions and operators on the lattice. (2) Solve the resulting lattice equations using iterative techniques. Step (1) is an interpolation problem for which we could take advantage of the techniques developed using the basis-spline functions [18,11]. . The use of the Basis-Spline Collocation Method (BSCM) leads to a matrix-vector representation on the collocation lattice with a metric describing the transformation properties of the collocation lattice.

3.1 Basis-Splines

Given a set of points or *knots* denoted by the set $\{x_i\}$ a basis-spline (B-spline denoted by B_i^M) function of order M is constructed from continuous piecewise polynomials of order $M-1$. B-splines have continuous derivatives up to $(M-2)nd$ derivative and a discontinuous $(M-1)st$ derivative. We only consider odd order splines or even order polynomials for reasons related to the choice of the collocation points. The ith B-spline is nonzero only in the interval (x_i, x_{i+M}). This property is commonly referred to as limited support. The knots are the points where polynomials that make up the B-spline join. In the interval containing the tail region B-splines fall off very rapidly to zero. The explicit construction of B-splines is explained elsewhere[11]. We can also construct exact derivatives of B-splines provided the order does not exceed $M-1$.

A continuous function $f(x)$, defined in the interval (x_{min}, x_{max}), can be expanded in terms of B-spline functions as

$$f(x) = \sum_i B_i^M(x) c^i \ , \tag{20}$$

where quantities c^i denote the expansion coefficients. We can solve for the expansion coefficients in terms of a given (or to be determined) set of function values evaluated at a set of data points more commonly known as *collocation points*. There are a number of ways to choose collocation points [11,18], however, for odd order B-splines an optimal and simple choice is to place one collocation point at the center of each knot interval within the physical boundaries

$$x_\alpha = \frac{x_{\alpha+M-1} + x_{\alpha+M}}{2} \quad , \quad \alpha = 1, \ldots, N. \tag{21}$$

Note that collocation points are denoted by greek subscripts. We can now write a linear system of equations by evaluating (20) at these collocation points

$$f_\alpha = \sum_i B_{\alpha i} c^i \,, \tag{22}$$

where $f_\alpha \equiv f(x_\alpha)$, and $B_{\alpha i} \equiv B_i^M(x_\alpha)$. In order to solve for the expansion coefficients the matrix $\mathbf{B}$ needs to be inverted. However, as it stands matrix $\mathbf{B}$ is not a square matrix. In order to achieve this inversion we need to introduce additional linear equations which represent the boundary conditions imposed on $f(x)$ at the two boundary points, x_M and x_{M+N}. The essence of the lattice method is to eliminate the expansion coefficients c^i using this inverse matrix. The details of using the boundary conditions and inverting the resulting square matrix will be discussed elsewhere[11]. After inversion the coefficients are given by

$$c^i = \sum_\alpha \left[\mathbf{B}^{-1}\right]^{i\alpha} f_\alpha \,. \tag{23}$$

One can trivially show that all local functions will have a local representation in the finite dimensional collocation space

$$f(x) \longrightarrow f_\alpha \,. \tag{24}$$

The collocation representation of the operators can be obtained by considering the action of an operator $\mathcal{O}$ onto a function $f(x)$

$$\mathcal{O}f(x) = \sum_i [\mathcal{O}B_i^M(x)]c^i \,. \tag{25}$$

If we evaluate the above expression at the collocation points x_α we can write

$$[\mathcal{O}f]_\alpha = \sum_i [\mathcal{O}B]_{\alpha i} c^i \,. \tag{26}$$

Substituting from Eq.(23) for the coefficients c^i we obtain

$$\begin{aligned} [\mathcal{O}f]_\alpha &= \sum_{i\beta} [\mathcal{O}B]_{\alpha i} \left[\mathbf{B}^{-1}\right]^{i\beta} f_\beta \\ &= \sum_\beta O_\alpha^\beta f_\beta \,, \end{aligned} \tag{27}$$

where we have defined the collocation space matrix representation of the operator $\mathcal{O}$ by

$$O_\alpha^\beta = \sum_i [\mathcal{O}B]_{\alpha i} [\tilde{C}]^{i\beta} \,. \tag{28}$$

Notice that the construction of the collocation space operators can be performed once and for all at the beginning of a calculation, using only the given knot sequence and collocation points. Due to the presence of the inverse in Eq.(28) the operator O is not a sparse matrix. In practice, operator $\mathcal{O}$ is chosen to be a differential operator such as d/dx or d^2/dx^2. By similar constructions it also possible to obtain the appropriate integration weights for the collocation lattice.

3.2 Discrete HF Equations

In order to obtain a a set of equations which preserve the conservation laws associated with the continuous equations we must apply the variation to the discretized form of a conserved quantity, in this case the total energy. In particular, we will show that this variation results in the BSCM representation of the HF equations. Since the detailed derivation involves many terms that are present in the energy functional, we will only demonstrate the method for one term.

As an example for a local term let us consider a part of the t_0 contribution to the total energy

$$\frac{t_0}{2}(1+\frac{x_0}{2})\int d^3r\rho^2 = \frac{t_0}{2}(1+\frac{x_0}{2})\sum_{\alpha\beta\gamma} w^\alpha w^\beta w^\gamma [\rho(\alpha\beta\gamma)]^2 \,, \tag{29}$$

where on the right-hand side we have written the discretized form on a collocation lattice with collocation weights, denoted by greek letters α, β, γ. In order to be able to perform the variation with respect to the single-particle states ψ^*_λ we rewrite equation (29) explicitly

$$\frac{t_0}{2}(1+\frac{x_0}{2})\sum_{\alpha\beta\gamma} w^\alpha w^\beta w^\gamma \sum_{\mu\nu} \psi^*_\mu\psi_\mu\psi^*_\nu\psi_\nu \,. \tag{30}$$

The general variation is defined as

$$\frac{\delta\psi^*_\mu(\alpha\beta\gamma)}{\delta\psi^*_\lambda(\alpha'\beta'\gamma')} = \delta_{\lambda\mu}\delta_{\alpha'\alpha}\delta_{\beta'\beta}\delta_{\gamma'\gamma} \,. \tag{31}$$

Using Eq.(31) in the variation of Eq.(30) we obtain (after replacing the primed indices with unprimed ones)

$$t_0(1+\frac{x_0}{2})\rho(\alpha\beta\gamma)\psi_\lambda(\alpha\beta\gamma) \,, \tag{32}$$

where we have rewritten a summation as the total density. Together with the expansion in terms of B-splines

$$\psi_\lambda(\alpha\beta\gamma) = \sum_{ijk} c^{ijk} B_{\alpha i} B_{\beta j} B_{\gamma k} \,, \tag{33}$$

this procedure is a way to obtain a stable collocation lattice representation of the HF equations. The same procedure can be carried out for the nonlocal terms in the energy density. A typical term is illustrated below

$$(\nabla\psi_\lambda^\pm)_{\alpha\beta\gamma}=\sum_{\alpha'} D_\alpha^{\alpha'}\psi_\lambda^\pm(\alpha'\beta\gamma)\,\hat{\imath}+\sum_{\beta'} D_\beta^{\beta'}\psi_\lambda^\pm(\alpha\beta'\gamma)\,\hat{\jmath}$$
$$+\sum_{\gamma'} D_\gamma^{\gamma'}\psi_\lambda^\pm(\alpha\beta\gamma')\,\hat{k}$$

where the matrices $\mathbf{D}$ denote the first derivative matrices in x, y, z directions (they can be different although the notation does not make this obvious), calculated as described in the previous section. Finally, the HF equations can be written as matrix-vector equations on the collocation lattice

$$h\psi_\lambda^\pm \longrightarrow \mathbf{h}\cdot\psi_\lambda^\pm \,. \tag{34}$$

3.3 Solution of the Discrete HF Equations

The solution of the HF equations (34) is found by using the damped relaxation method described in Refs.[19,3)]

$$\chi_\lambda^{k+1} = \mathcal{O}[\chi_\lambda^k - x_0\mathbf{D}(E_0)(\mathbf{h}^k - \varepsilon_\lambda^k)\chi_\lambda^k] \,, \tag{35}$$

where $\mathcal{O}$ stands for Gram-Schmidt orhonormalization. The damping operator $\mathbf{D}$ is chosen to be[19,3)]

$$\mathbf{D}(E_0)=\left[1+\frac{\mathbf{T}}{E_0}\right]^{-1}$$
$$\approx\left[1+\frac{\mathbf{T}_x}{E_0}\right]^{-1}\left[1+\frac{\mathbf{T}_y}{E_0}\right]^{-1}\left[1+\frac{\mathbf{T}_z}{E_0}\right]^{-1},$$

where $\mathbf{T}$ denotes the kinetic energy operator. As a convergence criteria we have required the fluctuations in energy

$$\Delta E^2 \equiv \sqrt{< H^2 > - < H >^2}\,, \tag{36}$$

to be less than 10^{-6}. This is a more stringent condition than the simple energy difference between two iterations, which is about 10^{-13} when the fluctuation accuracy is satisfied.

The calculation of the HF Hamiltonian also the requires the evaluation of Yukawa and direct Coulomb contributions given by Eqs.(18) and (19). The evaluation of the three-dimensional integrals is very costly, however, instead one could solve the corresponding differential equations

$$(\nabla^2 - \frac{1}{a^2})W_Y(\mathbf{r}) = -4\pi a\rho_q(\mathbf{r})$$

$$\nabla^2 U_C(\mathbf{r}) = -4\pi e^2 \rho_p(\mathbf{r})\ .$$

Details of solving the Helmholtz and Poisson equations using the BSCM is given in Ref.[11].

4. RESULTS

In Table 1 we have tabulated the results obtained from the solution of the three-dimensional HF equations together with the solution of the radial Schrödinger equation for the spherical ^{16}O nucleus, using the BKN force[20]. Three-dimensional calculations result in a spherical nucleus with a quadrupole moment on the order of $10^{-13} fm^2$. We have used 22 points in x, y, and z-directions in interval $(-10, +10) fm$. Radial calculations had converged to two significant digits. As we see from Table 1, increase in the B-spline order M leads to a significant improvement in all quantities. The error in binding energy for $M = 7$ is 0.03%. The energies were calculated using the Koopman's formula with corrections for the density dependent three-body term and the Coulomb exchange. The three-dimensional calculations involve 16 single-particle states each with a spin-up and a spin-down component. Due to the absence of the spin-orbit interaction in the BKN force, the 6 p-states for neutrons and protons are degenerate in energy. We have also repeated these calculations for the Skyrme M^* force using $M = 5$ with accuracies almost identical to the BKN case. In this case the numerical computation of the parity and the third component of the total angular momentum are within 1 part in 10^4 of the exact spherical value. Each HF calculation consumed 15 minutes on a Cray-2 supercomputer. The computer time for heavier systems will be directly proportional to their mass number.

At this time we have also completed one TDHF calculation for the $^{16}O + {}^{16}O$ system at $E_{lab} = 64\ MeV$ using the BKN force with $b = 0.0\ fm$ (head-on collision). We have used a time step of $\Delta t = 0.4\ fm/c$ and followed the reaction for 850 time steps. These numbers are commensurate with the few restricted three-dimensional calculations available in the literature[21,22,23,24]. At the end of the calculation the error in energy was 32 keV out of the total energy of 193 MeV (error of 0.02 %). During the collision norm was conserved to better than 1 part in 10^8.

Table 1
A comparative study of HF results for ^{16}O. The force used is the Bonche, Koonin, Negele (BKN) form with $t_0 = -497.726\,MeV \cdot fm^3$, $t_1 = t_2 = 0.0\,MeV \cdot fm^5$, $t_3 = 17270.0\,MeV \cdot fm^6$, $t_4 = 0.0\,MeV \cdot fm^5$, $x_0 = x_3 = 1.0$, $\alpha = 1.0$, $V_U = V_L = -363.044\,MeV$, and $a = 0.45979\,fm$. We also include the direct and exchange parts of the Coulomb interaction and consider separate neutron and proton densities. The HF energy is calculated via Koopman's formula with corrections.

	Radial $N = 500$ $\Delta = 0.025\,fm$	B-Spline, M=3 $N = 22^3$ $\Delta = 0.9\,fm$	B-Spline, M=5 $N = 22^3$ $\Delta = 0.9\,fm$	B-Spline, M=7 $N = 22^3$ $\Delta = 0.9\,fm$
E_{HF} (MeV)	-119.641	-113.849	-119.135	-119.674
$K.E.$ (MeV)	249.734	250.416	249.809	250.070
r_p (fm)	2.6263	2.6491	2.6284	2.6254
r_n (fm)	2.5970	2.6202	2.5993	2.5961
$E_s(n)$ (MeV)	-30.120	-30.181	-30.160	-30.148
$E_p(n)$ (MeV)	-16.949	-16.702	-16.954	-16.975
$E_s(p)$ (MeV)	-26.237	-26.351	-26.282	-26.265
$E_p(p)$ (MeV)	-13.361	-13.167	-13.372	-13.386

5. DISCUSSION

We have for the first time performed three-dimensional HF and TDHF calculations with no simplfying assumptions. The use of the BSCM method results in an unprecedented accuracy for relatively coarse meshes. The evolution of nuclear HF and TDHF calculations have closely paralled the advances made in computer technology. With these new numerical methods and the technology available today it is possible to perform static HF calculations with accuracies comparable to more phenomenological models. In near future we will incorporate constraints into the HF program which will enable us to perform energy surface calculations for heavier systems. Similarly, unrestricted dynamical calculations will help us correctly address the questions about the amount of mean-field dissipation present in heavy-ion collisions. But, perhaps more importantly, the development of such numerical technologies and programs will lead to spin-offs for addressing many other problems in physics.

6. ACKNOWLEDGMENTS

This research was sponsored in part by the U.S. Department of Energy under contract No. DE-FG05-87ER40376 with Vanderbilt University. The numerical calculations were carried out on CRAY-2 supercomputers at the NCSA, Illinois, and at the NMFECC Livermore.

REFERENCES

1. K. T. R. Davies, K. R. S. Devi, S. E. Koonin, and M. R. Strayer, in *Treatise on heavy ion Science*, edited by D. A. Bromley, (Plenum, New York, 1985), Vol.3, page 3.
2. J. W. Negele, Rev. Mod. Phys. 54, 913 (1982).
3. A. S. Umar, M. R. Strayer, R. Y. Cusson, P.-G. Reinhard, and D. A. Bromley, Phys. Rev. C32, 172 (1985).
4. A. S. Umar, M. R. Strayer, P. -G. Reinhard, K. T. R. Davies, and S. -J. Lee, Phys. Rev. C40, 706 (1989).
5. P. -G. Reinhard, A. S. Umar, K. T. R. Davies, M. R. Strayer, and S. -J. Lee, Phys. Rev. C37, 1026 (1988).
6. S. -J. Lee, A. S. Umar, K. T. R. Davies, M. R. Strayer, and P. -G. Reinhard, Phys. Lett. 196B, 419 (1987).
7. A. S. Umar, M. R. Strayer, and P.-G. Reinhard, Phys. Rev. Lett. 56, 2793 (1986).
8. K. T. R. Davies and S. E. Koonin, Phys. Rev. C23, 2042 (1981).
9. P. Hoodbhoy and J. W. Negele, Nucl. Phys. A288, 23 (1977).
10. S. E. Koonin, K. T. R. Davies, V. Maruhn-Rezwani, H. Feldmeier, S. J. Krieger, and J. W. Negele, Phys. Rev. C15, 1359 (1977).
11. A. S. Umar, J. Wu, M. R. Strayer, and C. Bottcher, (submitted to the Journal of Computational Physics).
12. A. K. Kerman and S. E. Koonin, Ann, Phys. (N.Y.) 100, 332 (1976).
13. D. Vautherin and D. M. Brink, Phys. Rev. C5, 626 (1972).
14. Y. M. Engel, D. M. Brink, K. Goeke, S. J. Krieger, and D. Vautherin, Nucl. Phys. A249, 215 (1975).
15. P. Quentin and H. Flocard, Annu. Rev. Nucl. Sci. 28, 523 (1978).
16. M. J. Giannoni and P. Quentin, Phys. Rev. C21, 2076 (1980).
17. J. Bartel, P. Quentin, M. Brack, C. Guet, and H. B. Hakansson, Nucl. Phys. A386, 79 (1982).
18. C. De Boor, *Practical Guide to Splines*, (Springer-Verlag, New York, 1978).

19. C. Bottcher, M. R. Strayer, A. S. Umar, and P. -G. Reinhard, Phys. Rev. A40, October (1989).

20. P. Bonche, S. E. Koonin, and J. W. Negele, Phys. Rev. C13, 1226 (1976).

21. H. Flocard, S. E. Koonin, and M. S. Weiss, Phys. Rev. C17, 1682 (1978).

22. P. Bonche, B. Grammaticos, and S. E. Koonin, Phys. Rev. C17, 1700 (1978).

23. K. T. R. Davies, H. T. Feldmeier, H. Flocard, and M. S. Weiss, Phys.Rev. C18, 2631 (1978).

24. K. R. Sandhya Devi, M. R. Strayer, J. M. Irvine, K. T. R. Davies, Phys. Rev. C23, 1064 (1981).

Optimization of FORTRAN programs

Niels R. Walet, Dep. of Physics, University of Pennsylvania,
Philadelphia PA 19104-6396

Abstract

In this contribution, based on one-and-a-half informal seminars given at the summer school, I shall discuss the optimization of FORTRAN programs on scalar and vector machines. The discussion is limited to the minimal subset of ideas needed for general problems. A number of references is given.

1 Introduction

Optimization of programs, in physics almost by definition written in FORTRAN, has become a very important part of writing a code. In these notes I want to give some hints about the general techniques that are used for this task. I will limit the discussion to so-called scalar and vector machines, thus by-passing the interesting, but complicated, parallel architectures.

I am certainly not the first to discuss the techniques useful for optimization, and the reader should certainly consult Refs. [1, 2] if he has any further interest in the subject. Anyhow, optimization of programs for vector machines is in general "*déja vu* all over again", trying to apply the same well-established tricks in every context possible. Part of the message these notes want to convey is that optimization may be tedious, but is not a hard job. In doing that I address myself to the novice (or inexperienced) optimizer. If you do not belong to this group this story may be boring!

At this point I would like to shortly discuss the reference list. There are two books I already referenced before, [1, 2], which discuss mainly optimization and some pertinent details of the architecture. If you would like to know more about the architecture of supercomputers, please consult [3, 4]. Some of the conferences listed in the references ([5, 6, 7, 8, 9, 10]) also contain a lot of information about architecture, albeit mostly for parallel computers. These proceedings, together with the other proceedings of conferences dedicated to applications ([11, 12, 13, 14]) give a good number of papers where algorithms specially designed for vector computers are discussed. These may give you some hints about the special numerical methods that lead to easy vectorization.

Before we bury ourselves in technical details, I would like to make one very straightforward, but reasonably profound, remark: The most dramatic speed-up of your program is generally not obtained from optimization, but is obtained through using a different algorithm or a different data structure. This is nothing but the old adagium "think before you begin".

As stated before FORTRAN is the language of choice in science. For this reason all my examples are in FORTRAN, the only freedom I allowed myself is to use a `DO ... END DO` construction instead of the old-fashioned `DO ... CONTINUE`.

2 What is a vector computer?

The meaning of the word vector computer is somewhat misleading, the name "pipeline computer" gives a better understanding of its capabilities. Whereas one tends to think of a multiplication, addition, division as a single process, this is not the way such an operation is performed on a computer. If we consider the smallest step of time on a computer, the "clock cycle", a computer needs a few of these clock cycles to perform one of the operations mentioned above.

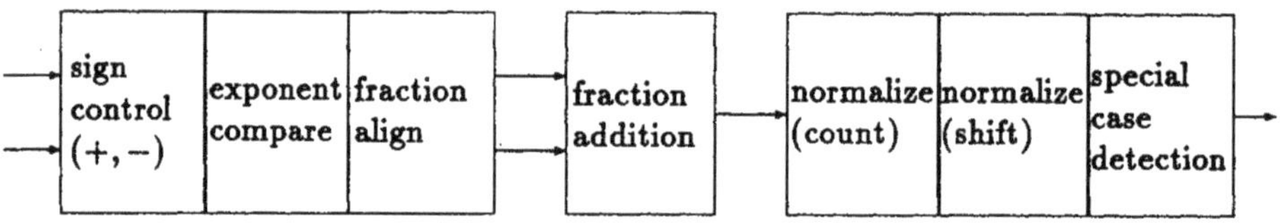

Figure 1: Pipelined addition unit (adapted from Ref. [3])

Consider for example the simplified version of the adding unit on the CYBER 205 as given in fig. 1, where we have indicated each step needed to perform the addition of two floating point numbers by a square block. I think most of the processes are self-explanatory once you realize that a real number is represented as a fraction and an exponent, where the exponents are not the same for the two real numbers.

We see that the process of addition takes seven steps. On a classic computer we would wait for one pair of operands to clear the addition unit before taking on the next pair. For a pipeline architecture we start to perform the first suboperation on the second pair of operands when we perform the second suboperation on the first pair, etc.. In view of the similarity with a fluid running through a pipeline this is called "pipelining". What we thus need is a continuous and contiguous stream of data (actually two streams) going into the addition unit to obtain one stream going out.

From now on we shall use the term vector computer for any computer with some form of pipelining. The usual computers, from PC to mainframes, that use the ordinary one-at-a-time approach, are referred to as "scalar" computers. As I shall mention again later all vector computers have also (separate) scalar processors. If you just look at the schematic vector unit above can you think of a reason why this may be profitable? I shall give the answer in section 5.

3 Compiler optimization

Before discussing your optimizations, let us discuss the switch "OPT" on your compiler. This is the easiest way to speed up your program and may, depending on the quality of the optimizer, be a very good start. The best compilers perform almost all the optimizations I shall discuss in the next two sections. Since it is an automatic process, it chooses the path that is the best in general, and may not be the best for your program. So even if you have the best optimizer available, you still may need to do some of your own optimizations, since it usually requires some understanding of your program to find the best.

4 Scalar optimization

4.1 Some standard tricks

Some of the most trivial examples of optimizations that almost all compilers will do for you are

- Removal of loop invariants from DO-loops, e.g.

  ```
  DO I = 1, N
    A = 99
    X(I) = A / I
  END DO
  ```

  where we have a unnecessarily repeated statement A=99 in the DO-loop, which can be taken outside off the loop.

- Replacing division by a constant by multiplication by its inverse, e.g.,

  ```
  DO I = 1, N
    X(I) = A(I) / N
  END DO
  ```

  is replaced by

  ```
  C = 1.0 / N
  DO I = 1, N
    X(I) = A(I) * C
  END DO
  ```

This is profitable since a division takes more time than a multiplication. It also introduces possible errors into the program; if N=0 the first program will run perfectly, but the second will crash due to a division by zero.

- removal of redundant machine instructions: After literally translating your program statement by statement into machine instructions there usually is some redundancy on the level of machine instructions. Many compilers are well designed to remove these redundancies, thus speeding-up your program.

4.2 Memory bank conflicts

A memory bank has to do with the way memory is organized on many computers. Consider the computer memory as schematically indicated in fig. 2. There we show

a_1	a_2	a_3	a_4	a_5	a_6	a_7	a_8
a_9	a_{10}	a_{11}	a_{12}	a_{13}	a_{14}	a_{15}	a_{16}
a_{17}	a_{18}	a_{19}	a_{20}	a_{21}	a_{22}	a_{23}	a_{24}
$\vdots$	$\vdots$	$\vdots$	$\vdots$	$\vdots$	$\vdots$	$\vdots$	$\vdots$

Figure 2: Schematic memory layout. The vertical columns indicate one bank, and the symbols a_i denote an array stored in memory.

that the memory is divided up in 8 slices (chosen for this example, but a number of 8 or 16 is not unrealistic), the banks. The subdivision of memory in these functionally independent parts derives its use from the fact that it takes much more time to get to elements out of the same bank than out of different banks. This is due to the fact that, due to physical constraints, the memory has a certain dead time during which you cannot access the same bank.

For the old CRAY 1 it takes 8 clock cycles to obtain `A(9)` after we get `A(1)` from memory. The specific example was chosen in such a way that if we have a do loop that uses subsequent elements of `A`,

```
DO I = 1, N
  X(I) = A(I) * ....
END DO
```

we get one element each clock cycle. If we change the loop to a stride 8 loop,

```
DO I = 1, N, 8
  X(I) = A(I) * ....
END DO
```

we find that we only get one element every eight clock cycles. This leads to a substantial increase in execution time. The message: be careful about bank conflicts when using stride $\neq 1$.

A well-known example of implicit bank conflicts is in the storage of complex arrays on a cray computer, where writing your own program using separate arrays for real and imaginary part is much faster than using complex arrays.

4.3 DO loop unrolling

One way to speed up the performance of scalar do-loops is loop unroling. In this case one makes use of compiler optimization of machine code. Consider the two loops in

```
      DO I = 1, N
        A(I) = C(I)*FDD(I)/I
      END DO
.....
      DO I = 1, N, 2
        A(I)   = C(I)*FDD(I)/I
        A(I+1) = C(I+1)*FDD(I+1)/(I+1)
      END DO
```

Figure 3: DO loop unrolling

Fig. 3. On most computers the second loop executes faster than the first one, due to the fact that the compiler can optimize the extra code generated by duplicating the instruction line by removing some redundant machine instructions. Usually a fourfold unroling (i.e., duplicating the instruction line four times) is as optimal as it gets.

5 Vector optimization

5.1 Basic operations

Basic vector operations are

```
DO I = 1, N
  A(I) = B(I) o C(I)
END DO
```

where o is any of the four standard operations $+, -, *, /$. Several of the standard functions are also vectorized, most using the building blocks discussed above, although some may be vectorized in hardware.

We have not yet discussed the way a vectorpipe is initialized from memory. There are two different ways to solve this. One in which we can only fill a vector pipe from a contiguous part of memory, so that only stride 1 do loops can be vectorized, and other strides have to be mimicked by performing the vector operation on the contiguous stream of data, and storing only a subset of the data in memory. Clearly in this case a stride 2 do-loop is twice as slow as a stride 1 do loop. Another option is the CRAY approach where a pipe can be randomly filled from memory and randomly stored to memory. These operations are called gather and scatter, and all computer producers include routines in their software libraries to implement these operations as closely as possible.

Now let us again consider the vector pipe discussed in section 2. If we perform a calculation, we can distinguish two phases, one in which the pipe is being filled, and one where the pipe delivers a result every clock-cycle. The first period is called the start-up time. Clearly it takes some time, and thus some minimal length of the vectors, to wash out the effects of the start-up time. This is due to the fact that the design consideration for a vector processor are not to minimize the total number of operations, but to get make the operations completely independent. Thus a scalar processor is able to treat a single operation much faster than a vector operation. There is a point where vector operations win over scalar operations. The minimal length of a vector to make vector operations profitable depends on the computer system used: It is small (in the order of 64) for a CRAY, and large (in the order of 1000) for CDC and ETA machines. This is the reason why it makes sense to use scalar arithmetic when either the vectors are to short, or the data is not contiguous.

5.2 Data dependency & recursion

One of the important no-no's in vectorization is recursion or data dependency. Consider the inner product in Fig. 4. If we vectorize the elementwise product as in the second

```
      SUM = 0.0
      DO I = 1, N
        SUM = SUM + A(I) * B(I)
      END DO
....
      DO I = 1, N
        C(I) = A(I) * B(I)
      END DO
      SUM = 0.0
      DO I = 1, N
        SUM = SUM + C(I)
      END DO
```

Figure 4: Inner product

half, we see that the sum can not be vectorized, since we need the previous value of

SUM to calculate the next value. Fortunately this operation occurs so often in programs that the manufacturers include an optimal routine to do this with their computers. Optimizing a recursive statement requires some ingenuity. In the case of an algorithm that contains some recursive statements, it usually pays to look for algorithms that do not have this property. A large number of such algorithms are discussed in the literature.

5.3 Some common linear algebra

One of the most common structures in physical computing is ordinary linear algebra. Consider the matrix vector multiplication $c_i = \sum_j A_{ij} b_j$. In Fig. 5 we give two possible

```
      DO I = 1, N
        C(I) = 0.0
        DO J = 1, N
          C(I) = C(I) + A(I,J)*B(J)
        END DO
      END DO
.....
      DO I = 1, N
        C(I) = 0.0
      END DO

      DO J = 1, N
        DO I = 1, N
          C(I) = C(I) + A(I,J)*B(J)
        END DO
      END DO
```

Figure 5: Matrix vector multiplication

ways to do this calculation. Again we prefer the second way over the first. The reason is, apart from the fact that we take giant strides through the array A in the first case, that in the second the innermost loop is just the multiplication of a contiguous vector times a constant, a standard vector operation.

In Fig. 6 we give a similar procedure for a matrix matrix multiplication. Try to figure out why (again) the second way is better.

5.4 IF statements

I shall try to illustrate the tricky concept of optimizing IF statements to the example in Fig. 7. I have supposed that RND() is a (pseudo-)random number generator which gives uniformly distributed numbers between zero and one. Y is some predetermined

```
DO I = 1, N
  DO J = 1, N
    C(I,J) = 0.0
    DO K = 1, N
      C(I,J) = C(I,J) + A(I,K)*B(K,J)
    END DO
  END DO
END DO
.....
DO I = 1, N
  DO J = 1, N
    C(J,I) = 0.0
  END DO
END DO
DO K = 1, N
  DO J = 1, N
    DO I = 1, N
      C(I,J) = C(I,J) + A(I,K)*B(K,J)
    END DO
  END DO
END DO
```

Figure 6: Matrix matrix multiplication

```
DO I = 1, N
   IF (RND() .GT. Y) THEN
     A(I) = A(I)/2.0
   END IF
END DO
```

Figure 7: Generic IF statement

number between zero and one. The way to optimize this statement depends strongly on the value of Y. One can distinguish three regions where different techniques play a role.

1. Y is close to zero. In this case only a very small number of divisions will actually be performed. The optimal solution is to force the compiler to translatee the loop into scalar code; most compilers have some way to include directives that allow you to do that.

2. Y is close to one. In this case it pays to first divide every number by 2 and correct this action for the small number where we were not allowed to do it. We again have to compile the second loop into scalar code.

```
DO I = 1, N
  A(I) = A(I) / 2.0
END DO
DO I = 1, N
  IF (RND() .GT. Y) THEN
    A(I) = A(I) * 2.0
  END IF
END DO
```

Figure 8: Y is close to 1

3. Other values of Y. In this case the most profitable way is to using the controled store capablities almost all vector computers have. The statements given above

```
DO I = 1, N
  COND(I) = RND() .GT. Y
END DO
DO I = 1, N
  IF (COND(I)) THEN
    A(I) = A(I) / 2.0
  END IF
END DO
```

Figure 9: Y is neither close to 1 nor 0

are compiled as follows: the computer evaluates the division for all elements, also the unwanted ones, but stores only the ones where COND(I) is true. This is faster than the scalar loops given above only if we are going to use a sizable portion of elements.

6 What should you optimize?

After this introduction to optimization you may have the impression that you should optimize your whole program. In general that would be a waste of your time since usually 50% of the computation time is spent in 15% of your program. There are many fancy utilities that tell you how much time is spent in each part of your program. Use them to find out the computationally intensive parts, and direct your attention to these parts. If you need more speed-up repeat the process, until you are either satisfied or do not see any further improvements. Good luck!

References

[1] W. Schönauer, *Scientific computing on vector computers* (North-Holland, Amsterdam, 1987)

[2] J. M. Levesque and J. W. Williamson, *A guide book to FORTRAN on supercomputers* (Academic, San Diego, 1989)

[3] P. B. Schneck, *Supercomputer architecture* (Kluwer Academic, Boston, 1987)

[4] K. Hwang (ed.), *Tutorial supercomputer design and applications* (IEEE Computer Society Press, Silverspring, 1984)

[5] P. B. Wilhelmson (ed.), *High speed computing: scientific applications and algorithm design* (University of Illinois Press, Urbana, 1988).

[6] F. A. Matsen and T. Tajima (eds.), *Supercomputers: algorithms, architecture and scientific computation* (University of Texas Press, Austin, 1986)

[7] *Proceedings of the first international conference on supercomputing systems* (IEEE Computer Society Press, 1985)

[8] *Proceedings of the supercomputing conference* (IEEE Computer Society Press, 1988)

[9] A. Lichnewsky and C. Saguez (eds.), *Supercomputing: state-of-the-art* (North-Holland, Amsterdam, 1987)

[10] N. Metropolis *et al* (eds.), *Frontiers of Supercomputing* (University of California Press, Berkeley, 1986)

[11] K. F. Jensen and D. Truhlar (eds.), *Supercomputing research in chemistry and chemical engineering* (Am. Chem. Soc., Washington, 1987)

[12] P. W. Numrich (ed.), *Supercomputer Applications* (Plenum, New York, 1985)

[13] D. T. Devreese and P. van Camp (eds.), *Supercomputers in theoretical and experimental science* (Plenum, New York, 1985)

[14] A. M. L. Emmen (ed.), *Supercomputer applications* (North-Holland, Amsterdam, 1985)

[15] S. Karni and N. P. Smith, *The supercomputer era* (Harcourt-Brace-Jovanovich, Boston, 1987)

[16] C. Lazou, *Supercomputers and their use* (Oxford University Press, Oxford, 1986)

QCD PHENOMENOLOGY ON HADRON FORM FACTORS*

Chueng-Ryong Ji
Department of Physics, North Carolina State University
Raleigh, North Carolina 27695-8202

ABSTRACT

A brief summary on perturbative QCD analyses on hadron form factors is presented and the problem in these analyses is pointed out. The improved method using "frozen" coupling constant is discussed and recent results obtained by this method for the pion, proton and neutron form factors and the cross section of the pion pair production in two photon collisions are presented.

1. INTRODUCTION

A simple picture on hadrons is that mesons are composed of quark and antiquark ($Q\bar{Q}$) and baryons are made of three quarks (QQQ). However, the underlying theory on this simple picture, Quantum Chromodynamics (QCD), predicts other types of meson such as glueballs (composite of two or more gluons), and exotic mesons and baryons such as $Q\bar{Q}Q\bar{Q}$ and $QQQQ\bar{Q}$. QCD is a theory of interactions among quark, antiquark and gluon. Thus an important question to be asked is whether experimental data on structures of hadron can be understood by QCD. So far, I believe, the answer is not yet known or at best the answer to this question is not yet complete. In this talk, I cannot answer this question as others but show some recent QCD analyses[1] whose results I think are rather amusing. Since this school is designed for graduate students or beginning post-docs, I will briefly explain how to measure the structure of hadrons and what the form factor means in the next section (Section 2). In Section 3, experimental data on pion and proton form factors and phenomenological understanding of these data based on perturbative QCD are presented. In Section 4, our recent QCD analyses based on the idea of "frozen" coupling constant are presented as well as the motivation of this idea. Conclusions and future directions are followed in Section 5.

---

*Invited Talk Presented at the Summer School of Computational Atomic and Nuclear Physics, The University of the South, Sewanee, Tennessee, June 26 - July 7, 1989.

2. FORM FACTORS IN ELECTRON SCATTERING EXPERIMENT

Let's consider an electron scattering experiment to show how to measure the structure of hadrons and what the form factor means: An electron beam which has the beam energy E scatters on a target atomic nucleus which has a size of order 10 fm (or 10^{-12} cm). In this experiment, if $E < \frac{1}{10\ \mathrm{fm}}$, then the target would be seen as a point by the beam and the differential cross section at a scattered angle θ would follow well-known Rutherford formula,

$$\left(\frac{d\sigma}{d\Omega}\right)_{\text{point}} \sim \frac{1}{E^2 \sin^4 \frac{\theta}{2}} . \tag{2.1}$$

The angular dependence, $\sin^{-4}\theta/2$, is smooth enough to give a violent large angle (or back angle) scattering, which is a characteristic of scattering on a point target. However, if $E > \frac{1}{10\ \mathrm{fm}}$, then the beam begins to see the structure of the target and the total charge Ze of the nucleus should be replaced by a charge density $\rho(\vec{r})$ due to the spatial extension of the target seen by the beam. In this case, the differential cross section is given by

$$\frac{d\sigma}{d\Omega} = \left(\frac{d\sigma}{d\Omega}\right)_{\text{point}} |F(Q^2)|^2 , \tag{2.2}$$

where the form factor $F(Q^2)$ is a Fourier transform of the charge density and Q^2 is the square of the momentum transfer of the scattered electron beam. For a static case, the form factor is given by

$$F(\vec{q}^{\,2}) = \int d^3\vec{r}\,\rho(\vec{r})e^{i\vec{q}\cdot\vec{r}} . \tag{2.3}$$

For example, if $\rho(\vec{r})$ is given by a gaussian form $e^{-m^2\vec{r}^{\,2}}$ then $F(\vec{q}^{\,2})$ is also given by a gaussian form $e^{-\vec{q}^{\,2}/4m^2}$. To show an actual experimental result[2], the differential cross section obtained in electron-carbon (C_{12}) scattering is shown in Fig. 1. The data are obtained at approximately one energy, 190 MeV, but at two angles, 80° and 135° (Figs. 1(a) and 1(b) respectively). Fig. 1(a) shows a clear elastic peak at the energy of the scattered electron, $E' \approx 185$ MeV. However, as shown in Fig. 1(b), this peak dramatically shrinks at the larger angle and correspondingly larger Q^2 than Fig. 1(a). This effect is caused by the form factor (see Eq. (2.2)) of the carbon. Thus, similarly, the form factor of the hadron can be measured by observing the change of the height of the elastic peak of the differential cross section in electron-hadron scattering at various angles and energies.

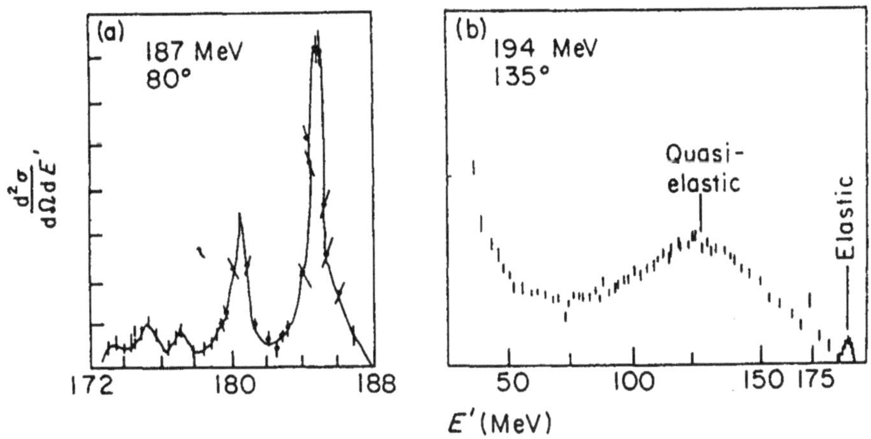

Fig. 1 Electron-carbon scattering: (a) through 80°, (b) through 135°. The results are shown in arbitrary units.

3. PION AND PROTON FORM FACTORS

The experimental data of the pion form factor[3] cannot be obtained by $e - \pi^{\pm}$ scattering because the pion target decays so fast (~ 10^{-8} sec). Instead, one can obtain the data in $ep \to e\pi^+n$ experiment. Anyway, the result for $F_\pi(Q^2)$ multiplied by Q^2 is shown in Fig. 2 as well as phenomenological fits given by

$$F_\pi(Q^2) = \frac{1}{\left(1 + \frac{Q^2}{m^2}\right)}, \tag{3.1}$$

where the corresponding static charge distribution is given by $\rho(r) \sim \frac{e^{-mr}}{r}$. In Fig. 2, the fits with $m^2 = 0.35$, 0.45 and 0.55 GeV^2 are shown and the corresponding pion rms radii are $\langle r_\pi \rangle = 0.81$, 0.72 and 0.66 fm respectively. However, an important fact is that the data can be described by a "monopole" form given by Eq. (3.1). To emphasize this fact, $Q^2F_\pi(Q^2)$ is shown in Fig. 2. The data for $Q^2F_\pi(Q^2)$ at $Q^2 > 1$ GeV^2 seems to be consistent with a constant even though more accurate data at high Q^2 are clearly needed.

More accurate data[4] were obtained for the proton Dirac form factor $F_1^P(Q^2)$. Interestingly enough, the proton data are described by a "dipole" form

$$F_1^P(Q^2) = \frac{1}{\left(1 + \frac{Q^2}{m^2}\right)^2}, \tag{3.2}$$

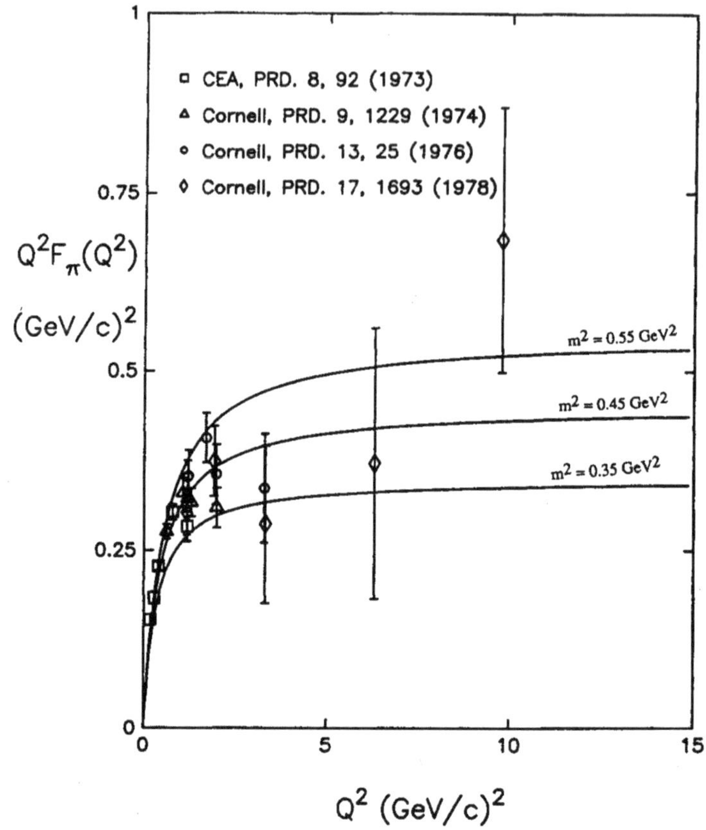

Fig. 2 The pion form factor mutiplied by Q^2, $\bar{Q}^2F_\pi(Q^2)$. Three curves plot Eq. (3.1) with $m^2 = 0.35$, 0.45 and 0.55 GeV^2.

where the corresponding static charge distribution is given by $\rho(r) \sim e^{-mr}$. To emphasize the asymptotic behavior $F_1^P(Q^2) \sim m^4/Q^4$ at $Q^2 >> m^2$, $Q^4F_1^P(Q^2)$ is plotted in Fig. 3 as well as the curves of Eq. (3.2) with $m^2 = 1.0$, 1.05 and 1.1 GeV^2 whose corresponding proton radii are $<r_p> = 0.76$, 0.75 and 0.74 fm, respectively.

Now, an important question is why the pion and the proton have the "monopole" and the "dipole" form factors respectively (i.e. why Eqs. (3.1) and (3.2) have the power "1" and "2" in the denominator respectively). Some people may try to explain this by vector meson dominance model or the similar. However, at least in the high Q^2 region, QCD can explain the reason more rigorously and intuitively than any

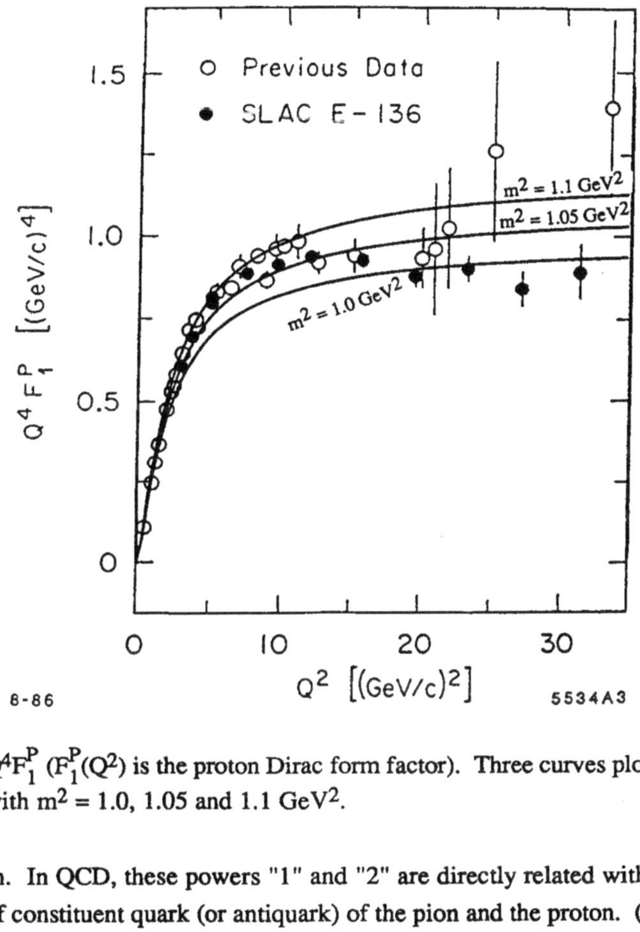

Fig. 3 $Q^4F_1^P$ ($F_1^P(Q^2)$ is the proton Dirac form factor). Three curves plot Eq. (3.2) with $m^2 = 1.0$, 1.05 and 1.1 GeV^2.

others can. In QCD, these powers "1" and "2" are directly related with the minimum number of constituent quark (or antiquark) of the pion and the proton. (The pion has 2 constituents and the proton has 3 constituents.) Therefore, one can make a prediction for any hadron form factor at asymptotic high Q^2. If the hadron has minimally n constituents, then $F_n(Q^2) \rightarrow \frac{1}{(Q^2)^{n-1}}$ as $Q^2 \rightarrow \infty$. This is called the quark counting rule[5]. The reason behind this quark counting rule can be rigorously proved[6] in QCD. However, the following simple explanation may be sufficient in this short talk.

If the momentum transfer carried by a virtual photon emitted by the beam electron is very large (i.e. $Q^2 >> \Lambda^2_{QCD}$, Λ_{QCD} is the QCD scale parameter.), then the struck quark inside the hadron is far off-shell so that it must transfer its momentum to other spectators in the hadron by exchanging gluons to make all constituents be collinear. Without the collinearity of all constituents, initial or final bound state (hadron) cannot be formed. Because of the asymptotic freedom in QCD, more gluons emission is more suppressed at high momentum transfer region. Therefore only the minimum number of gluon exchange which connects all constituents is the dominant contribution in the high Q^2 region, and this number determines the Q^2 behavior of hadron form factors. For example, in the pion and the proton, the minimum numbers of gluon exchange are "1" and "2" respectively. The above steps of explanation for the pion are shown in Fig. 4 and the form factor can be factorized by three amplitudes[6]:

$$F_\pi(Q^2) \approx \int_0^1 dx \int_0^1 dy\, \phi_\pi^*(y, Q) T_H(x, y, Q) \phi_\pi(x, Q)$$
$$\sim C \cdot \frac{\alpha_s(Q^2)}{Q^2} \left(\ln \frac{Q^2}{\Lambda^2} \right)^{-2\gamma_o}, \tag{3.3}$$

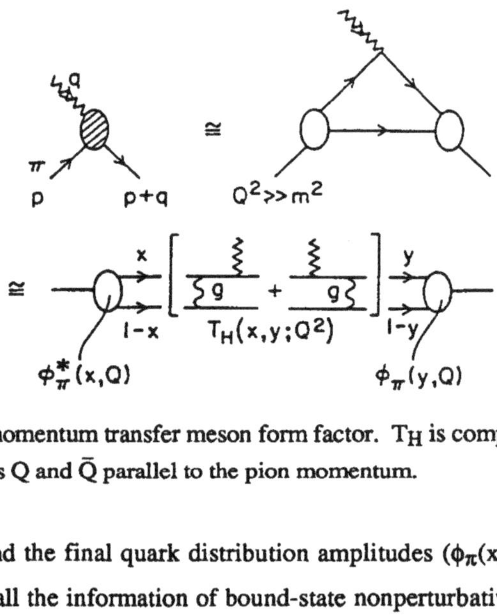

Fig. 4 The large momentum transfer meson form factor. T_H is computed for zero mass quarks Q and $\bar{Q}$ parallel to the pion momentum.

where the initial and the final quark distribution amplitudes ($\phi_\pi(x, Q)$ and $\phi_\pi^*(y, Q)$ respectively) have all the information of bound-state nonperturbative physics and the quark and gluon hard scattering amplitude $T_H(x, y, Q)$ can be expanded in the order of QCD running coupling constant $\alpha_s(Q^2)$ ($Q^2 = -q^2 >> \Lambda_{QCD}^2$). In Eq. (3.3), γ_o is the calculable leading anomalous dimension[7] of the quark distribution amplitude ϕ_π and the normalization C can be obtained after calculating hard scattering amplitude T_H.

4. QCD ANALYSES WITH FROZEN COUPLING CONSTANT

In the last section, it was claimed that the experimental data on the pion and proton form factors are consistent with the form $\frac{1}{(1+Q^2/m^2)^{n-1}}$, where n is the minimum number of constituent and this form can be predicted rigorously from QCD in the asymptotic high Q^2 region ($Q^2 >> \Lambda_{QCD}^2$ or m^2). However, it is rather surprising that experimental data are well fitted by small m^2 ($m^2 < 1$ GeV^2). Does it mean, then, that $Q^2 > 1$ GeV^2 for the pion and $Q^2 > 5$ GeV^2 for the proton are asymptotic free regions where the perturbative QCD works well? The answer seems no[8], because the leading order perturbative QCD calculation for hard scattering amplitude T_H shows that the dominant contribution of the form factor comes from small momentum transfer of the exchanged gluon. As an example, for the pion, the leading order T_H (see Fig. 4) is given by[1,6]

$$T_H(x, y, Q) = e_1 \frac{\alpha_s\left((1-x)(1-y)Q^2\right)}{(1-x)(1-y)Q^2} + e_2 \frac{\alpha_s\left(xyQ^2\right)}{xyQ^2}, \tag{4.1}$$

where e_1 and e_2 are the charges carried by particles 1 and 2 where longitudinal momentum fractions are $x(y)$ and $1-x$ $(1-y)$ respectively in the initial (final) quark distribution amplitude. Since the argument of the running coupling constant should be taken as the square of the momentum transfer of the exchanged gluon[9] (see Eq. (4.1)), the dominance in the small momentum transfer of the exchanged gluon precludes the possibility that $Q^2 > 1$ GeV^2 for the pion, for example, is the asymptotic free region.

Furthermore, one cannot use the "one-loop" QCD running coupling constant

$$\alpha_s(Q^2) = \frac{4\pi}{\left(11 - \frac{2}{3} n_f\right) \ell n\left(\frac{Q^2}{\Lambda_{QCD}^2}\right)} \tag{4.2}$$

to calculate the form factor because this formula blows up at small Q^2 ($Q^2 \sim \Lambda_{QCD}^2$). Fortunately, it has been suggested[10] to replace α_s of Eq. (4.2) by the "frozen" coupling constant which includes nonperturbative effects and is more natural to understand the color confinement problem[11]:

$$\alpha_s(Q^2) = \frac{4\pi}{\left(11 - \frac{2}{3} n_f\right) \ell n\left(\frac{Q^2 + 4m_g^2}{\Lambda_{QCD}^2}\right)}, \tag{4.3}$$

where the dimensional parameter m_g is numerically related with Λ_{QCD} by solving a special set of Schwinger-Dyson equations.[10] If $\Lambda_{QCD} \sim 100$ MeV to be consistent with various experimental results, then $m_g = 500 \pm 200$ MeV.

Therefore, even though the pion and proton form factor data are consistent with the quark counting rule obtained by the perturbative QCD, much more careful numerical analyses to calculate the normalization of the form factor are clearly necessary. This is the motivation why we reanalyze hadron form factors using the frozen coupling constant given by Eq. (4.3). To do this, we take the argument of α_s as the momentum transfer of the exchanged gluon (see e.g. Eq. (4.1)) and perform the integration (see Eq. (3.3) for the pion) after placing α_s inside the integration. Since we take $\Lambda_{QCD} \sim 100$ MeV in our analyses, the lower and upper bounds of m_g^2 are given by 0.1 GeV^2 and 0.5 GeV^2 respectively. Figs. 5 and 6 show the results on the pion and the proton form factors, respectively, using the quark distribution amplitude obtained by Chernyak and Zhitnitsky[12]. We caution that the sensitivity of m_g^2 shown in Figs. 5 and 6 are amplified by extra Q^2 factors. The actual sensitivity in $F_\pi(Q^2)$ and $F_1^P(Q^2)$ should be much more suppressed at high Q^2. Our calculation on the neutron Dirac form factor $F_1^n(Q^2)$ is also consistent with poorly known experimental data ($Q^2 \lesssim 10$ GeV^2). Finally, in Fig. 7, the most recent QCD analysis using frozen coupling constant for the pion pair production in two photon collisions is compared with the experimental data[13] which has the range of the two photon c.m. energy, 1.3 GeV/c^2 < M < 3.0 GeV/c^2. Our result is also compared with the previous QCD result which was obtained by Brodsky and Lepage[14] without using the frozen coupling constant.

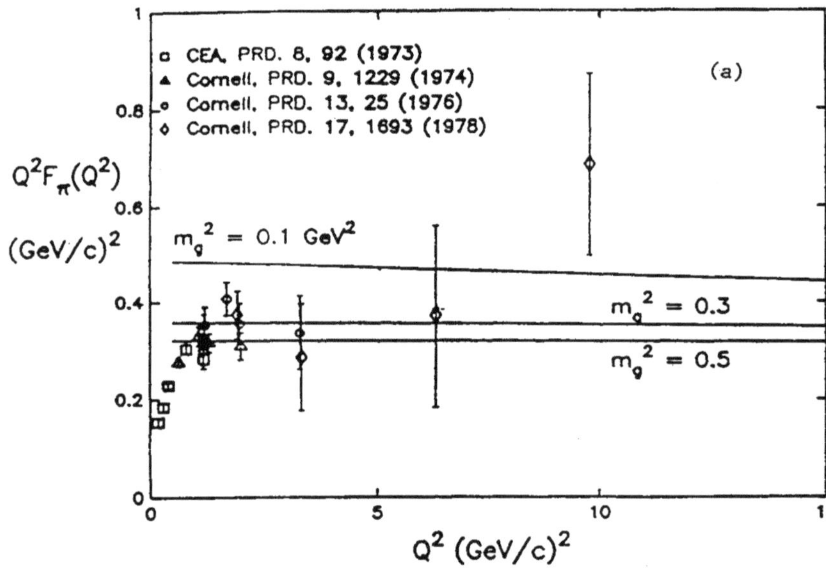

Fig. 5 Pion form factor calculation with Chernyak and Zhitniksky's amplitude (Ref. 12) and with the argument of $\alpha_s(Q^2)$ evaluated at correct value of exchanged gluon momentum for each diagram. $Q^2F_\pi(Q^2)$.

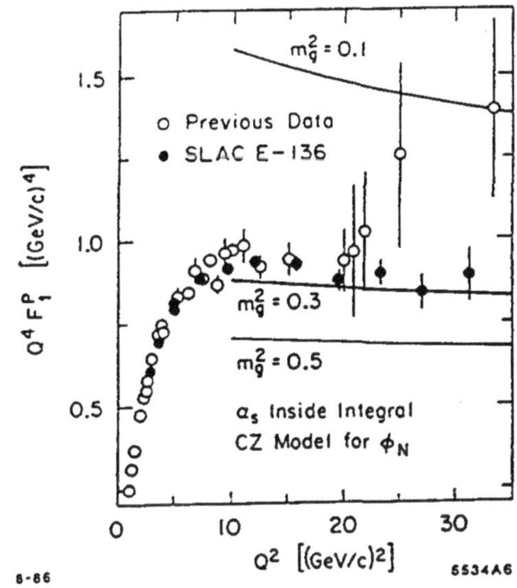

Fig. 6 Proton Dirac form factor calculation with the same procedure as Fig. 5. $Q^4F_1^P(Q^2)$.

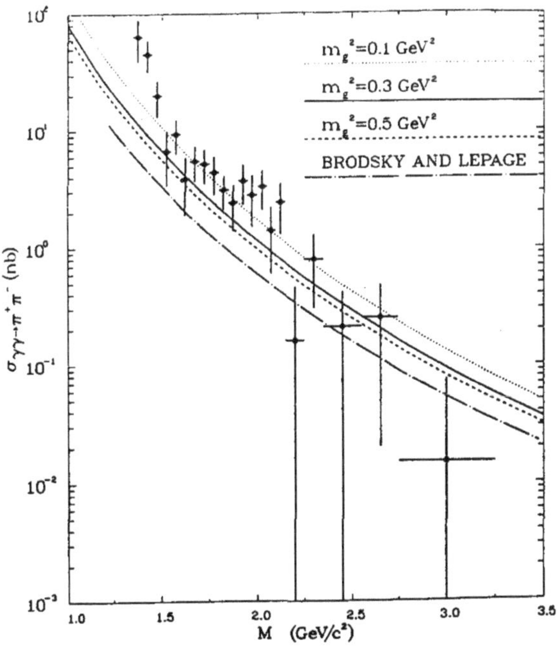

Fig. 7 The cross section for $\gamma\gamma \to \pi^+\pi^-$. The angular range is $|\cos\theta^*| \leq 0.3$. The calculation was done with the same procedure as mentioned in the text and Fig. 5.

5. CONCLUSIONS AND FUTURE DIRECTIONS

In this talk, I explained how the form factors are measured, how the hadron form factor can be understood by QCD especially at high Q^2 region and also what the problems in perturbative QCD calculations are. To remedy the problem caused by the divergence of the "one-loop" QCD running coupling constant $\alpha_s(Q^2)$ at $Q^2 \approx \Lambda^2_{QCD}$, the frozen coupling constant given by Eq. (4.3) is suggested. I showed some recent QCD analyses using the frozen coupling constant and the quark distribution obtained by the QCD sum rule[12]. It seems remarkable that the pion, proton and neutron form factors and the cross section for the pair production in two photon collisions are reasonably consistent with the theoretical predictions which do not have any ad hoc parameters.

It seems that the nonperturbative effect is important in the kinematic region where we have data. However, it is amusing that such complicated nonperturbative effect is simply summarized by a parameter m_g and furthermore this parameter is not arbitrary but related to Λ_{QCD} by a special set of Schwinger-Dyson equation[10]. Our method of using frozen coupling constant may not be the final one but, I think, at least provides an important phenomenology based on the first principle QCD.

The systematic development such as fermi motion of quarks and quark mass effects to the present analyses will be necessary to push Q^2 even to smaller values. Furthermore, applications of our method to other observables such as $N\Delta$ transition form factors and structure functions in deep inelastic lepton-hadron scattering etc. are necessary to further investigate the consistency of our method.

ACKNOWLEDGEMENTS

I would like to thank the organizers of this meeting C. Bottcher, J. B. McGrory and M. R. Strayer for their hospitality and for arranging this stimulating school. Parts of the results presented here are based on collaborations with F. Amiri, S. J. Brodsky, R. Lombard-Nelson and A. F. Sill. I would like to acknowledge S. R. Cotanch for many useful discussions. This work was supported by the U.S. Department of Energy.

REFERENCES

1. C.-R. Ji, A. F. Sill and R. M. Lombard-Nelson, Phys. Rev. D36, 165 (1987); C.-R. Ji and F. Amiri, SLAC-PUB-4869 (1989).

2. E. Bloom, in Proceedings of the Summer Institute on Particle Physics, Stanford (SLAC-Report 191 (1975)).

3. S. R. Amendolia et al., Nucl. Phys. B227, 168 (1986).

4. R. G. Arnold et al., Phys. Rev. Lett. 57, 174 (1986).

5. S. J. Brodsky and G. Farrar, Phys. Rev. Lett. 31, 1153 (1973); Phys. Rev. D11, 1309 (1975); D. Sivers, S. Brodsky and R. Blankenbecler, Phys. Rept. 23C, 1 (1976).

6. G. P. Lepage and S. J. Brodsky, Phys. Rev. D22, 2157 (1980).

7. S. J. Brodsky and C.-R. Ji, Phys. Rev. D33, 1951 (1986).

8. N. Isgur and C. H. Llewellyn Smith, Phys. Rev. Lett. 52, 1080 (1984).

9. S. J. Brodsky, G. P. Lepage and P. B. Mackenzie, Phys. Rev. D28, 228 (1983).

10. J. M. Cornwall, Phys. Rev. D26, 1453 (1982); G. Parisi and R. Petronzio, Phys. Lett. 94B, 51 (1980).

11. For more detailed discussion, see C.-R. Ji, in Proceedings of Nuclear and Particle Physics on the Light Cone, Los Alamos, edited by M. B. Johnson and L. S. Kisslinger, pp. 291-308 (1988).

12. V. L. Chernyak and I. R. Zhitnitsky, Phys. Rept. 112, 1973 (1984); Nucl. Phys. B246, 52 (1984).

13. M. Aihara et al., Phys. Rev. Lett. 57, 404 (1986).

14. S. J. Brodsky and G. P. Lepage, Phys. Rev. D24, 1808 (1981).

LIST OF PARTICIPANTS

C. J. Albert
Department of Physics
Texas A&M University

T. Appelquist
Department of Physics
Yale University

G. Arbanas
Department of Physics
MIT

H. Arellano
Department of Physics
University of Georgia

N. R. Badnell
AWE Aldermaston
Reading, UK

M. Baranger
Department of Physics
MIT

J. N. Bardsley
Lawrence Livermore National Laboratory
California

B. Bateson
Lawrence Livermore National Laboratory
California

P. M. Bergstrom
Department of Physics & Astronomy
University of Pittsburgh

R. Bijker
Bartol Research Institute
University of Delaware

C. Bottcher
Physics Division
Oak Ridge National Laboratory

J. Boyle
Department of Physics
University of Virginia

T. Brage
Department of Physics & Astronomy
Vanderbilt University

D. M. Brink
Department of Theoretical Physics
Oxford University

M. J. Buie
Naval Research Laboratory
Washington

C. Carlsund
Manne Siegbahninstitutet for Fysik
Stockholm, Sweden

Z. H. Chen
Department of Physics
New York University

F. E. Close
Department of Physics
University of Tennessee

I. L. Cordrey
Department of Physics
University of Tennessee

S. Crall
Department of Physics
Brown University

S. Dai
Department of Chemistry
University of Tennessee

D. J. Dean
Physics Division
Oak Ridge National Laboratory

M. Dekker
Utrecht
The Netherlands

X. S. Fang
Department of Physics
Texas A&M University

C. F. Fischer
Department of Computer Science
Vanderbilt University

D. C. Griffin
Bush Science Center
Rollins College

E. C. Halbert
Physics Division
Oak Ridge National Laboratory

J. S. Hyslop
Department of Physics
Virginia Polytechnic Institute & State University

Y. Ishikawa
Department of Chemistry
University of Puerto Rico

C. R. Ji
Department of Physics
North Carolina State University

Y. R. Jiang
Physics Division
Oak Ridge National Laboratory

W. R. Johnson
Department of Physics
Notre Dame University

M. H. Kalos
Center for Theoretical Physics
Cornell University

S. E. Koonin
California Institute of Technology

Y. Kuang
Department of Physics
University of Pittsburgh

S. Kumano
Department of Physics & Astronomy
University of Tennessee

M. Leonhard
Department of Physics
Texas A&M University

Z. J. Li
Department of Physics
Virginia Polytechnic Institute & State University

Z. P. Li
Department of Physics
University of Tennessee

Z. M. Lu
Department of Physics & Atmospheric Science
Drexel University

J. H. Macek
Department of Physics
University of Tennessee

A. Mahmoud
Department of Physics
Texas A&M University

E. J. Mansky
Department of Physics
Georgia Institute of Technology

J. B. McGrory
Physics Division
Oak Ridge National Laboratory

P. Mohr
Physics Building
National Institute of Standards & Technology
Maryland

D. L. Moores
Department of Physics
Auburn University

J. Morrison
Department of Physics
University of Louisville

D. Murdock
IU Nuclear Theory Center
Indiana

J. W. Negele
Department of Physics
MIT

V. E. Oberacker
Department of Physics & Astronomy
Vanderbilt University

M. S. Pindzola
Department of Physics
Auburn University

N. L. Roeth
Department of Physics
University of Wisconsin

M. E. Rosenkrantz
Air Force Astronautics Lab./LSX
Edwards Air Force Base
California

H. R. Sadeghpour
JILA
University of Colorado

B. Serot
Department of Physics
Indiana University

H. Sporn
Department of Physics
State University of New York

M. R. Strayer
Physics Division
Oak Ridge National Laboratory

M. Q. Sun
Department of Physics
Ohio State University

X. M. Tong
Department of Physics
University of Pittsburgh

A. S. Umar
Department of Physics & Astronomy
Vanderbilt University

M. Vallieres
Department of Physics & Atmospheric Science
Drexel University

J. D. Walecka
Research Division
CEBAF
Virginia

N. R. Walet
Department of Physics
University of Pennsylvania

S. Ward
Department of Physics
University of Tennessee

J. C. Wells
Department of Physics & Astronomy
Vanderbilt University

J. S. Wu
Physics Division
Oak Ridge National Laboratory

L. H. Xia
Cyclotron Institute
Texas A&M University

L. Xiong
Cyclotron Institute
Texas A&M University

N. Yoshinaga
Department of Theoretical Physics
Oxford University

B. Zhou
Department of Physics
University of Pittsburgh

R. S. Zighelboim
Department of Physics
Tulane University

T. Zuo
Department of Physics
York University
Ontario, Canada

www.ingramcontent.com/pod-product-compliance
Lightning Source LLC
LaVergne TN
LVHW071223011026
849280LV00007B/36
* 9 7 8 9 8 1 0 2 0 1 2 5 8 *